旅游规划篇（第二版）

梦游记

Dreamlike Traveling

中国建筑工业出版社

图书在版编目(CIP)数据

梦游记：旅游规划篇/伍新风编著. —2版.
北京：中国建筑工业出版社，2011.10
ISBN 978-7-112-13607-0

Ⅰ. ①梦… Ⅱ. ①伍… Ⅲ. ①旅游业—经济规划
Ⅳ. ①F590.1

中国版本图书馆CIP数据核字(2011)第199679号

责任编辑：常 燕 付 娇

梦游记：旅游规划篇(第二版)
伍新风 编著
*
中国建筑工业出版社出版、发行(北京西郊百万庄)
各地新华书店、建筑书店经销
贵州天海规划设计有限公司制版
广州市一丰印刷有限公司印制
*
开本：546×886毫米 1/12 印张：27 字数：813千字
2011年10月第二版 2011年10月第二次印刷
定价：230.00元
ISBN 978-7-112-13607-0
(21432)

编委 Compiling Members

作者简介——伍新凤

1958年生于湖南凤凰，毕业于贵州大学艺术学院美术系
贵州天海规划设计有限公司法人代表、董事长、总经理、首席策划、规划、设计师
贵州民族学院建筑工程学院副教授
中国景观设计专业委员会副主任
国际ICAD注册A级职业景观设计师
贵州省美术家协会理事
贵州省旅游规划评审专家组成员
贵州省设计艺术委员会副主任

经历风雨，刚毅顽强，立志高远，他做过知青，当过矿工，办过工厂，在深圳打过工，曾在贵州民族学院美术系任教，现任教于贵州民族学院建筑工程学院，2002 年创建了贵州天海规划设计有限公司……这些承载了他 50 载的经历。

十多年前，当刚接触到规划设计行业时，他突然感觉到这就是其人生理想之所在。天行健，君子以自强不息，地势坤，君子以厚德载物，在他的脑海里，世间有一种比山峦更高更远的风景，那就是天空；世间有一种比江河更宽更深的景象，那就是海洋，普天之下，四海之内，万物囊于其中。于是 2002 年他注册了贵州天海规划设计有限公司。他立志于通过天海规划这样一个平台在体现个人价值的同时，为贵州乃至全国的规划设计行业贡献自己的一份力量，为他的学生带去更鲜活更具竞争力的理念。

博观约取，厚积薄发，天海自成立以来正在以惊人的速度蓬勃发展，给予世人不断的震撼。

每个企业都需要有个领军人物，带来好的思路，先进的管理方法，与时俱进的开拓精神，引导这个企业的健康发展，显然，他便是这个赋予天海灵魂的人。做人当心志要苦，意趣要乐；气度要宏，言动要谨；修学不以诚，则学杂；行事不以诚，则事改；自谋不以诚，则是欺自心而自弃其忠；与人不以诚，则是丧其德而增人之怨，这便是他的人生信条。正是这样一个人，以他丰富的人生阅历，深刻的人生感悟，独到的眼光，诚信刚毅的品质，演绎着天海的完美绝伦，让天海之船驶向辉煌。

Introduction of the Author —Wu Xinfeng

He was born in Fenghuang Hunan 1958. As a design major, he graduated from the Art Institute of Guizhou University.
Wu Xinfeng—Legal representor, general manager, prime planner, designer of Guizhou Tianhai Planning and Design Co., Ltd.

A professor of Guizhou National University
Vice Director of Professional Committee of Landscape Design of China
Class A Professional Landscape Designer of International ICAD
Executive Member of the Artists Association of Guizhou Province
One of the Expert Group of Tourism Planning of Guizhou Province
Vice Director of Design and Art Committee of Guizhou Province

With massive life experience of working in the countryside as educated youth, in the mines as miner, in some corporations of Shenzhen, and running his our plants, he has become stronger in mind and more promising,And now he is a professor in Architecture Institute of Guizhou National Collage, he set up Tianhai Planning and Design Company in 2002, which is on the basis of his fifty years life experience.

Once he contacted the field of planning and design more than ten years ago, he was aware that it is the very point of his life expectation. Thus he registered the foundation of his company with the name of Tianhai (天海) Planning and Design Company with the hope to achieve his individual value, contribute to the planning and design profession of Guizhou province and even China, and bring his students more vivid and competitive concepts. The Chinese characters "天海" mean the universe and oceans. While the universe keeps rolling on orbit, a superior man makes untiring endeavour for advancement. In Wu xinfeng's mind, universe is the highest and most distant landscape, while ocean the broadest and deepest scenery, of the two which everything is contained.

Tianhai has been developing at the incredible speed since its foundation, which requires massive learning and accumulation, and of which people are shocked continually.

A leader, with great guidance, advanced managing method and pioneer spirit of keeping pace with the times, is needed in an enterprise. Obviously, he is the very person who endows Tianhai with soul. With his life credo of bearing hardship, pleasing mind, and extensive tolerance and cautious behavior, he believe that sincerity is the basic principle of doing everything and the guarantee of successful enterprise, without sincerity, one may fail in learning and lose his friends and even himself. It is he, with his rich life experience, profound life understanding, unique viewpoint and faithful and stouthearted quality, leads Tianhai going towards brilliance.

“一个行业游戏规则六十年未变，城市却已经变化巨大、沧海桑田，由此而带来灾难性的后果，那就是所有中国城市同质化现象严重，变得千城一面。”

千城一面何时止？！
——对话伍新凤
A THOUSAND OF THE SAME CITY

- 中国 30 年的变化让世界感到震撼，特别是北京、上海这样的大城市，现代化程度可以与纽约媲美。
- 中国一直在宣传自己五千年的历史文化，长期引以为豪。遗憾的是，在中国的城市里很难看到五千年历史文化的辉煌。
- 我走在中国的大城市，如果没有中文，街上行人不是中国人，我根本不知道自己身处于哪个国家。

这三句话，是一位美国专家访问中国，被问及关于中国城市的观感时所言。这三句话一直令伍新凤耿耿于怀。作为一个城市规划的圈内人，也许他更加有切肤之痛。

伍新凤是较早认识到这个问题严重性的业内人士之一，记得很多年前采访他时，提起这个问题，焦急、愤怒、心有不甘便溢于言表。不幸的是，几年过去了，“千城一面”的现象非但没有得到有效遏制，反而有愈演愈烈之势。有人戏称，中国 200 个城市一奶同胞。英国《卫报》说："中国是一个由一千座雷同的城市构成的国家。”不知道是调侃，还是“千城一面”的英文直译。作为每一个生活在其中的人，没有理由不感到压抑和悲哀。“高楼、高密度、大广场”，“呆板、毫无生气的、火柴盒般的钢筋水泥森林”——这便是当今中国大多数城市的真实写照。举个例子，中国县级城市的外环线无入口，无一例外地有一个圆形大转盘，盘中立着一个宏伟的不锈钢雕塑，多半为球形。全国城雕艺术委员会主任、中国雕塑院院长吴为山披露，近 30 年来，全国各地“托起的球”不下 100 万尊。许多城市沦为“无特点、无文化、无历史”的三无城市。

These three sentences, said by an American expert when he came over to China and was asked about his impression of Chinese cities, has been lingering over Wu Xinfeng. As a person in the occupation of urban design, he may be more suffered.

Wu Xinfeng is one of the professional insiders who is aware of the seriousness of this problem. I remember many years ago I interviewed him, he was worried, angry and enterprising when talking about this issue. Unfortunately, in stead of being effectively confined, the phenomenon of "singularity of thousands of cities" has become more and more intense. It is quipped that more than 200 cities of China are twins. The UK Guardian said:"China is a country contributed with one thousand similar cities". Is it ridicule, or the literal translation of the "singularity of thousands of cities"? Nobody has no reason to be pressed and sad living in such cities. "high buildings, high density, large squares ", "stiffness, no activeness, and the

"The rules hadn't been changed for sixty years , but the cities have transformed completely, which causes badly similarity of Chinese cities"

What time is the end of "singularity of all cities" An interview with Wu Xinfeng

A THOUSAND OF THE SAME CITY

- The world is shocked by the transformation of China in 30 years, especially the metropolises such as Beijing and Shanghai, of which the modernization can compete with New York.
- China has publicized and been proud of its five thousand years of historical culture. However, it is hardly to see the brilliance of the five thousand years of historical culture.
- When I walk through the metropolises of China, I even can't know which country I am in if without Chinese characters and pedestrians.

中国城市的“千城一面”广受诟病非今日始。英国皇家城市规划学会主席帕白森曾痛心疾首：“中国历史和文化的传统太珍贵了！不能允许它们被西方来的这些虚假的、肤浅的、标准化概念的洪水所淹没。我确信，你们遭到了这种威胁。”其实，这已经不是威胁，而是活生生的现实。上海世博会的总设计师，法国人马丁 · 罗班也说中国城市普遍缺乏特点。从他住的酒店望出去，城市风貌都差不多。著名建筑师库哈斯则这样调侃他的中国同行：“中国建筑师的数量是美国建筑师的十分之一，在五分之一的时间内设计了五倍数量的建筑。这就是说，中国建筑师的效率是美国同行的 2500 倍。”

在国内，城市的“千城一面”则更是受到不断质疑。作家冯骥才说：“近年来，我国的城市有非常迅猛的发展，但是也有不可挽回的损失，‘千城一面’失去了文化个性，失去了自己独有的文化灵魂，破坏了文化的多样性。”上海大学教授林少雄给博士生放幻灯片，讲述在国内不同城市的游历观感，放着放着就串了，切换了好多张，好像还没有走出一座城市，其实已经是好几座城市了。

matchbox-shape concrete forest"— this is the real image of most of the present Chinese city. For example, the outer races of county-level cities in China are without entrances, in stead, a large round platform with a stainless steel sculpture, usually in a ball shape which, revealed by Wu Weishan, director of national urban sculpture art committee and the president of sculpture institute of China, were no less than one million in China. Many cities turn down to be without specialty, culture and history.

In China, the "singularity of all cities" has drawn much attention. "In recent years, although the cities have developed rapidly in China, we got irretrievable loss. Singularity of thousands of cities' causes the loss of culture and the unique cultural soul, and the damage of cultural diversity", an author Feng Jicai said. "It seems to be in one city, but already run through many cities", displayed by Professor Lin Shaoxiong of Shanghai University to his doctoral students.

林少雄说："土耳其诗人希格梅有句名言，'人一生中有两样东西是永远不能忘却的，这就是母亲的面孔和城市的面貌。'可我们的城市只有同一副面孔。几乎百分之百的中国城市互相抄袭，三线看二线、二线看一线，只有楼房高矮、汽车多少等量的区别，没有质的区别，这就叫'同质化'。"许多历史、人文学者也表达了同样的观点。《中国青年报》甚至评论说："没有思想和个性的城市犹如没有灵魂的人。对于具有几千年城市史的中国来说。如果城市市区特点，将历史文化的底蕴破坏殆尽，'千城一面'都像一个模子里刻出来的，那么，不仅是中国建筑和中国城市的悲哀，也是对中国文明的糟蹋。"

悲哀，莫大的悲哀！

罗素先生说："须知参差多态，乃是幸福的本源。"亿万中国人生活在一模一样，毫无差别的城市里，生活的幸福感受，不说毁之殆尽，至少也大打折扣，这也难怪为什么官方统计的生活幸福指数，总和普罗大众的切身感受有着相当大的出入。

要知道，个人是无法脱离群体和环境生存的。人们之所以选择进入城市生活，就是要享受城市生活的品质。因为只有城市这种集聚形式，才集中了现代文明和先进科技的生活方式，为生活提供了优越的条件。但是，假如一个城市以建设为名，将历史的传承印记从城市上抹去，将优秀的文化遗产破坏殆尽，将值得继承和发扬的传统资源弃若敝屣的时候，城市无所凭借，变成无根之城；人心失去依托，灵魂漂浮半空，于是物欲横流、人情冷漠、道德沦丧。看上去毁掉的是城市，实质上泯灭的是世道人心。这也是为什么"千城一面"并不仅仅是一个城市发展路径的选择问题，也是一个很严峻的社会现实问题的重要原因。

可是，为什么"千城一面"饱受质疑与诟病并非今日始，却依然愈演愈烈？为什么这个长期为社会瞩目的问题，会迟迟没有办法解决，并逐步沦为茶余饭后的谈资？

2010 年末，新华网刊发了一篇题为《拒绝"千城一面"：中国西部给城市穿上"民族服装"》的报道，给解决城市的同质化现象带来了一丝新鲜气息。文章对民族地区运用来自生活的当地民族元素改造城市给予积极评价，其中特别提到了凯里，认为凯里的公共汽车站使用坡屋面和垂瓜，体育馆吸取了苗族的吊脚楼、侗族的鼓楼等特点，并使用了许多苗侗服饰上的元素，使得凯里的城市特点突出，给人印象深刻。同济大学规划六所所长，上海世博会规划顾问委员会成员，城市规划博士，《城市中国》执行副主编苏运升在文

Lin Shaoxiong said, "A turkey poet said 'In one's life,there are two things that are unforgettable for ever;one is his mother's face,the other is his town's image.' However, our cities are in the same image, almost all the Chinese cities copy each other. Small cities imitate medium cities, while medium cities imitate big city. The difference is not essentially, but the different level of constructions and different number of cars. This is so-called 'homogenization'". Many historical and human learners also express the same idea. The Youth Daily even presents comment that cities without thought and uniqueness are like people without soul. As a country with thousands of years of urban history, China will be sorrowful of its constructions and cities, and ravaged of its civilization, if the cities lost their features and were damaged totally on their historical culture.

Sorrow, badly sorrow!

Mr. Luosu says: "we should recognize that the diversity is the origin of happiness. " However, millions of Chinese are living in the cities which are in the same image. No wonder the happiness index by authority statistics is fair different from the real public experience.

We should know that an individual can not exist without the groups, so that people choose to live in the cities to enjoy the living quality of cities. Only this collective living style can provide the superior conditions with modern civilization and advanced technology. Nevertheless, if a city were constructed by erasing the historical inheritance, destroying the cultural heritage and deleting the valuable traditional resources, it will become a city without root, and which is the cause of shallow relationships and moral bankruptcy. In all, "singularity of thousands of cities" is not only an issue of the developing method of a city, but also a severe realistic social problem.

Why "singularity of thousands of cities" has become more and more intense since the query is not from today? Why such a hot spot hasn't been solved, but turned off to be conversation resources?

At the end of 2010, "Xinhua Net" issued a report named "Reject 'Singularity of Thousands of Cities': The West of China Put Ethnic Costume on the Cities" which introduced a fresh idea to figure out the problems of urban homogenization. The paper presented positive comments on the urban transformation using local ethnic elements of the ethnic regions. Kaili City was especially mentioned in the paper as a city with prominent features and will leave people deep impression

2011 年 5 月，一个阳光明媚的早晨，带着对中国城市"千城一面"的关注与困惑，我们造访了位于贵阳华坤大厦 16 楼的天海规划设计有限公司，在伍新凤的办公室进行了一场对话。

对公共空间可能提供的美、安慰和关怀，人们总是寄希望于规划。然而现实给人们的一个嘲讽是，大部分所谓的规划正在成为一种破坏。举目四望，"规划"所到之处，城市原有的魅力尽失。"规划正在变成城市魅力的杀手。"

中评价说："比如凯里的体育场，风雨廊和墙体的结合从场内和场外看都和水平线是平衡的，看起来很舒服。"

凯里的城市规划设计者，正是伍新凤。

多年来，伍新凤一直致力于解决"千城一面"的思路与模式的探索，并在九年前创办天海规划作为实践。天海成立 9 年来，坚持"民族现代化、现代民族化"的城市规划新理念，坚持城市发展的"可延续性"和挖掘城市的"核心价值"，在此基础上打造特色城市。

目前，经过九年探索，天海规划设计已经形成了一套规划设计方法和模式：在项目规划满足功能需求的同时更强调对规划地核心价值的深入挖掘，对规划地民族性、文化性和本土性的全力彰显并最终使之体现于项目中；整合策划、规划、建筑、创意、效果表现、动画等与规划设计密切相关的各个行业于一体，使得规划设计从资源评估、背景分析到战略定位、规划布局、建筑创意、效果表现一气呵成，最大限度地克服了传统规划设计模式中各个程序脱节的弊端。

多年来，伍新凤和天海规划以独特的规划设计理念、颠覆性的操作模式以及极富想象力的创意，承担和完成了 60 多项大中型规划设计项目，包括云南省丘北县县城整体设计及普者黑景区规划设计、贵州省凯里市城市整体形象设计、贵州省铜仁市整体设计、贵州省三都县县城整体设计、贵州省黔西县县城整体设计、贵州省威宁县县城整体设计、山东黄金主题公园规划设计、华藏世界多国佛教园规划设计、贵州省东风湖百里画廊休闲度假区规划设计、贵州省香纸沟温泉度假区规划设计、贵州省博物馆创意设计等。这些项目的实践，都是伍新凤为改变"千城一面"现状而作的可贵尝试。

2010 年 8 月，伍新凤根据多年思考和实践经历编著的《变城记——城市规划改造篇》、《筑魂记——原创建筑篇》、《梦游记——旅游规划篇》、《造景记——原创景观篇》四部专著由中国建筑工业出版社出版，这四部专著总结了伍新凤八年来从事规划设计行业的经验及独到见解，同时将天海的项目操作流程及规划设计理念向读者展示，试图对规划设计行业引入新的操作模式和表现手法，并为这行业带来一些新鲜观点供有关人士借鉴与参考。

as its bus stations were designed adopting pitched roof and gourd and the gymnasium was designed absorbing the features of the Diaojiao pavilion of Miao and the Drum pavilion of Dong and some ethnic elements of the costumes of Miao and Dong.

It is Wu Xinfeng who designed Kaili City.

Wu Xinfeng has devoted himself to the exploration of tackling "singularity of thousands of cities" and organized Tianhai Planning and Design to make practices for nine years. Tianhai has been focusing on the new concept of urban design "Modernization of Nationality, Nationalization of Modernity" and "sustainability" of city development and digging the "core value" of cities, so as to create special cities.

At present, Tianhai Planning and Design has concluded a certain method and mode of planning and design after nine years of exploration: In addition to meeting the needs of functions, the projects are stressed more on excavating the core value and highlighting the nationalities, culture and localities of the planning land, so as to express them in the design; The involved professions such as scheming, planning, architecture, design, 3D display and animation display are integrated with each other to complete a project from the resources assessment and background analysis to strategic orientation, planning layout, architectural design and image display, so as to avoid the processes apart.

During these years, Tianhai Planning and Design has undertaken and completed more than 60 large and medium projects with its unique planning and design ideas, subversive operation mode and innovative creation. The works, such as the project in Qiubei City of Yunnan, Kaili City, Tongren City, and Sandu City of Guizhou, and the Golden Theme Park of Shandong, the design of Sino Tibetan World of Qingzhen, the creative design of Guizhou Museum, planning and design of Xiangzhigou Resort of Guiyang, are the valuable practice that Wu Xinfeng engaged in to change the the current situation of "singularity of thousands of cities".

In August 2008, Wu Xinfeng completed his four monographs named "City Changing—Urban Design", "Spirit Shaping—Original Architecture", "Dreamlike Traveling—Tourism Planning", "Landscape Building—Original Landscape" which are issued by China Architecture and Industry Press, and conclude Wu Xinfeng's experience and unique ideas after eight years working on this profession.

A sunny morning in May, 2011, we visited Tianhai Planning in the 16th floor of Huakun Building and had a dialogue with Wu Xinfeng in his office.

People always expect that planning can make the public space beautiful, comfortable and agreeable. However, the reality is so ridiculous that most of the so called "planning" is turning out to be ruin. "Planning is becoming the assassin of urban charm"

对话 // Dialogue

Newtarget：中国城市“千城一面”的现象已被诟病多时，但到底是什么造成这种现象的原因众说纷纭，作为一个长期从事规划设计并在这个问题上有所建树的专家，您是怎么看待这个问题的?

伍新凤：造成中国城市“千城一面”的原因有三。

第一是城市规划设计行业的操作模式和游戏规则六十年来未变。六十年不变的城市规划行业，怎么能适应和跟上城市翻天覆地的变化?

第二是体制性的问题，行政主导学术，个别政府官员将城市建设作为升迁的政绩，忽略了这个专业的特殊性，不重视专家的意见，搞“政绩工程”，因此作出的规划往往不是出自科学和文化论断，而是出自个人判断和喜好。

第三是行业管理上的问题。由于行业管理上多年不变，思想僵化、缺乏创新，造成了对决策部门的误导。

正是这三个原因，造成了中国城市的“千城一面”。其中，游戏规则和操作模式六十年未变，是最核心和最根本的原因。

Newtarget：那么，应对这样的问题，有没有什么行之有效的解决办法?

伍新凤：一个城市规划项目，同时要面对六七个不同的学科和专业，现行体制下，涉及城市规划的各个领域之间互不相关，甚至相互拆台，这样的现实，怎么能做好城市规划?

改变“千城一面”的现实，关键在于发现潜入城市肌理的文脉。当一种资源成为一个城市的特色，不仅为世人所知，而且被挖掘出来，通过一定的手段展现，为这个城市打上深深的烙印，那么这种资源就构成了这个城市的核心价值。所以，在城市规划之前要首先对城市进行资源评估和定位，再围绕核心价值和定位进行功能布局，创造方便舒适内容丰富的新生活。最后，通过艺术创作手段，使城市建筑或景观的现实性升华为公众所能感受到的具象。

规划要在设计之前，城市规划师一定要具备人文素质和城市战略眼光，而不是一个只会规划设计的工匠。中国建筑大师梁思成，是梁启超先生的大公子，从小就攻读《史记》、《左传》，有很深的国学功底，念书时爱好美术和音乐，是校刊的美术编辑。在宾夕法尼亚大学学习建筑期间，自学中国古代历史。正是这样的历史积淀，才使得他的眼光超越了时代，对当时的北京城市建设提出许多真知灼见，当时没有被采纳。直到 1993 年，北京的城市规划方案，最终还是回到了梁思成的思路上。

Newtarget：In China, the phenomenon of "singularity of thousands of cities" has existed for a long time, and there are different versions of the cause of this phenomenon. As an expert working on planning and design and making contributions to this field, what do you think about this problem？

WU：There are three causes of "singularity of thousands of cities" in China. First of all, the operation mode and occupation rules never changes for sixty years. But the cities has experienced earth-shaking changes during the sixty years, so how can the old urban design concept adapt to the changing cities? Secondly, it is a systematic issue, the administration confines the academy. Some of the governors treat urban construction as his merit for his promotion, but ignore the specialty of this occupation and the experts' suggestions, so that the planning is generally not according to scientific and cultural conclusion, but the personal judgement and favor. Thirdly, it is the management.

These three points are exactly the cause of "singularity of thousands of cities", and the core and essential reason is the unchanged operational mode and rules for sixty years.

Newtarget：Then, do you have any methods to figure out these problems effectively？

WU：There are six or seven different objects and specialties involved when carry out a project of planning and design of a city, unfortunately, the involved departments don't cooperate with each other or even discriminate against each other due to the present system. Under this circumstance, how to make a good urban design.

The key to change the reality of "singularity of thousands of cities" is to discover the cultural context which permeates through the city. If a kind of resource is one of the specialties of a city, it will be known and displayed by certain methods to brand the city deeply, and it, therefore, will contribute to the core value of this city. Thus, the process of urban design is, firstly, assessing resources and making orientation, then working out the functional layout according to the core values and orientation to create convenient and comfortable new life, and lastly, transforming the urban architectures and landscape to be real constructions by art methods, so that the public can enjoy them.

Planning should be prior to design. Urban planners can't be a workman of planning and design, but should be of humanistic quality and urban strategy. The Chinese architectural master Liang Sicheng, the first son of Liang Qichao, accumulated wide Chinese ancient civilization as he had study "Historical Records" and "Zuo Zhuan" since his childhood, served as the art editor of the school journal due to his favor of art and music when he was in the school, and learned by himself the ancient Chinese history during his study in University of Pennsylvania. All the experiences endowed him with foresight, so

《中国新闻周刊》和新浪网曾经进行一项名为“你的城市理想”的网上调查，16877 人中，有 29.46% 的人们最希望自己所居住的城市建筑具有的品质是“外观很美，有艺术感的”。在“你希望居住的城市的历史年龄”一项中，38.38% 的人希望居住城市的历史年龄在 200 以上。一旦离开城市，多数人表示，除去经济因素，最受不了的一点是“生活缺少变化”。根据第十五次中国互联网报告发布的数据，69% 网民的年龄在 35 岁以下，可见，丰富多变、有历史内涵和审美情趣的城市“魅力”，已经成为当下年轻人城市理想的价值偏好。

我们渴望生活的富足与美好，但更渴望精神的充实与尊严，我们有权利希望更好的城市文明。从这一点上，我们对伍新凤与他的实践充满期待。

Newtarget：天海也是正在做着这方面的努力?

伍新凤：是的，我们认为城市创意设计是主题性、原创性和震撼性的综合体——以体现城市定位为核心和主题，抓住城市文脉特征，从人文背景中挖掘创意设计元素，从公众的视觉、听觉感受出发，以新颖独特为生命进行城市建筑和景观的创意设计。

同时，在进行创意设计时还要注重现代与民族融合。民族与时尚并不矛盾，民族并不代表落后，也不仅存在于历史中，只要合理挖掘，城市规划完全可以将民族的变成时尚的主体，带领人们穿越时空，去感受民族与现代的交融。

在操作层面，天海整合策划、规划、建筑、创意、效果表现、动画等与规划设计密切相关的行业于一体，使得规划设计从资源评估、背景分析到战略定位、规划布局、建筑创意、效果表现一气呵成，力图以全新理念与操作模式来改变当前行业的弊端。

Newtarget：在您眼里，一个美好的城市应该是什么样子?

伍新凤：马丁·罗班说："一百年前说到城市，我们讲得最多的是经济增长、工业发展，很少有人站在人的角度去评价城市。如今我们认识到，城市是为人而建，人才是城市里最重要的元素。衡量一个城市是否有魅力，最重要的标准是，当地人在当地文化历史背景下，对空间是否感觉舒适惬意。"

维也纳与音乐、鹿特丹与航海、哥本哈根与童话、威尼斯与水……一个成功的城市应该是有性格、有魅力的个性城市，是建基于自己文化传统、社会习惯和民风民俗基础上的集合。我到法国考察，法国议会大厦浮雕上的灰尘鸟粪很厚，据说一两百年没打扫，这是一种对自身文化的敬畏与自豪，这种态度很值得今天的城市规划者们学习。

一个美好的城市，不仅是 GDP 增长快速的城市，而应该是性格鲜明、文化积累深厚、历史传统清晰、生活方便舒适，并能有长时段可延续发展的城市。而绝不是高楼大厦密集，但却让人困惑、受到伤害、感到失望甚至绝望的城市。

that he presented many valuable suggestions to urban construction of Beijing City. However, his suggestions were not adopted until 1993 when the proposal of Beijing urban design came back to Liang Sicheng's ideas.

Newtarget：Is Tianhai making effort on this aspect?

WU：Yes, we believe that the creative urban design should be thematic, original and shocked. The proposal should be of the core and theme of describing urban orientation, and innovation and uniqueness of urban architectures and landscape with creative elements from cultural background and considering public visual and auditory sense.

At the same time, the integration of fashion and ethnicity should be focused on while make the design. Ethnicity is neither contradictory against fashion, nor the symbol of backwardness. Ethnicity can be absolutely transformed into fashionable mainstay if developed reasonably to lead people to experience the integration of fashion and ethnicity through time and space.

Tianhai unites the specialties of scheming, planning, architecture, design, 3D display and animation to complete a project entirely from the resources assessment, background analysis to strategic orientation, layout, architecture and animation display, with the hope to avoid the disadvantages of present profession with a brand-new operation concept and mode.

Newtarget：What is a nice city like in your mind?

WU："Martin Robain said:"We talked more about economic increase, industry development, but seldom gave comment from the views of human when it comes to cites hundred of years ago. Now we recognize that cities are constructed for human who is the most important elements of the cities. A city is charming or not mainly depending on if local people feel agreeable and comfortable "

Vienna and music, Rotterdam and voyage, Copenhagen and fairy tales, and Venice and water—a successful city should be of its character and charm, and the union on the basis of its own cultural tradition, social custom and folk styles. It is said that the sculptures of French Parliament building carry a lot of dust and bird droppingshas as they have not been cleaned for one or two hundred years, which I learned when I traveled in France. It is a kind of respect and pride of their own culture, and worth the planners learning.

A nice city should not be of rapid GDP increase only, but of striking characters, accumulated culture, clear historical tradition, convenient and comfortable environment, and sustainable development. It is absolutely not a nice city which is filled of confusion, damage, disappointment and desperation even with dense tall buildings.

The "China Newsweek" and Sina net have ever made a investigation named "Your Ideals of the City" on internet, among 16 877 informants, 29.46% of them hope their cities with "beautiful looks and art sense", and 38.38% of the informants hope that the cities they lived are with more than 200 years history when they are asked "how old do you hope your city is". Most of them said except the economic factor, what they can't stand is "life is lack of alternation" once they leave the cities. 69% netizen are no more than 35 years old according to the statistics issued by the Fifteenth China Internet Report, thus, the urban "charm" of diversity, historical connotation and aesthetic taste has become the modern youth's preference of value. We are eager for rich and beautiful life, but what we aspire more is substantial and dignity. We have the right to expect a better urban civilization, so that we are looking forward to Wu Xinfeng's practices.

自序 PREFACE

“中国的城市化与美国的高科技发展将是深刻影响 21 世纪人类发展的两大主题。”

诺贝尔经济学奖获得者——斯蒂格利茨

"China's urbanization and America's development of high-tech will be the two great themes to impact profoundly human development in 21st century".

The winner of the Nobel Prize on economy—Stiglitz

进入 21 世纪以来，中国经济持续快速发展，城市化进程不断加快，成为贯穿本世纪的主题。在此背景下，如何确保城市战略发展方向的准确性，如何实现城市的可持续发展，如何最大限度地构建城市核心竞争力成为城市可持续发展的核心。因此，在这个时代，规划设计行业肩负起此项重任。毫不夸张地说，当下是属于规划设计行业的，是规划设计行业展示自我的大好时机——在此过程中，责任与压力并存。

路漫漫其修远兮，吾将上下而求索。回首改革开放以来的 30 年，我国的城市规划虽说取得了很大进展，但是却陷入了“千城一面”的泥潭。城市是人类文明的主要组成部分，城市也是伴随人类文明与进步发展起来的。随着工商业的发展，城市还成为了传播文明的场所。遗憾的是，当下的中国城市，文化缺失、内涵不足、特色不明、协调不力，整体上略显浮躁，没有成为传播当地文明的有效载体——中国城市处于病态中，就像每个人从出生到成长都会不同程度地经历病痛一样，这是当今中国社会形态下所必经的阶段。缓解病情、治愈疾病，提高免疫力，规划设计师责无旁贷。

没有一味地批判，没有一味地指责，仅仅是作为一个规划人对我国规划设计现状进行反思。

反思一：规划设计行业体制

在计划经济体制下规划设计工作主要是政府行为，规划设计行业统一在事业编制下，规划设计任务由上级主管部门统一下发，行业竞争及压力有限。进入市场经济体制后，这一特殊行业现象在很大程度上延续下来，与此共同沿袭下来的还有几十年来一成不变的规划设计模式，缺乏与时俱进的创新思维，结果便是规划设计中规中矩，却缺乏特色，缺乏灵气，缺乏创意，更缺乏社会责任感。

反思二：规划设计模式

一个项目需要多个专业之间的配合才能完成。然而这些相关专业都是各自独立的，专业之间各自为政，本位主义较强。一个项目仅凭一个专业解决问题是远远不够的。现状是：从策划、规划到设计缺乏协调，严重脱节，导致规划逐步细化后与最初构想相去甚远。

By 21st century, with the rapid and continual development of economy in China, urbanization has speeded up and become the theme throughout this century. Under this circumstance, the core of sustainable development of cities has depended on how to make sure the right direction of strategic development of cities, accomplish sustainable development and form urban core competitions as strong as possible. And this is the responsibility that the planning and design profession should take. Thus, we dare say that this is the era of planning and design field and the best opportunity for the planning and design unities to express themselves with responsibility and stress.

The way stretches endless ahead, I shall search. What a long journey ahead!

Although urban planning in our country has made great progress during the thirty years since reform and opening up, it has been put in an awkward situation of "one thousand cities, one appearance". As the main components of human civilization, cities have developed along with the human civilization and progress. With the development of industry and business, cities have become the places spreading local civilization, too. Nevertheless, the current cities, with deficient culture and connotation, unobvious characteristics, and weak coordination abilities, are not successful to disseminate local civilization effectively. Chinese cities are in sick! This is the inevitable period for cities under the current social formation in China just like every people should go through lots of pain in their growth. Releasing sickness and enhancing immunity are the responsibilities of the planners and the planning and design profession.
As there is no absolute criticism, there is no absolute blame. Just as a planning person, I've made reassessment of the current status of the planning and design of our country.

First, system of planning and design
Planning and design is the governmental work during the planned economy that the field of planning and design is controlled by the governmental departments and the tasks of planning and design are arranged by the state, which caused little competition and press in this field. This phenomenon has been lasted after entering market economy, as well as the planning mode which has not changed for many decades and has been lack of innovative ideas. Thus, the planning works are regular and common, without characteristics and soul.

Second, mode of planning and design
Completing a project needs the cooperation of several separate specialties which, however, are leaded by different departments, and it's hard for them to reach a consensus on the project, as a result, the processes of scheme, planning and design are divorced seriously from each other and the initial image is different a lot from real effect, also working efficiency is not in control.

反思三：规划设计教育

规划设计师的职责是宏观地制定区域发展路线，定格区域发展基调，合理地进行区域功能布局。这需要宏观经济学、社会学、历史学等大量学科的知识积累，只有这样才能站在正确的战略高度上去计划区域未来的发展。在现今的教育制度下，更多的规划设计专业将大量精力放在了诸如行业规划标准，管网道路铺设，制图技术等微观战术层面上，对宏观战略的理解把握严重缺乏。除此之外，规划设计行业还需要艺术思维、创新思维的体现，但是现今“正统的”规划设计行业对审美、造型艺术等方面的教育缺乏甚至是空白。很难想象，一个缺乏大局观、缺乏审美眼光的规划设计师如何能使城市摆脱“千城一面”的窘境。

八年前，带着对行业的反思，怀揣着自己的信念，我创办了天海规划设计有限公司。从公司创立之初，我就不断吸取内外部经验教训，进行剖析比对，敢冒“天下之大不韪”，推行了一系列的改革措施：

规划理念方面。在项目规划时最大限度地满足功能需求的同时更强调对规划的核心价值的深入挖掘，对规划的民族性、文化性和本土性的全力彰显并最终使之体现于项目中。

规划模式方面。整合策划、规划、建筑、创意、效果表现、动画等与规划设计密切相关的各个行业于一体，使得规划设计从资源评估、背景分析到战略定位、规划布局、建筑创意、效果表现一气呵成，形成流水线规划模式，最大限度地克服了传统规划设计模式中各个程序脱节的弊端。

八年来，经过 60 多个项目的成功实践，在市场上赢得许多开发商和政府部门的认同和赞誉。事实证明我的改革与创新是有价值的。现在，我把近十年来的改革创新的规划设计成果分为城市设计、建筑设计、景观设计以及旅游规划四部分集结成册分别命名为“变城记”、“筑魂记”、“梦游记”、“造景记”，用我的挚热和真诚与大家共同交流探讨，同时更希望能与同行们携手为推动全国规划设计行业的改革与发展，贡献绵薄之力。

Third, education of planning and design

Urban planners, who are requested to be of large accumulation of knowledge in economics, sociology, and history and so on, take the responsibilities of working out the developing proposal of urban regions macroscopically, determining basic tone of cities and reasonably arranging urban functional districts. Only the ones who are competence enough can design the future development of the cities on a correct strategic view. Under the current educational system, many urban planners only know some about urban functional layout and distribution, various planning norms and setting of pipes and roads, but they are badly lack of macroscopic innovative consciousness. Besides, artistic and innovative thought are needed to be demonstrated in the planning and design works. But the education of aesthetic and art etc. are lost in the existing "orthodox" planning and design profession. It's hard to image how can a planner without overall sense and macro view help cities escape from the awkward situation of "one thousand cities, one appearance".

Eight years ago, I found Tianhai Planning and Design Company with rethinking about this sector and my strong conviction. We have totally discarded traditional operation mode, which was regarded "fighting a rearguard action against the mass of public opinion" at that time.

In the aspect of planning idea, besides meeting the needs of functions, our projects are stressed on excavating deeply the core value and highlighting the nationalities, culture and localities of the planned place so as to show them in the project eventually.

In the aspect of planning mode, the closely relative professions of scheme, planning, architecture, creativity, effect display, and animation etc. have been integrated in a unity, so that the planning and design work can be accomplished smoothly and completely from assessing resources and background analysis to strategic orientation, planning and layout, architectural innovation and displaying effects, so as to avoid maximally the disconnection of each step in the traditional planning and design mode.

More than 60 successful items within 8 years has won accept of many investors and relevant governments and proved that our reformation and innovation are significant. Now, we divide our works based on reform and innovation into four parts: urban design, architectural design, landscape design and tourism planning and named respectively "City Changing", "Spirit Shaping", "Dreamlike Traveling" and "Landscape Building". We hope to communicate with you sincerely and passionately and also make great efforts and contribution along with the people in the same field to the reformation and development of the planning and design field in our country.

导言 Introduction

规划是对未来事务的合理组织与安排。旅游规划是“在调查研究与评价的基础上寻求旅游业对人类福利及环境质量的最优贡献的过程”。在世界范围内，1959 年的夏威夷规划被看做是现代旅游规划的先驱。我国的旅游规划实践可追溯到改革开放之初，国家旅游局牵头组织编写了国家级旅游业发展规划——《关于 1980 年至 1985 年旅游事业发展规划(草案)》，这是我国第一部正式的旅游规划文本。此后，各地旅游规划实践逐渐兴起。进入 21 世纪，我国旅游规划遍地开花，全国几乎所有省市都做过旅游发展规划和重点景区总体规划。目前，地区、县乡级旅游规划、专项旅游规划以及对原有规划的修编成为规划实践的主要类型。经过 20 多年的发展，虽然我国旅游规划的参与主体逐渐多元、规划类型日益多样、管理不断规范，但是问题也随之产生，其中最显著的问题便是旅游规划总摆脱不了墙上挂挂，最终不了了之的命运。那么是什么原因导致了这些问题的存在呢?

Planning is to prepare and arrange for the future affairs, and tourism planning is "the process of seeking the best way tourism brings the most benefit to human beings and environment on the basis of investigating and evaluating. The planning of Hawaii in 1959 was treated as the pioneer of the modern tourism planning throughout the world. China's tourism planning can be traced back to the beginning of the Open and Reform. The National Tourism Bureau leaded to draw up the developing planning of national tourism named "Planning of Tourism Development from 1980 to 1985 (draft)", this is China's first official text of tourism planning. From that time on, the tourism planning in every area has become more and more popular. By the 21 century, tourism planning has been prevailing in our country that it happened in almost all the cities. At present, the tourism planning in prefectures and counties, the special tourism planning, and modifying the original planning are the main types of planning practice. Although with more and more multiplied involvement subjects of China's tourism planning, diverse planning types and improved management after more than 20 years development, many problems appeared, too. Among them, the most protruding problem is that tourism planning has been always said without done. Then what cause these problems?

首先，规划之间的脱节致使最终结果与最初构想千差万别。

各类旅游规划各行其是，缺乏必要的沟通与协作，总体规划制定的宏观战略无法在控制性详细规划、修建性详细规划中体现出来。目前我国的旅游规划确实得到了大力的发展，地区、县乡级旅游发展规划、区域总体规划、控制性详细规划、修建性详细规划以及各类规划的修编层出不穷。但是由于各个规划出自不同规划编制单位之手，其对区域旅游发展战略、规划布局、建筑理念等方面认识的不同导致规划逐步细化后与最初构想相去甚远。

另外，由于我国体制壁垒的存在，各级规划的委托单位不尽相同，单位之间各自为政的现象沿袭到规划中去，导致了规划编制单位在规划编制中缺乏必要的沟通，使得各级规划之间脱节，无法达到当初预想的结果。

其次，旅游规划城市化现象严重，重技术，轻人文。

科学技术的发展使得人们越来越依赖于计算机各类软件的使用，通过软件可以绘制出精美的图片，计算出精确的尺度，但是缺乏对规划区背景条件的认知，对地方资源的深入挖掘，对地区人文内涵的系统分析，导致了旅游规划局限于条条框框的束缚之下而缺乏灵性，让人觉得很规范、具有普遍性，但缺乏独特性。

第三，旅游规划“工具化”现象明显。

旅游规划成为“业绩工具”。旅游业经过数年发展，逐步显示出其综合效益，上到省、自治区、直辖市，下到地方政府都将旅游业作为区域、地区的支柱产业、特色产业、国民经济新的增长点来加以扶持，旅游规划作为旅游业发展的依据，成为各级领导首先涉足的领域。在此过程中领导为了树政绩往往对旅游规划极为重视，发表了大量意见，这种绝对的话语权导致旅游规划编制成为领导意图的单方面演绎。

旅游规划成为“要钱工具”。旅游业得到扶持，在很大程度上体现在国家的拨款力度上。拨款要有依据，这个时候旅游规划便成为了获取拨款的“工具”。这种情况下产生的旅游规划仅仅处在“被利用”的范畴，无法起到实质性的作用，一旦拨款下发，规划便会无限期的搁置。

Firstly, the gap of every process of planning led to the final results varies from the initial idea.

The necessary communication and collaboration is in need when the planners make the various types of tourism planning, as a result, the macro-strategy of the overall planning can't play a role in the control detailed planning and constructive detailed planning. China's tourism planning did improved a lot so far that variety of planning emerged, such as prefectural tourism development planning, regional over planning, regulatory detailed planning, construction detailed planning etc. which, however, are different from each other on the aspects of strategy of regional tourism development, planning and layout, constructional concept and so on due to the different compiling units, and cause great difference between segmental planning and the initial conceived.

In addition, because of the systematic barriers of our country and the lack of communication among different planning units, a planning project can't achieve the original conceived planning undertaken by different planning units.

Secondly, the phenomenon of urbanization of the tourism planning is severe, and people pay more attention to technology, but ignore culture.

People depend more and more on computers and the variety of softwares as the scientific technology developed. With software, people can draw beautiful pictures and calculate accurate scales. Nevertheless, it has the deficiency of the cognition of the background conditions of the planning area, the deep excavation of local resources, and the systematic analysis of regional cultural connotation, which cause the tourism planning lack originality confined by the rules that they are regular and common with any uniqueness.

Thirdly, the "Tool" in Tourism Planning.

Tourism planning becomes" the tool of making merit". The synthetic benefit of tourism has displayed after several years of development. The government at all levels take tourism as the regional pillar industry, characteristic industry and the new increasing point of national economy to develop. As the basis for tourism growth, tourism planning becomes the main field that leaders at all levels enter. In this process, leaders put emphases on the tourism planning and give a lot of suggestion in order to make some merit.

Tourism becomes "the tool of asking for money". The country's efforts on fund for tourism mostly show the country's decision to develop tourism. It needs reasons to asking for money. In this case, tourism planning plays the "role" of getting the funding. However, tourism planning made in this circumstance is only a "tool" without any functions it should have. Once the purpose of getting funding is achieved, the planning will be delayed indefinitely.

旅游规划成为“攀比工具”。规划编制的过程中背离规划地的现状及发展条件，一味地进行体量规模上的攀比，过分地强调“全国第一”、“世界第一”的概念，致使规划缺乏可操作性。

天海规划数年来一直从事旅游规划编制的相关工作，并且积累了一些经验，我们认为，要实现旅游规划的健康发展，应该做到以下几方面的协调：

旅游规划应该是各级规划之间的协调

从旅游策划到旅游总体规划再到旅游区详细规划一步到位，一体化完成是天海规划经过多年实践证实了的行之有效的旅游规划编制方法，只有这样才能保持各个部门之间的沟通，才能确保对整体宏观战略的明确理解和有效贯彻，才能保证旅游区在最后的实施阶段与最初构想保持一致，才能保障理念构思的可操作性，可落地性。

当然，以上的方法并不是每一家旅游规划公司都能实现的，在这种情况下就需要以项目委托方为桥梁，加强各级规划编制单位之间的沟通协作，加强各级部门之间的协作意识，力求实现旅游规划编制的有效实用。

旅游规划应该是宏观战略与微观战术的协调

如果没有旅游区战略发展方向的指导，各类技术的运用只能是无头苍蝇到处乱窜，找不到方向。因此在旅游规划编制过程中首先应该根据对规划区宏观背景的切实分析，对地方资源的深入剖析，对客源市场的调查分析制定宏观的战略发展方向，完成规划区的整体规划构想，继而运用微观技术层层推进，围绕战略方向与整体构想将规划细化到各个技术层面上来，从而实现规划的延续性、完整性。

这里需要注意的是旅游本身就是一个涉及多个部门的综合性行业，因此旅游规划也具有了这种辐射性，旅游规划将会涉及多个部门，包括道路交通、城市规划、园林、历史、建筑等，因此在这个过程中也需要全面协调各个专业之间的关系。

旅游规划应该是本源与创新的协调

“本源”简言之就是规划区的资源现状，创新是指人为加工。在旅游规划编制的过程中，二者是相辅相成，缺一不可的。本源是规划区未经雕琢的自然痕迹、人文色彩，反映了规划区自古以来的地质变迁、社会生活、意识形态等方方面面，是规划区特有的，无法复制的，因此是规划区的核心吸引力，而这一核心吸引力需要通过创新手段表现出来。本源就像是等待加工的原材料，需要一定的工艺加工才具有使用价值，成为商品。具备开发价值的商品要通过怎样的加工才能别具一格，形成自身特色，怎样把当今潮流甚至未来潮流发展趋势与规划区资源本身结合起来，这些都是需要具有锐意的创新思维来完成的。

Tourism becomes "the tool of comparison". People always compare the scale and content with others when they make the planning, but without considering the current situation and developing conditions of the target area. They emphasis too much on the connotation of "the best of the country", and "the best of the world", which makes the planning beyond operation.

Tianhai Planning has engaged with the tourism planning, and accumulated some experience. We believe that the achievement of the healthy development should be with the following aspects of coordination:

Tourism planning should be the coordination of planning department at all levels.

Tianhai Planning has worked out a way of compiling tourism planning after many years of practice, which finish the primary tourism planning, the overall planning and the detailed planning of the tourism area on a assembly line process. Only in this way that the necessary communication can be kept between the departments, the overall macro-strategy can be clearly understood and effectively implemented, the appearance of tourism area can be ideal as the original image, and the design idea can be operational and realizable.

Of course, not all the tourism planning company is able to adopt the above way. It is indispensable to strengthen the consciousness of communication and cooperation of the department at all levels with the bridge of the client, so as to make the tourism planning effective and practical.

Tourism planning should be the coordination of grand and micro strategy.

The application of the various technologies won't have the developing direction without the guidance of the strategy of tourism planning. Therefore, the process of making tourism planning is: Firstly, we should finish the overall idea of planning for the planned area and macroscopical strategic direction based on the precise analysis of the macro background, the deep excavation of the local resource, and the investigation and analysis of the target market; secondly, we should segment the planning on the basis of the strategic direction and overall idea to every technical perspective by the means of micro-technique so as to get a sustainable and completed planning.

We need to note that tourism is a synthetic field containing many departments, so is tourism planning. So to make tourism planning we should consider the other departments such as road traffic, urban planning, garden, history, architecture and so on, which requires us to comprehensively compromise the relations of each specific.

Tourism planning should be the coordination of origin and innovation.

In a short word, origin is the current situation of the planned area, and innovation is the man-made process. They depend on each other in the process of compiling tourism planning. Origin, the natural traces and cultures without polished in planned area, reflects the ancient geological changes, social life, and ideology and so on. As origin is special and cannot be copied in the planned area, it is the core attraction which, however, needs innovative means to display. Only with certain technological processing that origin, like raw material waiting for processing, could be with use value and become commercial products. How the commercial products with developing value can be unique by man-made process, and how to combine the current trend even future trend with the resources in planned area? These are which need innovative creative idea to complete.

旅游规划应该是专家意见与本土意见的协调

这里所说的专家意见与本土意见的协调主要表现在两个方面，一个方面是规划编制单位与委托方意见的协调，委托方因为站在“顾客”的立场上，所以在规划编制的过程中希望自己的意见能够被规划方接受，这本身无可厚非，但是因为无法拒绝委托方的要求而背离自身专业的角度，导致委托方一些不成熟的意见被采纳的话就会影响到整个规划的质量。因此在规划编制的过程中，委托方与编制单位应该随时保持沟通渠道的畅通，为规划高质量地顺利完成打下基础。另一个方面是专家意见与区域利益方意见的协调，也就是说规划的编制方不是规划单位与地方政府自说自话的行为，而是要考虑规划编制地区所涉及的各个利益相关者，特别是社区民众的立场及意见，只有实现了各方利益的协调才能保证规划的科学性、有效性。

天海规划在认识到以上问题后，将在此思想指导下编制的旅游规划，分为不同专题向大家展示出来，与大家共同体会天海规划多年来在旅游规划道路上走过的历程。

Tourism planning should be the coordination of opinions of experts and local people

We will elaborate the coordination of the advice of experts and customers in two aspects. First, it is the coordination of the planning corporation and the client. As the customer, it's reasonable for him to require the company to accept his suggestion in the planning. But if the planning company accept unwilling the suggestions of the customers without considering the specific factors, the quality of the whole proposal will be decreased, as some of the suggestions are not proper. Therefore, the corporations and customers should keep their communication effective and smooth to lay the foundation for the high quality planning. Second, it is the coordination of the advice of the experts and local public. In the other words, it is not the planning company and the local government decides all aspects of the planning, but they should consider the relevant interest, especially the situation and suggestion of the residential public. Only achieve the interest coordination in every aspect that we can ensure effectiveness and scientificalness of the planning.

After realizing the above problems, Tianhai Planning would like to display our planning works based on this thought.

梦游记
Dreamlike Traveling

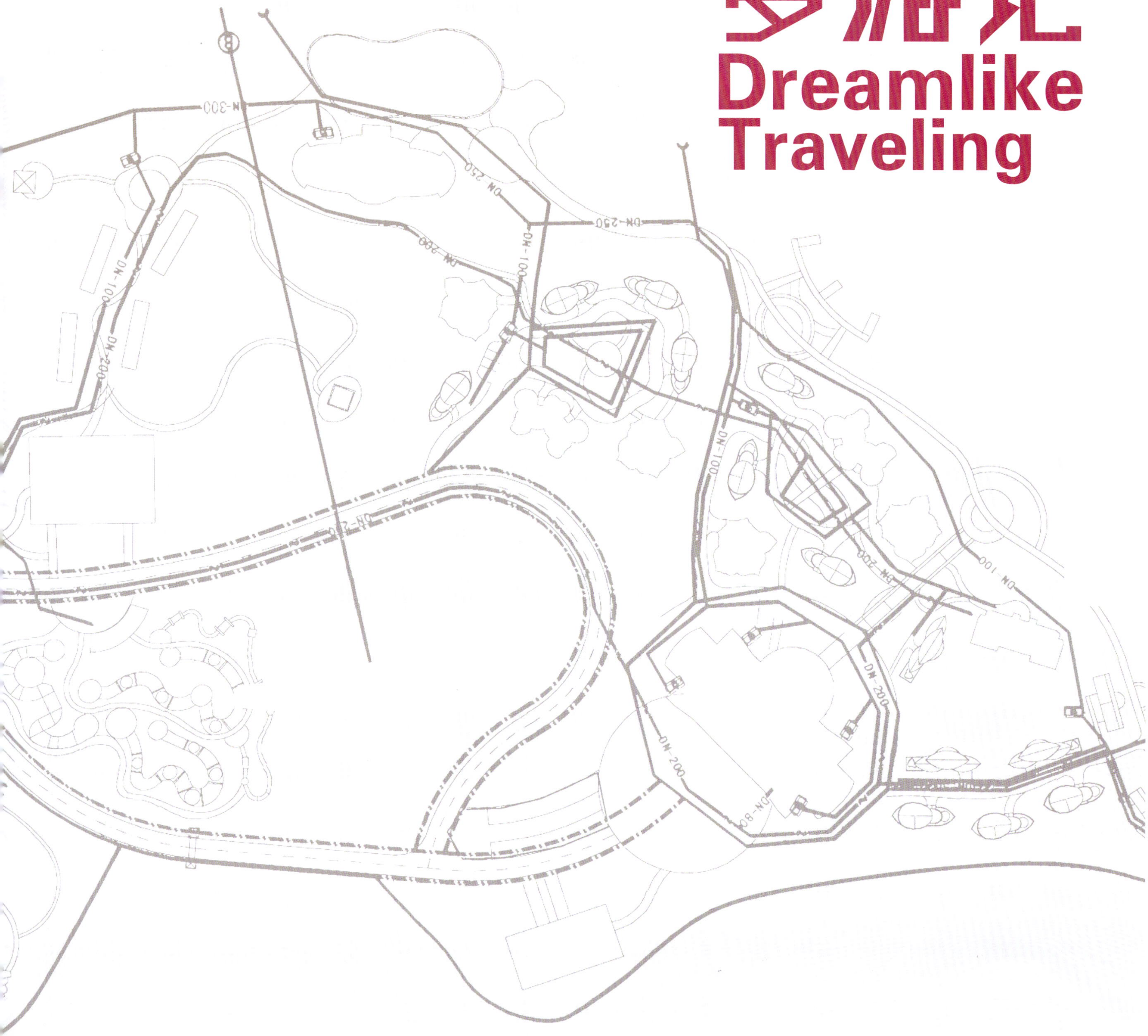

旅游规划篇（第二版）

梦游记

Dreamlike Traveling

Contents 目录

Theme

Park

题公园 ↘

中国旅游业发展30余年的经验证实发展旅游业能够带来经济效益、社会效益、生态效益，因此发展旅游业成为国内新的经济增长点。几乎全国所有的省、直辖市、自治区、地区市县均将发展旅游业作为促进经济发展的主要手段。随着旅游业的快速发展，旅游需求不断丰富，旅游手段不断更新，在这样的大背景下，主题公园应运而生。

Tourism has been proved to bring benefit of economy, society and ecology after 30 years of development, so it is a new economic growth source to develop tourism. Almost all the provinces or counties in the country have taken tourism as the main method to accelerate economic development. With the background of rapid tourism development, various tourism requirement, and up-dated tourism means, the theme park has come out.

1 Outline of Theme Park 了解主题公园

自 1955 年在美国诞生了世界上第一座主题公园——迪斯尼乐园以来，经过几十年的发展，主题公园这种新的旅游目的地形式，已经逐步为许多国家所接受。20 世纪 70 年代以后，主题公园发展迅速，逐渐在世界上一些国家，如美国、法国、日本，形成产业规模。

Theme park, as a new kind of tourism spot, has be accepted by people gradually after several decades since the first theme park has been built in the USA in 1955. After 1970s, theme park has developed rapidly, and gradually formed industrial scale in some countries, such as the USA, France, Japan, and so on.

1.1 主题公园开发的意义 The Significance of Developing Theme Park

主题公园是围绕一个或多个特定的主题，由模拟景观和园林环境为载体的人造休闲娱乐活动空间，是一种多属性的旅游产业。它在类别上属于现代公园的一种，除具备公园的一般要素外，主题公园的园区环境还突出地表现一个或多个主题意图。

Theme park is a man-made space for recreational activity, which is based on one or two specific theme with simulated landscape and garden environment. It is a multi-attribute tourism industry, and belongs to the modern park in the category. Besides the general factors of park, theme park is also with one or more special theme purpose in the garden environment.

主题公园被世界旅游组织称为当前和未来国际旅游发展的三大趋势之一。经过近十几年的发展，我国主题公园的发展已形成相当规模，涌现出一批具有中国特色和符合市场需求的大型主题公园。然而大多主题公园却在利益的追求中迷失了，忽略了历史背景及文化传统内涵的承载，规划设计和经营管理也多粗糙和缺陷，渐渐失去其应有的魅力。

Theme park is named one of the three trends of tourism development at present or in future by the World Tourism Organization. With more than ten years of development, theme park in China has expanded certain scales. And they are with Chinese characteristics and meet the market requirement. Nevertheless, most of the theme parks have lost the charm they should have in pursuing the benefit, overlooking the historic background and traditional culture connotation and ignoring planning and management.

然而近年来随着中国经济的持续发展，迪斯尼乐园、环球影城等世界级娱乐公司纷纷来华投资，说明我国已经成为世界上最大的主题公园潜在市场，主题公园在中国市场必定大有作为。

However, with the sustained economic development in the recent years, China has been proved to be the largest market of theme park in the world as many world recreational companies, such as Disneyland and Universal Studio, have come and invested in China. Theme park must take a good position in market share.

1.1.1 主题公园是现代旅游业发展的必然产物

改革开放30余年来，中国发生了翻天覆地的变化，这种变化适用于中国的各个产业，旅游产业当然也不例外，旅游学界一致公认中国现代旅游业的大发展从改革开放开始。改革开放以来，旅游，从可望而不可即的奢侈品逐渐成为中国居民家庭生活的一部分。随着旅游者外出旅游机会的增多，旅游者的见识越来越广，对旅游产品的要求也越来越高，一方面被动的参观已经无法满足旅游者的需求，他们希望自己能够成为主角，能够亲身参与、体验旅游产品；另一方面旅游业的发展也需要推陈出新，需要适应时代的旅游产品丰盈旅游业的发展。在这两方面原因的共同作用下，主题公园这一参与体验性强，时代特征明显，并且关联带动作用显著的旅游产品蔚然兴起。

1.1.2 主题公园是实现产业整合的有力武器

主题公园的建设可以带动一批与此相关的产业的全面发展，从而形成一个以主题公园为核心竞争力，相关产业共同发展的多赢局面。迪斯尼品牌涉及的产业既包括书刊、印刷、收藏品、纪念品、餐饮业、住宿业、房地产业、商业，还包括迪斯尼创意、构造、功能咨询和输出管理等不同的产业体系[1]。中国主题公园的翘楚深圳华侨城不仅局限于开发主题公园，还在中国的旅游行业中首创开发旅游城。这个旅游城里不仅有主题公园、主题酒店、主题地产、主题商业、主题创业园区，还有居民和文化教育设施。事实上对于主题公园来说，门票收入应该只是旅游者对当地经济所作贡献的较小部分，旅游者对其相关行业，如交通、住宿、饮食、商业、金融服务等带来的收入才占大部分。对于一个城市来说，主题公园在带动相关第三产业的发展方面，其作用最为明显，即通过主题公园带来的大量人流，带动当地的物流、信息流、资金流和商品流的运转，使区域经济的发展更具活力、更具竞争力、更具生命力[2]。

1.1.3 主题公园是推动产业结构调整的有效手段

目前我国的现状是

(1) 农村产生大量的剩余劳动力，急需要城镇化建设解决其就业问题；

(2) 我国是一个资源短缺的国家，以石油储备来说如果以人均消耗4t的美国标准计算的话，我国已探明的石油可采储备量只够消耗一年零三个月；

(3) 我国在新中国成立初期大规模的“工业基地建设”耗费了过多的资源，致使现在许多以矿产为支撑的城市陷入了资源枯竭的窘境。在认识到这些问题后，在新一轮的建设中需要调整产业结构，找到节约资源的、新的经济增长支柱。

以旅游业为龙头的第三产业日益成为国民经济发展的主导部门。第三产业的增加值和就业人数占国民生产总值和全部劳动力的比重各国都呈上升趋势。20世纪60年代以后，发达国家的第三产业发展更为迅速，所占比重都超过了60%。由此可见旅游业能够有效地带动产业结构的调整。而作为现代旅游业发展的新兴产物，主题公园在整合利用资源、协调各产业部门、促进就业等方面都能够最大限度地发挥作用。

1.1.1 Theme Park is the Inevitable Product of Modern Tourism Development

China has changed a lot since the reform and opening up, this change is adapted to all the industries of China, including tourism. The tourism scholars unanimously recognized the great development of China's modern tourism industry from the beginning of reform and opening up. Tourism, used to be regarded as the luxury, has gradually become the part of people's life in China. As the tourists have more opportunities to travel, and more and more experience and knowledge, they will have more requirements to the tourism product. For one thing, it can not meet the tourists requirement to only get sightseeing passively, while they hope to be the actors to experience and participate tourism products, and adapted to the current tourism products to accelerate tourism development. Thus, theme park, with experienced property. obvious time features and outstanding influence, has grown rapidly due to the above aspects.

1.1.2 Theme Park is the Powerful Means to Achieve Industrial Integration

The construction of theme park can accelerate the development of relevant industries, so as to form a system of mutual development of relevant industries with the theme park as the core competitiveness. The different industries attached to Disneyland Brand including not only books, printing, collections, souvenirs, food, accommodation, real estate, and business, but also innovation, structure, functional consulting, and consignment management. Chinatown, the outstanding theme park in China, has broken the regular way of just developing theme park, but also pioneered tourism city in tourism field. There are not only theme park, theme hotel, theme estate, theme business, theme entrepreneur park, but also residents and cultural and educational facilities. In fact, the income of admission tickets is just a small part of the tourist's contribution to the local economy, but the other incomes from the relevant industries such as traffic, accommodation, food, business, and financial service brought by the tourists take a big percentage.

1.1.3 Theme Park is the Effective Means to Accelerate Adjustment of Industrial Structure

The current situation of our country is

(1) There is plenty of rural surplus labor force, the employment problem needed to be tackled by urbanization construction;

(2) China is short of resource, take petrol for example, if the consumption is 4tons per person in terms of U.S. standards, the petrol storage that had been explored can last just one year and three months;

(3) Too many resources have been wasted in the "Construction of Industrial Bases" at the beginning of foundation of our country, which caused resource exhaustion of many mineral resource-based cities. Realized these problems, we need to adjust industrial structure and find out a resource saving and new pillar of economic growth.

Theme park can play an important role in the regional adjustment of industrial structure, which has an outstanding example of the construction of Hong Kong Disneyland Park. The pillar industry of Hong Kong is entrepot trade, which, however, will be impacted as the relationship of Chinese mainland and Taiwan developing and "Three Direct Links" (Shorthand for direct links in post, transportation and trade between the Chinese mainland and Taiwan) coming true, so the adjustment of industrial structure has become the key point of the sustainable development of economy of Hong Kong. Under this circumstance, Hong Kong government actively invested to construct a gigantic theme park, which was an excellent measure to tackle the probable unemployment problem, reintegrate the resources of Hong Kong, and improve the citizens' benefit.

[1] 李永文．论主题公园的区域经济影响、建设与发展[J]. 经济地理 .2005年9月．
[2] 张整魁．“锦绣中华现象”与中国主题公园发展[A]. 华侨城集团公司 .21世纪中国主题公园发展论坛[C]. 北京：中国旅游出版社 ,2003.39-41.

1.1.4 主题公园是实现社会全面发展的有力推手

首先，从经济效益来说，主题公园的开发不仅其自身能实现经济收益，还能带动相关产业的获益，同时还能为地方政府带来投资与税收。

其次，从社会效益来说，主题公园这样的劳动密集型产业能够为当地带来大量的工作岗位。以香港迪士尼乐园为例，在政府进行填海和基建工程时，就为劳工市场带来近万个职位。而第一期启用时，又可提供 18400 个职位。其后二十年，预期可增加至 35800 个新职位。而在建造主题公园期间，则另外会创造约 6000 个职位。并且乐园的所有雇员几乎都会是香港人。

再次，主题公园往往是用高品位文化、美学理念在特定地域上塑造艺术化、浓缩化的园林生态、水生生态或电影场景等综合景观[1]。其通过城市美景的创造，丰富城市文化，进而提高整个城市和城区的文化品位[2]，对城市文化留存、展示、提升等方面具有很大的积极影响。

1.1.4 Theme Park is the Advantageous Conditions to Realize the Comprehensive Development of Society

First of all, developing theme park can not only get incomes by itself, but also bring benefit to the relevant industries, and investment and tax revenue to the local government.

Secondly, as a labor-intensive industry, theme park can serve many positions for the local people. For instance, when constructed Disney Land (Hong Kong), almost ten thousand of positions were needed to fill sea and construct the foundation, and 18,400 positions provided in the run of the first phase. Also, 35.8 thousand new jobs will be expected in the next two decades, while additional 6,000 workers will be needed during constructing theme park. Almost all employees of the park are local people of Hong Kong.

Thirdly, theme park, consisting of the comprehensive landscape of artistic and concentrated garden ecology, aquatic ecology or film scene, enriches urban culture with aesthetic by creating beautiful scenery in city and then promotes cultural taste of the whole city, which has great positive impact to remaining, displaying and enhancing urban culture.

1.2 主题公园的特征 Features of Theme Park

1.2.1 创新性

创新是主题公园的精髓，是始终贯穿主题公园的关键词。从主题公园的策划到主题公园的经营管理都离不开“创新”二字。

建设初期开创主题的创新性。主题的创新是主题公园独特性与差异性的具体体现，是影响旅游者旅游决策的重要因素，但是这种创新也必须是建立在公众正常的审美范围、旅游业的发展态势之内的。

运营期间保持公园的新鲜度。主题确定的创新性只是主题公园成功的第一步，保持主题公园在游客心目中长期的新鲜度才是主题公园真正成功的途径。在主题公园运营管理中需要通过淘汰旧产品、改善旧产品、优化旧产品、开发新产品来达到创新从而保持公园长盛不衰的竞争力。

1.2.2 文化性

原香港中旅集团总经理、深圳华侨城建设指挥部主任马志民先生曾经在世界之窗开业 5 周年的大会上说“世界之窗是个旅游项目，但它又不仅仅是个旅游项目，它还是个文化项目，世界之窗最大的意义不在于它赚了多少钱，而在于它向人们介绍了世界”，明确阐述了主题公园的文化功能。 由此可见文化性是主题公园的价值和灵魂所在。主题公园通过“主题”来解释文化、传递文化，而且旅游者把主题表现的内容作为文化来理解和接受。因此，文化性是影响主题公园存在与引导其持续发展的主要标准之一。

主题公园的文化品位一是体现在整体的主题创意上，主题创意本身要具有浓郁的文化气息。二是体现在如何表现这个主题上，也就是园区建设的品质上，要通过丰富、生动、全面的造景手段将文化这一抽象概念具象、直观地向旅游者呈现出来。

1.2.1 Innovation

Innovation is the essence of theme park, and the key word throughout theme park-from the planning of theme park to the management of theme park.

Creating the theme innovation at the beginning of construction: theme innovation embodies specifically the uniqueness and difference of the theme park, is the important factor to affect the tourists' tourism decision, of course, this innovation must be based on the normal aesthetic that the public can accept and the developing situation of tourism.

The novelty of theme determination is only the first step to realize success, while keeping the park fresh in tourists' mind is the way to achieve true success. It is necessary to eliminate, improve and optimize the old products and create new one in the operation and management of theme park, so as to keep the sustainable competitiveness of the theme park.

1.2.2 Culture

On the conference of the fifth anniversary of "Windows of the World", Mr. Ma Zhimin, the general manager of China National Travel Service (HK) Group, the director of the construction headquarters of Chinatown (Shenzhen), has said: "'Windows of the World' is a tourism item, but it is not only a tourism item, but also a cultural item. The legal significance of 'Windows of the World' is not on how much money it can make, but the introduction of the world it gives", which stated clearly the cultural function of theme park. Hence we can see that culture is the value and soul of theme park. Theme park is to explain and transmit culture through "theme", and tourists understand and accept the content that the "theme" shows as culture. Therefore, culture is one of the main standards of affecting the existence and orienting the development of theme park.

The cultural taste of theme park reflects on the two aspects: 1. the overall thematic innovation, which should be with thick cultural atmosphere; 2. the method of embodying this theme, in other words, it is the constructed quality of the park, which illustrates culture, this abstract concept, to tourists specifically, directly by the means of constructing the landscape richly, vividly, comprehensively.

[1] 李永文．论主题公园的区域经济影响、建设与发展 [J]. 经济地理 .2005 年 9 月 .
[2] 池雄标．从城市旅游系统的角度看主题公园的发展 [A]. 华侨城集团公司 .21 世纪中国主题公园发展论坛 [C]. 北京：中国旅游出版社，2003.123-125.

例如苏州的福禄贝尔游乐园仅仅以科技为唯一主题，开业之初就遭游客冷落，这其中最重要的原因就是公园缺乏文化内涵。苏州、上海、杭州三座城市自古为江南重镇，历史文化名城，人间天堂，人文气息相当浓重，三地的人文和自然资源都很丰富，这种没有文化切入点的主题公园靠单纯的娱乐形式根本无法满足旅游者追求有意义、文化含量大、高层次的精神文化需求，不值得旅游者前往。

1.2.3 本土性

主题公园的本土性是由几个方面的原因决定的，首先是体现主题的差异性，为了逃脱主题公园“主题”千篇一律的桎梏，在主题上体现本土性，深入挖掘本土元素是形成主题特色的有效途径之一。其次是满足旅游者的需求。这里所说的旅游者包括外地旅游者和本地旅游者，外地旅游者大都对旅游目的地的本土文化情有独钟，因此主题公园的本土性满足了外来旅游者的需求；事实上主题公园的外地客源市场重游率极低，主题公园的主要客源还是本地游客，因此保持主题公园的本土性会使本地游客在文化上产生共鸣，产生认同。

1.2.4 综合性

主题公园的综合性主要从两个方面体现出来，首先是功能的综合性。主题公园的存在绝对不是单一只体现旅游六要素中的某一个方面，而是“吃、住、行、游、购、娱”六大要素的综合统一，只有做到六大要素相互带动、相互推进才能实现主题公园的可持续发展。其次是内容的综合性。主题公园呈现在旅游者眼前的既有静态的建筑物、园林景观、小品雕塑，也有动态的高科技娱乐设施、演员表演等动态景观。可以说旅游者在同一时间，同一地点内就可以体验层次鲜明，内容丰富的旅游活动，因此说主题公园具有极强的综合性。

1.2.5 时代性

主题公园的开发不是建完了就一了百了，一成不变的，它也具有其特定的生命周期，甚至可以说生命周期在主题公园中的体现更加明显。

主题公园是人为创造的产物，在有意无意间都会反映出当前时期内的经济社会发展水平、科学技术水平、文化理念、审美眼光以及思维层次等各个方面。因此，主题公园也会如抛物线一样达到顶峰，然后呈现出下坡的态势。当主题公园的生命走向衰竭的时候，不代表主题公园本身无药可救，而是宣告了这一个时期的终结，它需要紧随时代的步伐，呈现出另一个时期的特色，以获得新的生命，这便是主题公园时代性的体现。

1.2.6 体验性

主题公园的兴起是基于旅游者旅游需求的多样化，是建立在旅游者不能满足于静态的景物观光上的，因此注定了主题公园具备更高的的旅游功能——体验功能。这里所说的体验不仅仅是表现在游客亲身对某个活动项目的参与上而是一种身临其境、融入场景的错觉。首先是生理上的，也就是现在常常强调的参与性，即旅游者参与到某项活动中去，通过自己的实际体验感受生理上的反应，比

For example, the Fulu Belle Amusement Park (Hangzhou), only relied on scientific technology, couldn't attract the tourists at the beginning of its opening as the park was lack of cultural connotation. Suzhou, Shanghai, and Hangzhou have been the three important cities of the South of Yangtze River from the ancient time. They are the famous historic cultural cities, and enjoy the fame of paradise on earth. So the theme parks just depending on entertainment, but without any cultural can't meet the tourists' requirement of rich significance and culture, and high-grade spiritual culture in such places with rich cultural atmosphere and rich natural resources.

1.2.3 Locality

The locality of theme park is based on two factors. Firstly, it is the difference of the theme. To excavate deeply the local elements is one of the effective ways to form thematic characteristics, so as to escape from the mold of the "similar theme" of theme park. Secondly, it is the satisfaction to tourists' requirement. "Tourists" here refer to both foreign tourists and local tourists. Most of foreign tourists have special preference to the local culture of the tourism places, so the locality of theme can meet the foreign tourists' needs; actually, the rate of the tourists revisiting a scenic spot is fair low, and the main customers of theme park are local visitors. Therefore, to keep the locality of theme park will make local visitors agree on the sense of culture.

1.2.4 Comprehensiveness

There are two aspects of comprehensiveness of theme park. Firstly, it is the functional comprehensiveness. Among the functions of a theme park, the six parts of "food, accommodation, traffic, travel, shopping, and entertainment" are indispensable, any of them can't be separated. The theme park couldn't achieve sustainable development unless the six parts were in unity to move and accelerate mutually. Secondly, it is the comprehensiveness of content. The landscapes that theme parks show are not only static buildings, view of gardens, and sculpture elements, but also dynamic landscapes, such as high-tech recreational facilities, actors' performances, and so on. We can say that the visitors can taste tourism activities with clear grades and rich content. Above all, it is unproved that theme parks are with intense comprehensiveness.

1.2.5 Time

Theme parks are not immutable after finished, but have specific life cycle which can be even regarded as much obvious manifestation in theme parks.

As theme parks are creation of man, they will reflect intentionally or unintentionally the development of economy and society, the science and technology levels, the cultural concept, the aesthetic and thought level of the current period. Thus the development of theme parks will be the same as the paraboia. When it reaches the peak, it will tend to decline. However, it's not to prove that theme parks are incurable theirselves when their life become weak, but the declaration of the end of this period. And they need to keep pace with times to present features of another period, so as to obtain a new life, which is the manifestation of times of theme parks.

1.2.6 Experience

The prevail of theme parks is based on the tourists' various tourism requirement. As the tourists are not satisfied with the viewing of static scenery, theme parks should be of superior tourism function—experience function. "Experience" here not just refers to the participation of tourists to a certain activity, but a kind of illusion of taking part in the scene. First, on the physical aspect, it is the participation that we often stress. Specifically, it is the physical reflect of tourists' actual experience by taking part in a certain activity. i.e. taking all kinds of recreational facilities, taking part in the

如乘坐各类游乐设施，与演员共同参加歌舞表演等。其次便进入到心理层面上，游客不是被动地参观，而是主动进入主题公园设定的以某个主题为特色的模拟场景，亲身体验场景所表达的主题含义，使游客在心理上产生身临其境的感觉，并在视觉感受和心灵体验两个方面对园区环境达到共鸣，从而产生很强的体验感。

performance of singing and dancing with the actors, and so on. Then, on the psychological aspect, rather than viewing passively, tourists come into a simulated scene set in theme parks with the characteristics of a certain theme to feel by themselves the thematic connotation that the scenes express, so that tourists get immersive feelings and echoes of the environment of parks on the two visual feelings and spiritual experience.

2 Perspective on Theme Park Planning of Our Country

透视我国主题公园规划

2.1 前期定位不准确 Inaccurate Pre-positioning

主题公园投资大，风险高，在开发建设前期需要根据市场需求，进行精密的策划，找准定位，以确保主题公园开发的方向性。深圳华侨城“锦绣中华”诞生于 1989 年，当时的社会背景是中国改革开放 10 年，广大民众游客有饱览祖国河山的需求，却苦于时间、经济条件的限制，无法一一成行，在此背景下，华侨城集团瞄准这一人群，策划了一处浓缩了中国各处景观的主题公园，并且一炮而红；1994 年，华侨城旗下的“世界之窗”又在国民想要一览世界美景的需求下诞生了。反面的例子是苏州的福禄贝尔，其将主要客源市场锁定大上海，摆下了每天接待五千人的摊子。但是上海人天生对人造景观缺乏兴趣，加之交通不便，门票太贵，同类景观分流，公园门庭冷落车马稀，勉强维持一年后在本园员工哄抢中毁于一旦。由此可见，华侨城集团的成功便在于能够抓住当下或者是未来一段时间内旅游者的需求进行定位，从而构筑品牌，最终实现引领消费。

Theme park is a high investment and risk, thus it is necessary to make accurate planning and orientation according to the market needs in the preparing period of development, so as to ensure the orientation of development of theme park. "Splendid China Folk Culture Village" of "Overseas Chinese Town" was completed in 1989 with the social background of the tenth year since the reform and opening when the public had the desire to tourism around our country but without enough money and time. Under this circumstance, the group had schemed for a theme park with miniatures of the typical landscape around China, which had been popular immediately; In 1994, "The Window of the World", which is attached to the "Overseas China Town", came out to meet the public need of enjoying the landscape of the world. The example of failure is Froebel of Suzhou which had planned to have Shanghai as its main target market, and declared to welcome 5 000 visitors everyday. Unfortunately, with the Shanghai People's disinteresting in man-made landscape, inconvenient traffic and expensive entrance tickets, this park run hard, and finally broke down in a year. In conclude, the success of the group of "Overseas China Town" is that the orientation is on the basis of the present or future needs of tourists to create brand and finally accomplish leading consumption.

2.2 缺乏创新性 Lack of Innovation

缺乏创新性主要从两个方面体现出来，一是主题选择的重复，二是产品缺乏更新。

首先，是主题选择的重复。据不完全统计，我国有 2500 余处主题公园，但是主题公园的主题往往局限于中西微缩景观、民俗风情、影视城、器械娱乐、动物世界、仿古建筑几大类上，主题的选择趋于雷同。

其次，是产品缺乏更新。成功的主题公园通常采用的是“三三制”的经营法则，即每年淘汰三分之一的硬件设备，保留三分之一娱乐项目，新建三分之一的新概念项目。而大多数中国的主题公园是“一朝开发，终身受益”，开业的时候什么样子，数年后还是什么样子，缺乏创新性。

根据消费心理学中边际效用递减规律，同种消费产品重复使用次数越多，消费者对其产生的满意程度越低，吸引力也越低，这样导致重游率降低。另外社会在不断发展，人们的消费兴趣也在发生改变，因此主题公园的产品也应该随着市场的变化而有所变化。一成不变缺乏创新性的产品注定逃不脱失败的下场。

There are two aspects that theme parks are lack of innovation, one is the duplicate theme, and the other is the product without adapted.

First of all, it is the replicate of the choice of theme. According to incomplete statistics, there are more than 2 500 theme parks in China, but the themes of theme parks are confined to several categories, such as miniature landscapes of China and the Western, folk customs, movie and television production town, equipment entertainment, animal world, and antique construction, the choice of theme trend to be the same.

Secondly, the products are lack of innovation. Generally, the successful theme parks adopt a kind of running regulation "double three rules": to eliminate one of three equipments, remain one of three entertainment items, and build one of three projects of new concept. However, most of the Chinese theme parks are in the way of "Once completed, it won't be changed", which is lack of innovation.

According to the law of diminishing marginal utility of the consumer psychology, the more times a product was used, the lower satisfaction and attractiveness it has to a consumer, which leads to low rate of re-visiting; in addition, with the continuous development of society, people change their consumer taste, the products of theme parks should change, too.

2.3 本土文化挖掘不深入 Failing to Excavate Local Culture Deeply

本土文化的挖掘不深入主要体现在以下几个方面：首先，多数主题公园在主题的选择上侧重于模仿现有成功主题公园的主题，而忽视了这些主题能不能与当地文化相适应。这方面可以举北京的“世界公园”为例：北京是我国的政治、文化中心，具有悠久的历史，名胜古迹不胜枚举。从万里长城到故宫、天坛、颐和园再到胡同四合院，这些都是旅游者的必游之地，在此背景下，没有任何本土文化内涵的“世界公园”显然无法融入旅游者的旅游圈中。

其次，主题公园打着某某文化的幌子，但是这样的文化只是以口号的形式存在，无法对旅游者产生视觉上的文化主题冲击。文化是一个抽象的概念，主题公园要做的便是将这种无形的精神世界有形化，让旅游者通过实物感受精神。

It reflects mainly on three aspects that the local culture is not excavated deeply: first, most of the theme parks tend to choose themes of these succeed without considering their local culture. Take "The World Park" of Beijing for example, Beijing is the center of politics and culture, and with long history and many places of historic interest. From the Great Wall to the Imperial Palace, the Temple of Heaven, the Summer Palace, and then to the Siheyuan in Hutong, where the tourists must visit, in this circumstance, "The World Park", without any local cultural connotation, obviously can't be listed in the itinerary of the tourists.

Secondly, theme parks are covered with certain culture which, in fact, just exists in the form of slogan and can't have any visual impact of the cultural theme to the tourists. As culture is an abstract concept, theme parks should turn such invisible spiritual world into visible, so that the tourists can get spiritual experience through the material objects.

2.4 规划设计脱节 The Difference Between Planning and Design

规划设计脱节主要表现在主题公园被认同的主题与最终的实物出现较大的偏差。由于主题公园的策划、概念性规划、总体规划、修建性规划分别由不同的规划设计单位承担，规划设计人员对同一项目的理解不同，原本被大家认同的主题在变成实物展现出来时却不尽如人意。

It mainly refers to the difference between the agreed themes of theme parks and final actual items. As the planning, conceptual planning, overall planning, and construction planning of a theme park are in charge by different entities, and different planners and designers have different understanding to the same item, the acceptable theme won't be as satisfactory as it should be when it comes out as actual item.

3 Put Spirit into Theme Park
让主题公园活起来

3.1 确定主题 Determine the Theme

主题公园，顾名思义，具有明确的主题指向性，主题公园开发的一切程序都必须围绕主题展开，因此主题是核心，是决定主题公园走向的战略性方针，主题的确定对主题公园的成败起着至关重要的作用，特色鲜明、创意独特、文化内涵丰富的主题是主题公园成功的保障。

Theme park, as its name showed, has its definite thematic direction, and all procedures of developing theme park should be fulfilled with the basis of the theme, thus theme is the core, and the strategic direction of the development of theme park. The determining of theme plans a vital role in the running of theme park, and the theme with distinctive features, unique creation, and rich cultural connotation is the guarantee of success for a theme park.

3.1.1 资源是基础

天海规划一直强调主题的确立不是凭空想象的，更不是一味抄袭模仿其他主题公园的，而是来源于当地现有资源基础上的。一个地区有其固有的自然生态以及历史文化资源，经过无数年的积累和沉淀，会在当地居民以及外来旅游者心目中形成印象，而在此基础上确定的主题会自然而然的被游客接受，并且被认为是理所当然，从而使旅游者对品牌的认知更加容易。例如，提到西安，我们就会想到十三朝古都，想到繁盛一时的长安景象，因此在西安建立“大唐芙蓉园”就再适合不过，如果换一个地方建立“唐城”，缺乏了这种资源底蕴便达不到这种效果；又如我国主题公园的开创者——深圳华侨城旗下的“世界之窗”，提到深圳大家会想到经济之窗，对外开放的窗口，是中国了解世界的最初途径，因此在这里建立“世界之窗”便顺理成章。

除此之外，以当地资源为内涵的主题更易与其他主题公园确定的主题区分开来，是实现主题公园差异化的有效途径。以天海规划编制的《山东乳山黄金公园旅游规划》来说，该规划是受山东金洲矿业集团委托编制的，在考察了当地资源后我们认为，当地的核心资源在于现役的黄金活矿体，以及由此衍生出来的黄金开采、锻造以及黄金与生俱来的魅力，因此我们在规划的编制过程中考虑以黄金作为主题在现役活矿体的基础上规划一座黄金主题公园。

3.1.1 Resource is the Fundament

Tianhai Planning has always stressed that the establishment of the theme is neither a visionary scheme, nor a replication or imitation of other theme park, but on the basis of local current resources. Every region has its inherent resources of natural ecology and historic culture, and they will leave impression to the local residents and foreign tourists after many years of accumulation and precipitation. In this case, the theme which is determined on this basis will be accepted by tourists naturally and normally, and be taken for granted, so that tourists can grasp recognition of the brand more easily. For example, we will think about the 13 ancient dynasties and the short thriving scene of the ancient Chang'an city when it comes to Xi'an, so it can't be more suitable to build the "Datang Furong Garden" there; "Window of the World", attached to Shenzhen Overseas Chinese Town and the pioneer of theme parks of China, is another good example. Shenzhen is known as the window of economy, the window open to the overseas, and the first way for Chinese to understand the world, so it will be a matter of course to construct "Window of the World" here.

Besides, the themes with content of local resources are easier identified from the ones of other theme parks, which is an effective way to achieve difference of theme parks. Taking one works of Tianhai Planning—"Tourism Planning of Shandong Rushan Golden Park" as example, it is planned to be an gold theme park based on the current living mine.

3.1.2 文化是灵魂

主题公园不仅仅是一种游乐方式和商业手段，它还是一种文化形态，因为它具有解释文化的功能[1]。主题公园具有浓重的人工痕迹，是人类智慧的结晶，注定了其具有强烈的人文色彩，因此只有充分表现地域文化的主题公园才是活的，才是有生命力的。

我国5000多年的悠久历史，成就了博大精深的广袤文化，这种文化具有地域性、多元性，这些文化为中国主题公园的开发建设提供了无穷的灵感思维源泉，是具有中国特色的主题公园建设的不二之选。“民族的就是世界的”，本土文化的挖掘既可以与外地文化区分开来，形成自身独特的吸引力，又能够提升当地居民的文化认同感和自信心，另外本土文化是根植于当地社会生活的方方面面的，是当地最亲近熟悉最了解的，是最能代表当地的，是当地的核心竞争力所在，是当地的精神所在。因此在主题定位中体现本土文化内涵，是主题公园开发的重要手段。

天海规划在主题公园规划的过程中一直致力于文化的挖掘，这种文化不仅仅体现在显而易见的文化资源上，而更多的是强调对文化资源本身延伸出去的更深层次的文化。例如在《山东乳山黄金公园旅游规划》的编制过程中，在确定了以“黄金”为主题后，便是将黄金文化注入其中，这一文化不仅包括了黄金公园所在地——现役金矿本身表现出来的黄金采掘、选冶、成金流程，更重要的是抓住人们“黄金崇拜”的情结，展现黄金散发出来的那种与生俱来的财富与富有、尊贵与高雅、权力与强势的气质，以至光明、正义、美丽、和谐的象征，从而将“黄金文化”由表及里发挥得淋漓尽致。通过“黄金公园”这一载体实现黄金文化的传播及黄金精神这一理念的昭示，无疑会让旅游者感悟和领略到博大精深、高贵典雅的黄金文化意蕴，无疑会让游人从中获得自己所需要、所认同的内心体验和价值取向，从而对旅游者产生难以抗拒的诱惑力。

3.1.3 创意是命脉

对本土文化的挖掘并不意味着一成不变的“拿来”，而要体现文化继承与文化创新的结合，这其中就需要无穷无尽的“创意”。在确定主题的过程中怎样将本土文化与现代思维有机地结合起来，在主题公园的建设中怎样将主题相似的公园区分开来，怎样将主题公园的文化内涵向旅游者表现出来，这些都是需要通过创意来实现的。

创意是由始至终贯穿于主题公园开发全过程的，通过创意可以实现对主题的准确把握，通过创意可以实现主题公园文化的传播，通过创意可以保持主题公园新鲜度，从而实现主题公园生命周期的延续。总之创意就是延续主题公园命脉的特效药。

在《山东乳山黄金公园旅游规划》的编制过程中我们认识到将“黄金”这一主题用简单的矿山设备、矿山建筑硬生生地展现出来显然会使主题毫无新意，因此我们大胆创意，将黄金文化通过艺术手段进行再创造，将与“黄金”有关的成语、传说故事甚至金饰变成具象的建筑、雕塑等物体展现在旅游者面前，以创意表现文化、深化主题。

[1] 周向频．主题园建设与文化精致原则 [J] 城市规划汇刊 1995 年第 4 期

3.1.2 Culture is the Soul

Theme park is not only a kind of means of amusement and commerce, but also a form of culture as it has the function of explaining culture. With thick manual traces, theme park is the crystallization of human wisdom and with thick culture atmosphere, thus only the ones with full local culture are with life, and the alive theme parks.

With more than 5 000 years history, the culture in China is profound and extensive, which is the inspiration and resource of development of China's theme parks, and the best choice for construction of the theme parks with Chinese style. The development of local culture can be used to identify from the foreign culture and form its own unique attraction, and also can enhance the cultural identity and self-confidence of local residents; in addition, rooting in all aspects of life of local community, local culture is best choice of representative of local core competitiveness and spirit. Therefore, it is an important means of embodying local cultural connotation in thematic orientation to developing theme parks.

Tianhai Planning has been always engaging in cultural excavation in the planning of theme park. The culture we mentioned here is not only the obvious cultural resources, but more about the culture in deeper level. After determining "gold" as the theme in the process of compiling "Tourism Planning of Shandong Rushan Golden Park", what we need to do is to inject the planning with this gold culture which contains not only the located place of the Golden Park, but also the "golden worshiping" of human to manifest the innate quality of wealth and richness, nobleness and elegance, and rights and strength of gold, so as to express thoroughly the "gold culture". It's no doubt that the tourists can feel and understand the extensive and elegant charm of gold culture and obtain inner experience and value orientation of their needs and requirement with the carrier of "Golden Park" to realize the transportation of gold culture and popularity of gold spirit, so as to attract the tourists.

3.1.3 Creation is the Vital

Excavating local culture does not mean "getting" simply, but should reflect cultural heritage and innovation, which require lots of "creative". How to combine local culture with modern thought organically in the process of determining theme? How to identify the theme parks with similar themes, and how to display the cultural connotation of theme parks? These need creative idea to fulfill.

Creativity crosses the whole process of development of theme parks, with which the themes can be grasped adequately, the culture of theme parks can be transported, and the theme parks can be refreshed, eventually the life cycle of theme parks can be realized. In a word, Creativity is the miracle drug to continue the lifeblood of theme parks.

We have recognized in the compiling of "Tourism Planning of Shangdong Rushan Golden Park" that the design will be of no innovation if presented the theme of "gold" simply with equipments and buildings, thus we boldly used creativity to recreate the gold culture with the artistic means, turning the "gold"-related idioms, legends and even gold ornaments into actual buildings, sculptures, expressing culture with creativity, and deepening themes.

3.2 表现主题 Expressing the Theme

抽象的主题旅游者是感受不到的，换句话说旅游者到访主题公园是为了身心上的愉悦及放松，而不是费尽脑汁地去发掘主题，因此他们需要一进入公园就能感受，并且融入到主题中去。因此，主题确定后，需要将这一主题通过物质载体展示在旅游者面前，这一过程便是表现主题的过程。主题公园需要围绕既定主题从建筑、景观以及活动的安排等方面营造出一个供游客融入的模拟环境。

3.2.1 建筑

建筑作为功能性物质兴起的时候就注定了其“凝固的历史”的身份，也就是说建筑是某个地区某个时间段文化的体现。维克多・雨果说过“最伟大的建筑作品更多地是由社会而不是个人所创造的；它们是各民族辛勤劳动的产物，不是天才的灵感所致；它们是一个民族的遗产，是长期的积累财富，是人类社会不断升华的积淀，它们是人类结构”。由此推断，主题公园的建筑也应该是由其主题创造的，是主题公园文化的积淀，是主题功能的结构。如果主题公园的各个单体建筑之间背离主题，与整个主题之间没有联系，就无法表现出建筑之于整个主题公园的价值。因此建筑是主题表现的物质载体，营造氛围的重要内容。

主题公园的建筑与城市建筑最大的差别在于其除了追求功能性之外，更加强调的是建筑的景观性、观赏性、艺术性。只要建筑造型是契合主题的，就能够任凭设计师发挥天马行空的想象。主题公园除了建筑群的整体构思之外需要别具特色又迎合主题的主体建筑来起点睛的作用，这样既能够带出主题公园的亮点，又能使主题公园的建筑视觉做到张弛有度，有抑有扬。

天海规划在做黄金公园这一项目时充分考虑了这一点，整个园区的建筑色彩以黄金的金色为主，公园的建筑分为两条主线，一条主线为较为厚重的矿山建筑，与金矿这一背景契合；另一条主线则以黄金器具、首饰为原型结合现代建筑营造出一种梦幻浪漫的氛围。总之整个黄金公园的建筑虽然表现形式各异，但是都体现出了“黄金”这一主题。

3.2.2 景观

景观营造环境，环境表现主题。主题公园需要营造的就是一种与旅游者日常生活有巨大差异的虚拟氛围，因此需要根据公园的主题建造出一个模拟环境，使游客通过对这一模拟环境的视觉感官、心理刺激产生对主题的感受及认同。

主题的体现是与周围的环境密不可分的，旅游者置身于怎样的情境中就能体会怎样的主题，因此要想营造出一种真实的主题氛围就需要做到景观打造的真实细致，做到旅游者步入园区就能感受到景观设计营造出以假乱真的氛围，要做到主题公园内从亭台楼阁到路灯、雕塑、休息凳、垃圾桶再到植物配置都是在主题的范围内构思设计出来的。

Tourists can't feel the abstract themes, in other words, tourists visiting the theme park is to get physical and psychological pleasure and relaxation, rather than struggling to discover the themes, so they need to feel the themes easily as soon as they enter the parks. Therefore, the process of displaying theme is the one of displaying the thematic carrier after determining the theme. Theme parks need to create a simulation environment for tourists based on the aspects of buildings, landscape and the arrangement of activities.

3.2.1 Architecture

Architecture has gotten its status as the "solidified history" since the rise of functional substances, which means that architecture is cultural reflection certain area and time. Victor Hugo has said: "The greatest architectural works were created more by society rather than individuals; they are the products of hard work of all nations rather than the inspiration of genius; they are national heritages, the accumulated wealth of long time, the accumulation of promotion of human society, in a word, they are the human structure". Thereafter, architectures of theme park should be created by its theme, and the cultural accumulation of theme park and the structure of thematic function. If each individual architecture of theme park deviated from the theme, the significance of architecture to the whole theme park won't be demonstrated. Therefore, architecture is the material carrier of thematic demonstration, and important content of creating atmosphere.

Besides the pursuit of function, the main difference between the architecture of theme park and urban architecture is more about landscape, viewing, and artistry. The designers can give any imagination as long as the architectural shape fit to the theme. And besides the overall idea of the building groups, theme parks need main buildings with special features and catering to the theme as the finishing touch, in order to bring theme parks highlights and the architectural vision propose impact.

Tianhai Planning has fully considered this point. The main color of whole buildings in the park is golden. Architectures in the park are built on the basis of two lines: one is grand buildings of mining mountain, which is fit for the background of gold mine; the other is a kind of romantic atmosphere created by the gold tools and ornaments combined with modern architecture. In a word, although buildings of the whole Golden Park are in different shapes, they reflect the same theme "gold".

3.2.2 Landscape

Landscape creates environment, and environment embodies theme. What theme park needs to create is a virtual atmosphere which is great different from the tourists' daily life. Thus a simulated environment needs to be constructed according to the theme of park, so that tourists will feel and agree the theme through the visual sensory and mental stimulation of this simulated environment.

The embodiment of theme can't be segregated from the surrounding environment, and what theme the tourists can experience depends on what environment they are involved in. Therefore, only if created the landscape actually and detailedly that the real thematic atmosphere can be created.

“黄金公园”在景观的营造上同样是紧扣主题的：以“金牛下初”、“金乌负日”等传说为题的主题雕塑，以金矿破碎设备——球磨机剖面为主体造型的黄金舞台表演广场，以金石为开、金鸡报晓、一寸光阴一寸金、金榜题名、金印紫绶、金盆洗手、求得真金等以金字为题的独立景观，以及矿山风格的休息凳、垃圾桶都无处不在地表现了“黄金公园”的主题。

The layout of landscape of the "Golden Park" is also close to the theme: theme sculptures designed with the legends of "Golden Bird Bearing the Sun"; "Gold Bull Going Down" etc., and the Gold Performance Square with the shape of the profile of a gold crushing machine, and many single landscapes with the imagination of "gold", all of them demonstrate absolutely the theme of "Golden Park".

3.2.3 活动 Activity

通过活动的设置诱导旅游者主动参与到主题中去，从而使得主题得以实现。上面提到主题公园是一种与现实生活迥异的模拟场景，在这样的环境氛围中仅仅作为一个看客显然无法满足旅游者的需求，同时主题公园的体验性也决定了活动的设置是吸引旅游者的有效途径。旅游者在参与活动的过程中如果对活动本身印象深刻，就会认同园区主题，从而实现对主题公园的认同。迪斯尼乐园的大游行，让旅游者与电视荧屏上的米老鼠、唐老鸭、白雪公主大狂欢，从而体会到了迪斯尼童话世界的主题，使得迪斯尼乐园在旅游者的追捧中风靡全球，这便是活动表现主题的有力佐证。

The function of activity is to induce tourists to join in the theme in order to realize the theme. The theme park mentioned above is a kind of stimulated scene different from real life, in which tourists won't be satisfied only with watching; meanwhile, the function of experience of theme park proves that the setting of activities is an effective way to attract tourists. If tourists have deep impression of the activities resulted by their participation, they will agree on the theme, and finally the recognition of the theme park will be achieved. The parade of Disneyland has offered tourists a big carnival with the Mickey Mouse, Donald Duck, and Snow White which were in the TV programs, so that tourists experienced the theme of Disney fairy-tale world, and made Disneyland become popular throughout the world, which is a good example of activities expressing theme.

在黄金公园内，着金黄色工装，坐金黄色电机车进入黄金开采洞口，再坐木滑梯滑到“井下”，观看金矿开采现场，目睹金矿的开采过程，然后，又从竖井坐“罐笼”回到地面，再乘金黄色小火车游览复杂的选矿过程和神秘的冶炼出金过程，最后再到金库观赏，抚摸黄澄澄、沉甸甸的成品金块，这种旅游体验一定会强烈地刺激游客，增加旅游者对主题的了解。而通过参与原始淘金、黄金制品制作（简易性的）等活动，让游客获得动手参与的快意和欢悦，感悟和领略采金人为人类奉献宝贵财富的幸福感和成就感，使旅游者获得亲自参与其中乐趣的同时，更加真切地感受到蕴含其中的黄金文化。

In the Golden Park, tourists will put on golden overalls, take golden electric locomotive to the entrance of the cave, and slide into the "underground" by wooden slide to watch the process of gold mining in the worksite; then, take "cage" back to the ground, and view the complicate process of choosing mine and the mysterious process of refining gold on the moving small golden train; last, go to the gold storehouse to see and touch the glistening yellow and heavy gold finish, which must stimulate the tourists interest strongly and enlarge tourists' understanding about the theme.

Theme Park 主题公园

（山东黄金公园）

（多国佛教园）

Shandong Gold Park
山东黄金公园

Theme Park 主题公园

Shandong Gold Park

前期策划 Prophase Planning

1 前言 /PREFACE

中国经济的高速增长，必将带来中国旅游产业的空前繁荣和兴旺，人们的消费观念也必将随之发生深刻的变化，得到大幅度提升。毋庸置疑，中国已经进入到一个黄金般的旅游鼎盛时期，一个全民出游的大旅游时代已经不期而至，休闲度假、出门旅游已经愈来愈成为人们生活之必需，成为人们提高生存质量、追求生活品位之必然。

With the fast growth of China's economy, the tourism industry in China will definitely enjoy the unprecedented prosperity and thriving. As a result, the consumption concept of Chinese people will undergo far-reaching change with significant enhancement. No doubt, China is stepping into a prime time of tourism development and an "all people outgoing" travel era has knocked the door without invitation. Vacation tourism and outgoing travel increasingly becomes not only the necessary part of people's life, but also the trend of quality living improvement and life style pursuit.

面对这千载难逢的旅游开发大好时机，金洲矿业集团趁势而上，倾力打造乳山黄金文化主题公园，立志为自己所在的城市打造一处足以提升城市品位、构筑区域旅游强势的品牌旅游景区；立志另辟蹊径，在现役活矿体上打造一座以矿产资源为依托的黄金文化主题公园，在矿产资源产业的基础上衍生出旅游资源产业，将矿产资源产业转化为旅游资源产业，为资源性产业探求新的生存空间和发展空间，无疑是顺乎“天时地利人和”的精明之举。

In face with the rare opportunity of tourism development, the Jimson Mine Group goes ahead with the advantages and spares no effort in forging a theme park of Rushan gold culture. It aims: to establish in its own city a brand scenic spot which is good enough to promote the city status and shape the regional tourism advantage; to forge a theme park of Rushan gold culture on the basis of mine resource just at the existing mine site by seeking a new path. Therefore, it's, without dispute, a smart resolution to meet the great trend of "right time, advantageous place and harmonious society" that derives the tourism resource industry from the mine resource industry and change the mine resource industry into the tourism resource industry so as to explore the living and developing space for the mine resource industry. It indeed deserves our admiration of not only the super-normal strategist vision and the powerful ability of trend grasping and controlling, but also the strong social responsibility, the sincere aspiration of society return and the intensive spirit of pursuing exploration and innovation of the Jinzhou Mine Group.

作为一个以现役活矿体为依托而构建的黄金文化主题公园，无论是其外在的表现形式还是内在的文化内涵都必须是其他主题公园所不可比拟、无法企及的，它必须具有吸人眼球、撼人心魄的视觉冲击力和心灵震撼力，它所带给人们的绝不仅仅只是感官上的刺激，而更多的是带给人们精神上的享受、内心世界的深切体验和博大精深的黄金文化感悟。贵州天海规划设计公司正是秉承这一规划设计理念，在金洲黄金文化主题公园设计方案中注满了灵光四射、充满奇思妙想的创意设计，赋予了丰富深厚的、令人多有感悟的黄金文化内涵。

Undoubtedly, as a theme park of Rushan gold culture on the basis of mine resource just at the existing mine site, it must be incomparable, unsurpassable and far away from any other theme park no matter in external form of expression or in internal connotation of culture; must have an attractive and soul-stirring visual impact and heart-touching effect. What it brings may not only give tourists a sensory stimulus, but also offer them a spiritual enjoyment, deep experience in the heart and intensive perception of the extensive and profound gold culture. It's the right design concept that is adhered to by Guizhou Tianhai Planning and Design Company, which invented such design plan with great intelligence and creative ideas for the Jinzhou Gold Culture Theme Park and bestowed it a rich, profound and perception-induced gold culture connotation.

2 规划定位 /Orientation

以黄金文化旅游为主，生态休闲旅游为辅，集黄金生产展示、黄金制品观赏、黄金文化彰显，以及生态景观、人文景观游览和旅游地产开发于一体的黄金文化主题公园。

It shall be a one-stop integrative theme park of gold culture with major in gold culture tourism and minor in eco-fallow tourism covering gold processing layout, gold item appreciation, gold culture exhibition as well as ecological landscape, human landscape and tourist estate development.

3 目标愿景 /Target

1. 将园区打造为中国乃至亚洲唯一的、最具旅游价值的在役黄金矿黄金文化主题公园，使之与“银滩海景区”、“岠嵎山风景区”等域内景区互为映衬、相得益彰，促使“一轴、两带、多节点、一大片”旅游格局的尽快形成，开启域内旅游资源最大化利用、旅游景观南北呼应、东西辉映、珠联璧合的多彩局面，从而真正形成域内旅游强势，为域内旅游发展作出突出贡献。

Make a No.1 gold culture theme park in China even in Asia with the most tourism value, which matches and complement each other with Yintan Seascape and Juyu Scenic Spot in same area so as to promote a tourism frame of "one ax, two belts, many spots, entire scenery", open a splendid stage of full use in regional tourism resource and well complementarities of all-round scenery until a regional tourism advantage in shape to make due contribution to the local tourism industry.

2. 将园区建设成为资源性企业摆脱资源枯竭的窘境，注入新的活力，延续企业生命，寻求企业新的发展空间的成功案例，使之成为在全国都具有影响和示范性意义的矿产资源产业向旅游资源产业过渡和转移的典范。

Make the park a successful case, of which an enterprise relying much on natural resource has gone through the trap, absorbed new vigor, extended its life and seeked new development space, and make it a good example with national influence and significance, of which an enterprise relying much on natural resource has transformed and shifted to an enterprise focusing on tourism resource.

4 初步构想 /Preliminary Conception

1. 体验板块：通过深入金矿采掘现场、黄金选冶现场、金品制作现场等，让游客既充分感受黄金生产的神秘和不易，满足游客身临其境、亲眼目睹黄金生产过程的好奇心；又让游客深刻体验采金人的艰辛和采金人那种动人心魄的无畏和坚韧，从而使之对采金人有一个全新的认识，并产生由衷的敬意。

2. 展示板块：通过黄金文化博物馆（内陈设各类高贵、精美、独特的黄金制品，如金印、皇冠、金面具、金花、金砖、金条、金链、金鞋、金龟、奥斯卡金像奖杯等，并辅以相应的黄金科普知识）、太阳神鸟、黄金台、金銮殿、金字塔等实物实景和科普知识，淋漓尽致地展示黄金所特有的高贵、富有和权威等文化内涵，让游客感悟和领略博大精深、高贵典雅的黄金文化的意蕴。

3. 参与板块：通过参与原始淘金、黄金制品制作（简易性的）等活动，让游客获得动手参与的快意和欢悦，从而使其感悟和领略到采金人亲手为人类奉献宝贵财富的欢乐、喜悦的幸福感和骄傲、自豪的成就感；通过参与“金盆洗手”（制金盆洗手台，供游客净手）、“金口玉言”（制精美留言簿请游客留言）、“金榜题名”（制皇榜拜祭台，供学子祈求日后金榜题名）、“金印紫绶”（制金印紫绶台，供仕途之人祈求官运亨通）以及“金童玉女”、“金戈铁马”、“金鸡独立”（扮相、造型）等活动，体会黄金文化的博大精深。

1. Experiencing part: By means of on-site view of gold mining, gold refining and gold field production, make the tourist not only experience the mystery and hard-won property of the gold manufacture so as to satisfy their curiosity, but also experience the hardship during gold mining and the strong bravery and fortitude spirit of gold miner so as to shape a brand-new understanding on gold miner with a well uprising admiration.

2. Displaying part: By means of displaying real objects in the gold culture museum (where there are some luxurious, delicate and unique gold items like gold seal, gold crown, gold mask, gold flower, gold brick, gold bar, gold chain, gold shoe, gold turtle, the Oscar Statuette, etc.) including the Apollo Bird, the Gold Platform, the Gold Palace, the Pyramid and popular science, show tourist the culture connotation of nobility, richness and authority only featured by the gold so as to make them feel and get aware of the implied content of the comprehensive, profound, noble and graceful gold culture.

3. Participating part: By means of inviting the tourist to participate in gold mining and gold manufacturing, make the tourist get the happiness and joyment from the participation so as to feel and get aware of gold miner's joyment, happiness and pride from their effort in dedicating the treasure; by means of inviting the tourist to participate in the activity of "washing hands in gold basin" (provide gold basin for hands cleaning), "golden saying" (provide delicate visitor's book for valuable advice), "score signboard" (provide imperial altar to student praying for the future success in study), "gold seal platform" (provide gold seal platform to office-career pursuer praying for the future success in official circle), "golden boy and jade maiden" shining spears and armored horses" and "golden rooster standing on one leg", etc.

4. Ecological Part: Take avenges of water resource to establish a unique sightseeing water landscape, like lake, stream, waterfall, island, water step on rock, gallery bridge, pavilion, waterfall, windmill, waterwheel, fresh-water bathing spot, artificial torrent, hill surfing, etc.; take advantages of natural vegetation and artificial afforestation, bamboo garden, flower bed, orchard, vegetable bed to establish an ecological and botanical garden for sightseeing, appreciation and enjoyment, which makes a carefree and happy tourist with no thought of leaving, and to spare no effort in establishing in the north seaside an indispensable green land featured by the southern charm so as to make full use of the amazing natural resource.

5. Entertainment Part: With new sophisticated recreation facility of small train around the park, the Apollo cart, the Trojan horse, space shuttle, power-centralized aircraft, flaming chariot, UFO, discoverer, tornado to increase the entertainment item, makes the tourist enjoy the solely adventurous and stimulative experience triggered by modern recreation facilities.

6. Service Part: Construct some service facility like hotel, restaurant, tea house, gold shop, supermarket, KTV, dancing hall, gym with novel and unique architectural style to match the connotation of the theme park. All roofs will be decorated in yellow and all walls decorated in yellow with interspersed green so that it can imply the perfect integration of gold culture and ecological culture.

7. Real Estate Part: Satisfy the vacation travel need from VIP, expand the travel volume and enhance the park status by developing tourism real estate, forging property villa and creating charming gold village. The real estate shall be constructed in yellow with green interspersed in two forms: single house and combined house.

4. 生态板块：利用水资源制作风韵独具的观赏游乐水景，诸如湖泊、溪涧、瀑布、岛屿、水磴、廊桥、亭台、水榭、风车、水车、淡水浴场、人造山洪、山地冲浪等；利用天然植被和人工培植林带、竹苑、花圃、果园、莱畦，建成可游乐、可观赏、可享用的令人心旷神怡、流连忘返的生态植物乐园，着力营造北方滨海地区不可多得的极富南方风韵的山水绿地情调，最大限度地开发利用好令人惊叹的自然生态资源。

5. 娱乐板块：安置环园小火车、太阳神车、特洛伊木马、太空梭、聚能飞船、风火轮、UFO、发现者、龙卷风、过山车等高科技游乐设施，以丰富景区游乐项目，让游客享受现代游乐设施所带给人们的那份独有的惊险和刺激。

6. 服务板块：构建宾馆、酒楼、茶肆、金店、超市、歌厅、舞池、体育健身场馆等园区服务场馆，建筑风格力求新颖别致，并含公园主题意蕴。建筑顶部一律使用黄色，墙体则以黄色调为主，间以绿色调，以此暗示园区黄金文化与生态文化的完美融合。

7. 地产板块：开发旅游地产，构建产权式别墅，打造别有韵味的黄金村落，既满足高端游客旅游度假要求，又丰盈景区体量，提升园区品位。地产可按单体度假别墅和连体度假别墅形制构建，色调以黄色调为主，绿色调为辅。

5 规划建议 /Suggestion on Planning

1. 园区名称以企业名称冠之，一则更与主题相扣，二则更益彰显企业非常之举措，更有利于提升企业形象。

The park shall be named after the enterprise to: 1. match the theme of park; 2. ensure the enterprise an enhanced image with such super-normal resolution.

2. 项目分二期投建，一期为体验板块、展示板块、参与板块、生态板块、娱乐板块和服务板块，应同期投建，同期开放；二期为地产板块，可在一期工程建成后，相宜而建。这样，既可保证园区产生横空出世、不同凡响的轰动性效应，又可缓解资金压力，减少投资风险。

The whole project will have two phases: Phase 1: Experiencing part, displaying part, participating part, ecological part, entertainment part and service part shall be constructed and opened simultaneously. Phase 2: Real estate part shall be constructed based on the real condition after the phased 1. and the benefit shall not only ensure a great sensation caused by the park emergence without any notice, but also ease the capital shortage and reduce the risk of investment.

Shandong Gold Park

项目规划 Item Planning

1 项目开发背景 /Developing Background

第一节 中国旅游已经进入前所未有的鼎盛时期
Section Ⅰ China Has Entered the Unprecedented Prosperous Period in Tourism

一、中国已经开始步入全民旅游时代

随着中国经济的高速发展，人们生活水平和消费观念的不断提升，以旅游为主要载体的休闲生活方式已经愈来愈为人们所推崇和追捧，旅游已经不再是少数人的生活奢侈品，它已经不可遏制地开始成为大众需求，成为寻常百姓提高生活质量和生活品位的追求。

Ⅰ China Has Entered the Era of Universal Tour

With the rapid development of Chinese economy and the improvement of people's life and the promotion of people's consumptive concept, relaxing life with tourism has been admired and pursued by people that tour has not been the luxury of the minority, but become the public demands and the pursuit of people promoting their life quality and taste.

二、中国旅游已经昂首挺进世界前列

中国旅游业的繁荣已经开始备受国际市场青睐，据有关资料统计，至2006年，中国入境旅游人次已经跃居世界第三。又据世界有关旅游组织预测：到2020年，中国将成为仅次于美国的全球旅游主要消费市场，届时将有1.37亿国外游客涌入中国观光旅游，中国必将从此成为世界旅游强国，成为各国游客争相游览的旅游胜地。

Ⅱ China Has Leaped into the First Rank of the World in Tourism Field

The prosperity of China tourism has begun to be favored by the world market. By 2006, according to the relative document, the number of tourists from foreign countries has leaped to the third, almost the second of the world. By the year 2020, some relative tourism organization of the world has predicted that China will be the main consumption market of the world tourism, just behind the United States, when about 137 million visitors will stream into China to travel, and China will become a powerful tourism country and the tourist resort which visitors of every country prefer to come.

区位分析图

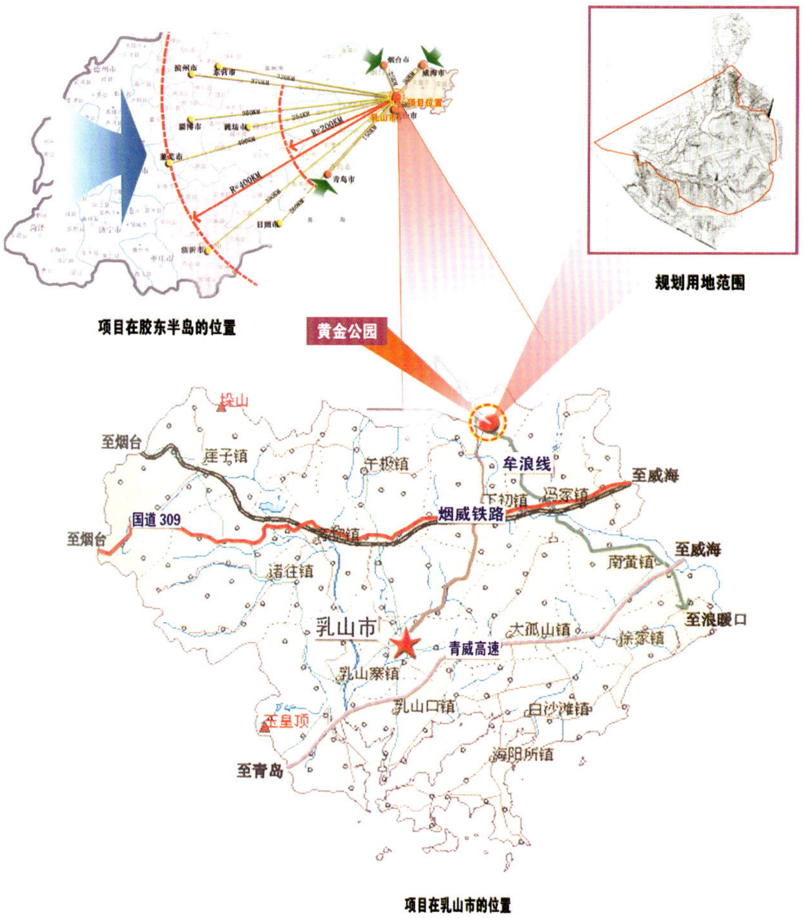

第二节 发展旅游——构建乳山旅游城市的必然选择

一、乳山市拥有丰富的旅游资源

乳山市拥有诸如“银滩”，“岠嵎山”，“天下三大道院”之一，有“道教圣地”之美誉的“圣水宫”，被誉为“海上仙境”的乳山口景区，“避暑胜地”月牙湾等美妙丰富的旅游资源，作为地处胶东半岛中心腹地、区位优势十分明显的乳山市，没有理由不把目光聚集于无烟工业——旅游产业。

Section Ⅱ Developing Tourism—The Inevitable Choice of Construct Rushan Tourism City

Ⅰ Rich Tourism Resources of Rushan

There are beautiful and rich tourism resources in Rushan city, such as "Silver Beach", "Juyushan Mountain", "Holy-Water Palace" which enjoys the fame of "One of the three great Taoist Park" and "Holy Land of Taoism", and so on. Lying on the center of Jiaodong Peninsula, Rushan city occupies an obvious position advantage, which is the good conditions to develop tourism industry, the no-smoking industry.

高程分析图

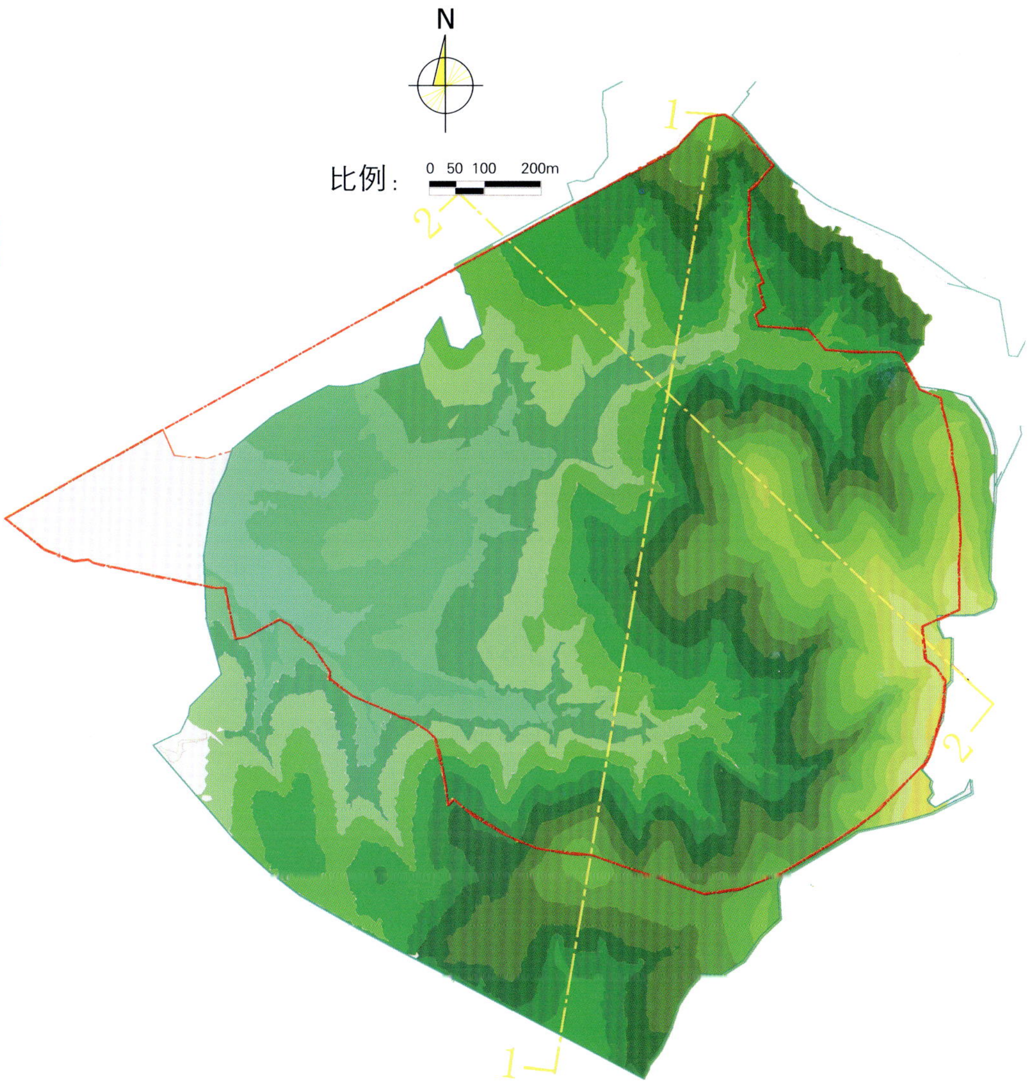

1-1剖面图

2-2剖面图

图例

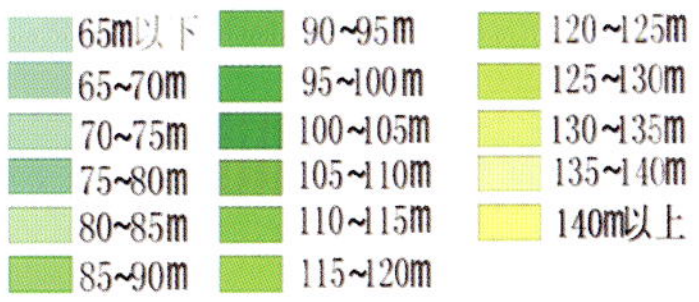

二、构建乳山旅游城市的时机已经成熟

乳山市委书记傅广照也在乳山新一轮的城市发展战略中提出了“旅游兴市，借旅兴乳”的战略方针。而近年来，乳山市以建设“生态型花园式滨海旅游城市”为目标，坚持“高起点规划、高标准建设、高质量管理”的方针，加快了景区的开发建设，乳山市的旅游事业已呈蓬勃发展态势。其构建旅游城市的条件已经具备，时机已经成熟，乳山实施“旅游兴市、借旅兴乳”、实现旅游产业历史性跨越的时期已经到来。

II The Right Moment of Construct Tourism City in Rushan

Fu Guangzhao, the Secretary of Rushan Municipal Committee, has pointed out the strategic principle of "Better Tourism, Better City" in the new strategy of urban development. Also, with the goal of constructing "Ecological Seashore Tourism Park City" and the principle of "High-Level Planning, High-Standard Construction and High-Quality Management", Rushan has developed prosperously in tourism industry. This is the right moment to create tourism city in Rushan.

坡度分析图

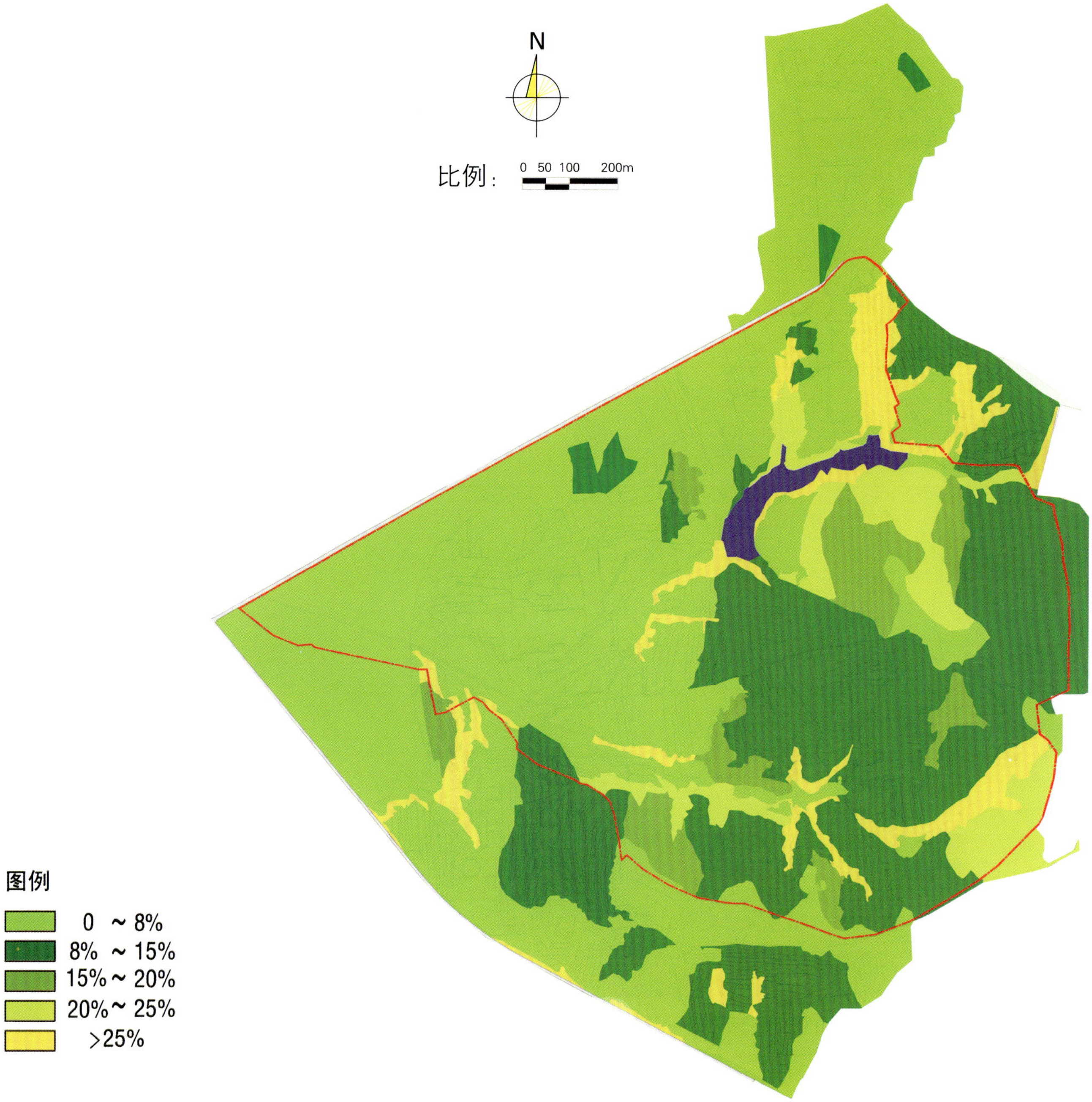

第三节 黄金文化主题公园在乳山市旅游资源大盘中的特殊地位

一、胶东半岛旅游产业的一张新名片

乳山黄金文化主题公园作为一座在现役活矿体上打造的旅游景区，在胶东半岛旅游圈所处的特殊位置不言而喻，公园融合采金体验、黄金文化、自然风光、休闲娱乐于一体，集工业旅游、生态观光、现代游乐以及参与性、趣味性、知识性于一身的独有的个性特色，必将成为胶东半岛旅游圈的新热点和新亮点，成为乳山旅游产业的一张新名片。

二、矿产资源向旅游资源过渡和转化的典范

金洲矿业集团未雨绸缪，倾力打造黄金文化主题公园，为资源性企业摆脱资源枯竭的窘境，注入新的活力，延续企业生命，寻求企业新的发展空间和生存空间进行有益尝试和大胆探索，必然会为工业企业如何实现可持续发展探索出一条新的路子，必然会成为矿产资源产业向旅游资源产业过渡和转移的成功案例，在胶东半岛乃至全国产生深远影响和具有广泛的示范意义，并成为工业旅游的典范。

第四节 地方政府的高度重视和金洲矿业集团的坚定信念

一、乳山政府的高度重视

作为地方政府，乳山市党政领导向来对于金洲矿业集团关爱备至，呵护有加。黄金文化主题公园更是得到了威海及乳山两级政府的重点扶持，明确指出旅游产业已成为乳山的重点发展产业。项目于 2004 年就纳入威海市旅游产业发展总体规划，并列入近期优先开发的 20 个重点项目工程之中。而在乳山市新一轮的旅游规划中，黄金文化主题公园已经被列为乳山市五大重点旅游项目之一，其重视程度显而易见。

二、金洲矿业集团的坚定信念

金洲矿业集团坚信打造黄金文化主题公园一定能够为资源性企业摆脱资源枯竭、寻求新的发展空间和生存空间找到一条有效的可持续发展的路子，一定会使黄金文化主题公园成为乳山、成为胶东半岛、成为山东乃至成为全国工业旅游的新亮点，成为资源性企业摆脱困境、走向辉煌的范本。

第五节 综述

一、项目开发占尽"天时、地利、人和"之先机

1. 项目开发正赶上中国进入全民旅游的"黄金"时期，恰遇世界大旅游时代所带来的千载难逢的旅游发展机遇，无论国内国外，旅游产业已经进入前所未有的鼎盛时期，可谓适逢其会，天公作美，已占难能可贵之"天时"。

2. 项目开发正赶上旅游发展的绝佳时期，恰遇乳山市把旅游产业作为重点发展产业，立志构建旅游大市的绝佳机遇，可谓左右逢源，万事俱备，又握必不可少之"地利"。

3. 项目开发正赶上乳山市政府和 59 万勤劳智慧的乳山人民立志发展旅游产业，实施"旅游兴市、借旅兴乳"战略的大好时机，可谓上下同德，万众一心，更得不可或缺之"人和"。

二、结论："天时地利人和"均占，项目开发势在必行

Section Ⅲ Special Position of Golden Cultural Theme Park in Tourism Resource of Rushan

Ⅰ A New Name Card of Jiaodong Peninsula

As a scenic spot created on an alive mine, Rushan Gold Cultural Theme Park has taken a special position undoubtedly. With the integration of experience of mining gold, gold culture, natural scenery, and relaxing and entertainment, Rushan Gold Cultural Theme Park will be a new popular and outstanding scenery attraction and a new name card of Rushan and Weihai Tourism industry, along with its features of participating, interesting and knowledge.

Ⅱ A good Example of Transformation From Mine Resource to Tourism Resource

With a predictable mind, we Tianhai people devote to create Gold Cultural Theme Park to help the enterprises depending on resources extricate from a difficult position of resources deficiency, and inject new blood to them to last their life, which will explore a new route for enterprises to achieve sustainable development and be a successful case of transition and transformation from mining resources industry to tourism resources industry.

Section Ⅳ Great Attention of Local Government and Strong Convictions of Jinzhou Mining Group

Ⅰ Great Attention

As local government, Rushan municipal committee has always pay great attention to Jinzhou Mining Group. With a definite instruction of determining tourism as the primary developing industry, the Gold Cultural Theme Park has even been supported extremely by the two levels government of Weihai and Rushan. This project has been in the overall planning of tourism industry development of Weihai Prefecture, and one of the twenty project prior to develop.

Ⅱ Strong Conviction of Jinzhou Mining Group

Jinzhou Mining Group believed that the creation of Gold Cultural Theme Park must seek an effective sustainable route for the enterprises depending on resources, and become a new highlight and a successful example of resource enterprises extricate from difficulty and win glory.

Section Ⅴ Overview

Ⅰ Right Moment

1. The project was developed at the very moment when tourism has become common in China and the world. Tourism has entered an unprecedented period no matter in China or abroad, which is a right moment to develop tourism.

2. The project was developed when tourism was determined as the most important industry by Rushan government, which is really an invaluable opportunity.

The project was developed when Rushan government and 0.59 million people aimed to develop tourism industry and fulfilled the principle of "Better tourism, better city", which is an inevitable conditions of concerted effort.

Ⅱ Conclusion: With All Good Conditions, the Development of This Project Must Be Enforced.

2 项目开发条件 /Developing Conditions

第一节 项目开发优势

一、经济发达、交通便捷的区位优势

黄金文化主题公园所在地乳山市位于横贯欧亚大陆交通大动脉“欧亚大陆桥”黄金经济走廊上的重要节点——胶东半岛（青岛、烟台、威海），与欧盟经济区、中西亚资源富集区、中国高增长经济区和日、韩经济区紧密相连，区域经济非常发达，大区位优势十分明显。不仅如此，黄金公园所在地乳山市地处胶东半岛腹地，位居青岛、烟台、威海三大旅游城市中心区域，距青岛、烟台和威海分别只有一个半小时左右的车程，青威高速公路、济威铁路、309国道均从境内经过，区域经济实力雄厚，交通便捷，旅游人气旺盛，小区位优势亦甚为明显。

Section Ⅰ Advantages

Ⅰ Position Advantage

Rushan Prefecture where the Gold Cultural Theme Park is located is an important node of the "Eurasian Continental Bridge"—Jiaodong Peninsula (including Qingdao, Yantai and Weihai), which connects with European Economic Area, the rich resources area of Middle and West Asia, the economic region in rapid growth of China, and the Japanese and Korea economic regions, is in a very developed economy state. Moreover, Rushan, the place where the Gold Park locates, lies on the Jiaodong Peninsula which is the center of the three tourism cities of Qingdao, Yantai and Weihai.

现状分析图

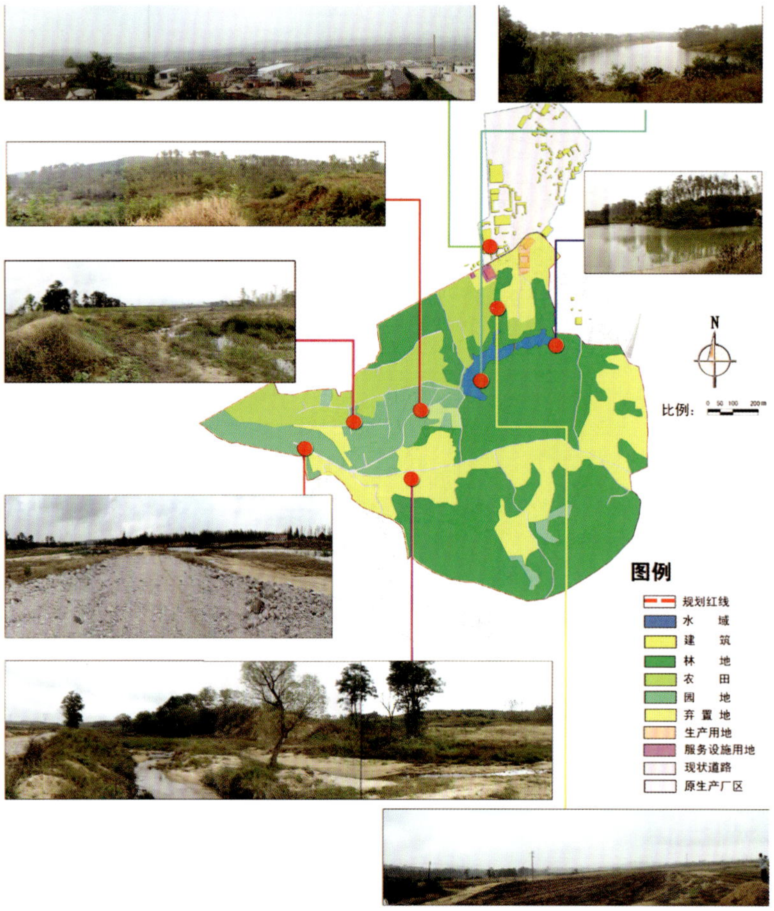

矿区现状图

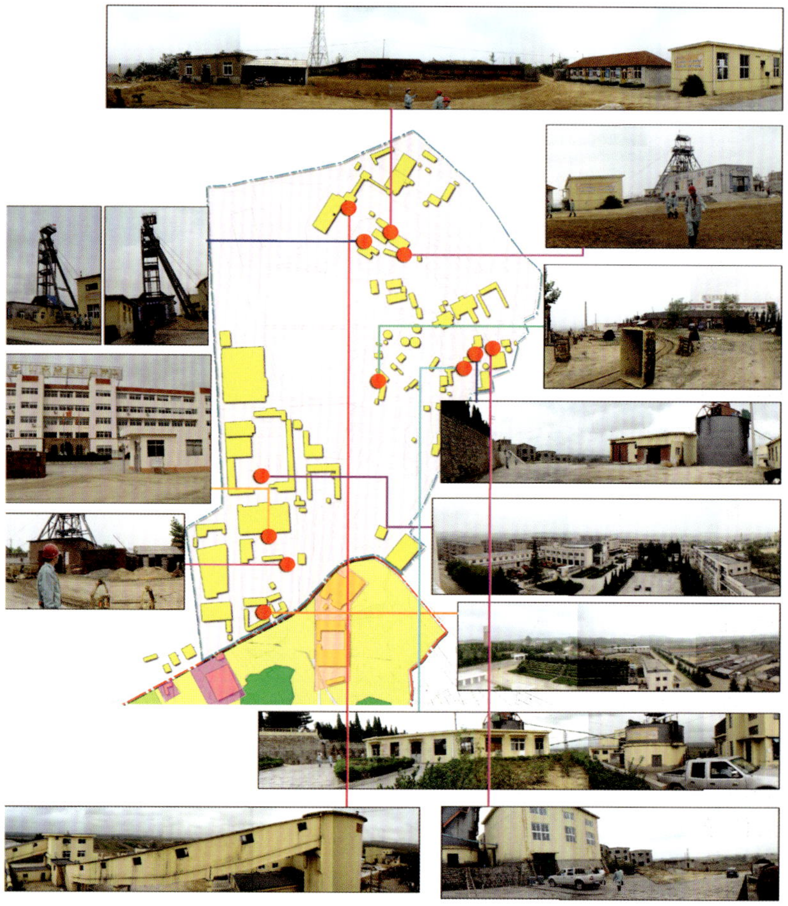

二、活矿成园的资源拓展优势

在还在服役的黄金矿产活矿体上建造一座黄金文化主题公园，把矿产资源拓展为旅游资源，以正在开采的黄金矿为依托，向游人展示黄金采掘、选冶、成金等一系列黄金生产过程，让游人身临其境、亲眼目睹黄金生产过程，充分感受黄金生产的神秘性，深刻体验采金的艰辛和采金人的豁达和无畏，并让游人参与其中，使其获得身心体验的愉悦，给人以其他景区、公园所不能得到的体验和快乐，这不能不说是黄金文化主题公园吸人眼球、撩人游兴的一大优势。

三、厚重、隽永的黄金文化优势

作为财富与富有、尊贵与高雅、权力与强势，以至光明、正义、美丽、和谐的象征，黄金原本就对人有着一种难以遏制的吸引力，无论是达官显贵，还是黎民百姓都不可避免地存在着神圣的“黄金崇拜”情结。以黄金为主题的乳山黄金文化主题公园，无疑具有黄金所特有的高贵、富有、权威和光明、正义、美丽、和谐等文化内涵，无疑会让游人感悟和领略到博大精深、高贵典雅的黄金文化意蕴，让游人从中获得自己所需要、所认同的内心体验和价值取向，从而对游人产生难以抗拒的诱惑力。

四、富足、优美的地块优势

黄金文化主题公园占地四千余亩，可用空间富足，极适宜天马行空、肆意发挥、尽情打造，极有利于丰盈景区体量。不仅如此，景区地貌条件甚优，林木茂密，植被葱绿，有丘陵、低谷，有水体、湿地，有果园、瓜地，自然环境甚是优美，极有利于打造项目游乐情趣的多样性，极有利于为游人营造出一园多种风景、一园多种情调、一园多种趣味的综合性大型主题文化园。

第二节 项目开发机遇

一、政策鼓励的机遇

“旅游促进发展”不仅是乳山市地方发展的战略决策，更是引起了中央的足够重视。不仅如此，中央关于带薪休假、调整节日假期的政策也必然会对旅游业起到极大的拉动作用。基于此，我们完全有理由说，乳山黄金文化主题公园正好赶上了旅游发展的又一个黄金时期。

二、全民旅游时代到来的机遇

随着经济的高速发展和人民生活水平的日益提高，中国业已开始步入全民旅游时代，旅游热情高涨，旅游盛况空前，必将极大地刺激旅游市场多元化发展，黄金文化主题公园实可谓应运而生，生逢其时，遇上了好时光。

三、中国步入世界旅游大国的机遇

据世界有关旅游组织预测：到 2020 年，中国将完全可以与世界最大的旅游市场——美国争雄，成为名副其实的旅游大国。黄金文化主题公园所在地乳山市位于横贯欧亚大陆交通大动脉“欧亚大陆桥”黄金经济走廊上的重要节点——胶东半岛腹地，优先接纳境外游客旅游观光势所必然。

Ⅱ Resource Transfer Advantage

Constructing a Gold Cultural Theme Park on an alive gold mine to transfer mining resources to tourism resources is a self-evident advantage of itself, which will show the visitors the processes of producing gold like mining, smelting and shaping to let them fell and see the mystery of gold producing and deeply experience the fearless and tolerance of miners that the other scenic spots or parks can't give them.

Ⅲ Advantages of Gold Culture

As the mark of wealth and richness, honor and elegance, power and strength, as well as brightness, justice, beauty and harmony, gold has attractions that people can't resist that no matter the nobility and gentry or the ordinary people are with the mind of holy "gold worship". Rushan Gold Cultural Theme Park, with the theme of gold, is undoubtedly of the cultural connotation of gold like honor, richness, authority, brightness, justice, and harmony, which will give the visitors the profound and extensive and elegant cultural charm of gold, as well as inner experience and value orientation.

Ⅳ Position Advantage

The Gold Cultural Theme Park occupies more than four thousand mu land, which is enough for us to create it with full imagination. What's more, this region is in good geographic conditions. With hills, valleys, streams, wetland, orchard, and melon field, the scenery is very beautiful, which is very beneficial to construct diversified relaxing project and a large comprehensive theme culture park with diversity of scenery and tastes.

Section Ⅱ Opportunities

Ⅰ The Policy

The principle "tourism accelerates development" not only takes the position of strategic decision, but also draws attention of the central government. In addition, the policies of adjusting holidays must accelerate tourism development. In conclusion, we dare say that Rushan Gold Cultural Theme Park will be born at the prime moment of tourism development.

Ⅱ Tour Becomes Common

With the rapid development of economy and the improvement of people's life, tourism in China has become common that the tourist industry has developed prosperous, which will stimulate the diversity tourism market. And this is the background with which Rushan Gold Cultural Theme Park grows.

Ⅲ The Great Tourism Country

According to the prediction of the relative world tourism organization, China will compete with the United State—the largest tourism market by the year 2020. The Gold Cultural Theme Park is located Jiaodong Peninsula, whish is the important node of the "Eurasian Continental Bridge" enjoying the fame the prime economic corridor.

第三节 项目开发面临的挑战

一、如何与招远黄金乐园 PK 的挑战

与相距不到 100 公里的招远"黄金乐园"PK，如何于不利中寻求有利，如何在共性中谋求差异，如何突出乳山黄金文化主题公园的个性特质，如何在策划、创意、设计、构筑等方面出奇招，亮绝活，使出高人一筹的手段，寻找制胜突破口，无疑将是乳山黄金文化主题公园成败的关键。

二、如何在胶东半岛抢占市场的挑战

黄金文化主题公园也面临来自胶东半岛早已名声在外的其他旅游景区压力，必将面临激烈的旅游市场竞争。如何对乳山黄金文化主题公园进行高规格、高品位的打造，如何将乳山黄金文化主题公园打造成乳山市独具魅力的品牌旅游景区，使其在早已声名远扬、人气旺盛的青岛、烟台、威海这些旅游城市中脱颖而出，在激烈的市场竞争中崭露头角也是当前面临的严峻挑战。

Section Ⅲ Challenge

Ⅰ How to Compete with Zhaoyuan Gold Paradise

Rushan Gold Cultural Theme Park is less than 100km from the "Gold Paradise" of Zhaoyuan, so how to seek benefit in the disadvantages and difference in the common features to highlight the specialty of Rushan Gold Park, and how to scheme, create, design and construct innovatively is undoubtedly the key of success.

Ⅱ How to Win Market Share in Jiaodong Peninsula

The Gold Cultural Theme Park also faces up press from other famous scenic spots of Jiaodong Peninsula. So how to create Rushan Gold Cultural Theme Park in large size and great taste and how to endow it with unique charm to emerge from the tourism cities of Qingdao, Yantai and Weihai are the challenge we are facing.

第四节 项目开发的制约因素

一、道路交通的制约

乳山黄金文化主题公园地处乳山市近郊的下初镇，与市内相距 25 公里，距离虽不远，但乳山市至黄金文化主题公园的交通尚欠发达，为三级公路，较为狭窄，难以承载日益增大的客流运载需要，更难以满足景区发展需要。

二、资金制约

高起点、高品位、大体量地打造黄金公园，就必须有雄厚的资金作保证，必须具有强大的经济实力，因而，能否保证足够的资金投入，能否筹措、吸纳和引进足够的资金同样是乳山黄金文化主题公园需要解决的问题。

Section Ⅳ Confined Factors

Ⅰ Confined by Traffic

Rushan Gold Cultural Theme Park is in Xiachu Town of Rushan Prefecture. Although it is not far away from Rushan city (about 25km), the traffic systems are not good as they should be that they can't meet the needs of the development of this scenic spot.

Ⅱ Confined by Capital

Abundant fund should be needed if we want to construct the Gold Park at high level and taste and large size. Therefore, whether we can collect, draw and introduce enough capital is the problem we have to salve.

3 规划定位及目标愿景 /Planning Orientation and Object

第一节 规划定位

以黄金文化旅游为主，生态旅游、休闲娱乐为辅，集黄金生产展示、黄金制品观赏、黄金文化彰显，以及生态景观、人文景观游览、现代游乐和旅游地产开发于一体的黄金文化主题公园。

Section Ⅰ Orientation

The Gold Park is major in golden cultural tourism, along with ecological tourism and relaxing and entertainment, gold producing, exhibition of golden articles and culture, ecological and humanistic landscape, and modern entertainment.

第二节 目标愿景

一、将项目打造成乳山市旅游产业的一张新名片，使之成为乳山市的旅游品牌景区，与域内景区互为映衬、相得益彰，促使"一轴、两带、多节点、一大片"旅游格局的尽快形成，开启域内旅游资源最大化利用、旅游景观南北呼应、东西辉映、珠联璧合的多彩局面，从而真正形成域内旅游强势，为域内旅游发展作出突出贡献。

二、将项目打造成中国乃至世界最有文化内涵、最具文化品位、最具旅游价值的在役金矿黄金文化主题公园。

三、将项目打造成矿产资源企业摆脱资源枯竭窘境，注入新的活力，延续企业生命，寻求新的生存空间和发展空间的成功案例，使之成为全国最有影响力和示范性意义的矿产资源产业向旅游资源产业过渡和转移的典范。

Section Ⅱ Object

Ⅰ We aim to create the Gold Park to be a new name card of Rushan Prefecture so that it can form its tourism brand and accelerate the pattern of "one axle, two lines, several nodes and a large space" reflecting with the other scenic spots of this region, which will surely form regional tourism strength and contribute to the development of regional tourism.

Ⅱ We aim to create this Gold Park to be a gold cultural theme park on an active gold mine with the most cultural connotation and taste and tourism value.

Ⅲ We aim to complete a successful example that the enterprises depending on mine resources escape from the difficulty of resource exhausting and seek new living and developing space with injection of new vigor.

4 规划原则与依据 /Planning Principle and Basis

第一节 规划原则

一、开发与开采和谐统一原则

黄金文化主题公园是一座以现役黄金生产矿区为依托开发、建构的工业旅游观光景区，因此，必须把握好园区开发与矿区开采和谐统一的原则，既要通过旅游开发提升企业的管理水平，促使企业矿业开采更加有序、更加规范；又要通过规范有序的开采和科学的管理为项目开发打下坚实的基础，提供良好的条件。

二、凸现和彰显黄金文化原则

黄金文化主题公园是一座集采金体验、黄金文化、自然风光、休闲娱乐于一体的现代工业旅游园，因而，必须最大限度地凸现和彰显园区独有的文化——黄金文化，最大限度地挖掘和提炼丰厚、深邃的黄金文化，让黄金文化主题公园无处不浸润和充盈着隽永诱人的黄金文化，让游人无论是在感官视觉上，还是在内心体验上，抑或是价值认同上，都强烈地感受到黄金文化耀眼夺目的光辉，感悟和领略到黄金文化不可替代的、永恒的魅力。

三、主体景观与配套设施有机结合原则

黄金文化主题公园以现役黄金矿区为依托，以博大精深、高贵典雅的黄金文化为主题，但必须配套相关游乐设施，以丰盈体量，增加看点，避免单一。园区必须注重主体景观与配套设施合理搭配，必须尽可能多层面地满足游客需要。既要注重主体景观的主导性，又要注重配套设施的补充性，并使之形成有机的整体，产生遥相呼应、相互辉映的效果。

四、人文景观与生态环境完美融合原则

黄金文化主题公园自然生态环境非常优美，因而，在构建黄金文化主题公园的时候，必须注重人文景观和自然景观的完美融合，既要充分展示园区人文景观的丰富性和趣味性，又要着力展现自然天成、景色宜人的生态环境，使自然景观和人文景观相互辉映，全力营造人义景观和自然景观水乳交融的景区氛围，确保人文景观与自然景观相映成趣，水乳交融，充满活力和灵性。

第二节 规划依据

一、国家、省、地（市）相关法律法规

二、项目规划设计说明书

Section Ⅰ Planning Principles

Ⅰ Development and Exploit

The Gold Cultural Theme Park is an industrial scenery attraction constructing on an existing gold mine. Therefore, we must emphasize the harmony between developing park and exploiting mine, that is we must improve enterprise management to form more regular and ordered mining exploiting by developing tourism, and also build up solid foundations and good conditions for this project by regular and ordered exploiting mine and scientific management.

Ⅱ Highlighting Golden Culture

The Gold Cultural Theme Park is a modern industrial tourist park concentrating on mining experience, golden culture, natural landscape and recreation. Therefore, we must highlight and extract as deeply as possible the unique culture of the park—golden culture, so that the visitors could feel strongly the brightness and the irreplaceable and endless charm of golden culture.

Ⅲ Organic Combination of Major Landscape and Assistant Facilities

Constructing on the existing gold mine and with the theme of profound and extending and elegant golden culture, the Gold Cultural Theme Park should be matched with relative facilities to enrich content and add scenery without singleness. We must pay attention to the reasonable facilities to go with the major landscape that we must meet the visitors' need as much as possible. We have to emphasize both on the leading position of major landscape and the complementarity of the assistant facilities, so as to form an organic entirety and reflecting effect.

Ⅳ Cultural Landscape and Ecologic Environment

The Gold Cultural Theme Park is of very beautiful natural ecological environment, thus, we must put emphasis on the perfect integration of cultural landscape and natural landscape, and demonstrate fully both the diversity and interesting of cultural landscape and the nature and beauty of ecological landscape, so as to form a harmonious and vigorous environment for the visitors.

Section Ⅱ Regulars of the Planning

Ⅰ Relative National, Provincial, Prefectural Laws and Regulars

Ⅱ Specification of the Planning and Design

5 规划范围及期限 /Planning of Scale and Deadline

第一节 规划范围

乳山黄金文化主题公园位于冯家镇西北、金洲矿业集团一带，规划用地位于矿区内，占地约4000余亩。

第二节 规划期限

一期投建内容：主体园区、现代游乐区、部分园林景观。
建设时间:2008年4月投建，2009年10月试运营，2010年3月开放。
建设周期：17个月。

二期投建内容：园林景观区、自然生态区、地产开发区（黄金度假别墅区和黄金住宅小区）。

建设时间：2009年10月投建，2011年10月建成发售（黄金度假别墅可用1年时间先期建成投入运营）。

Section I Planning Scope

Rushan Gold Cultural Theme Park lies on the northwest of Fengjia Town, in the region of Jinzhou Mine Group. The planning land is in the mine region and more than 4000mu.

Section II Planning Term

Phase I: major area, modern playing gardens and some landscape.
Constructing time: it will begin building in April 2008, be put into trial operation in October 2009, and open in March 2010.
Construction Term: 17 months.

Phase II: landscape area, natural ecological area and residential area

Construction Term: it will begin building in October 2009 and finish in October 2011.

水系规划图

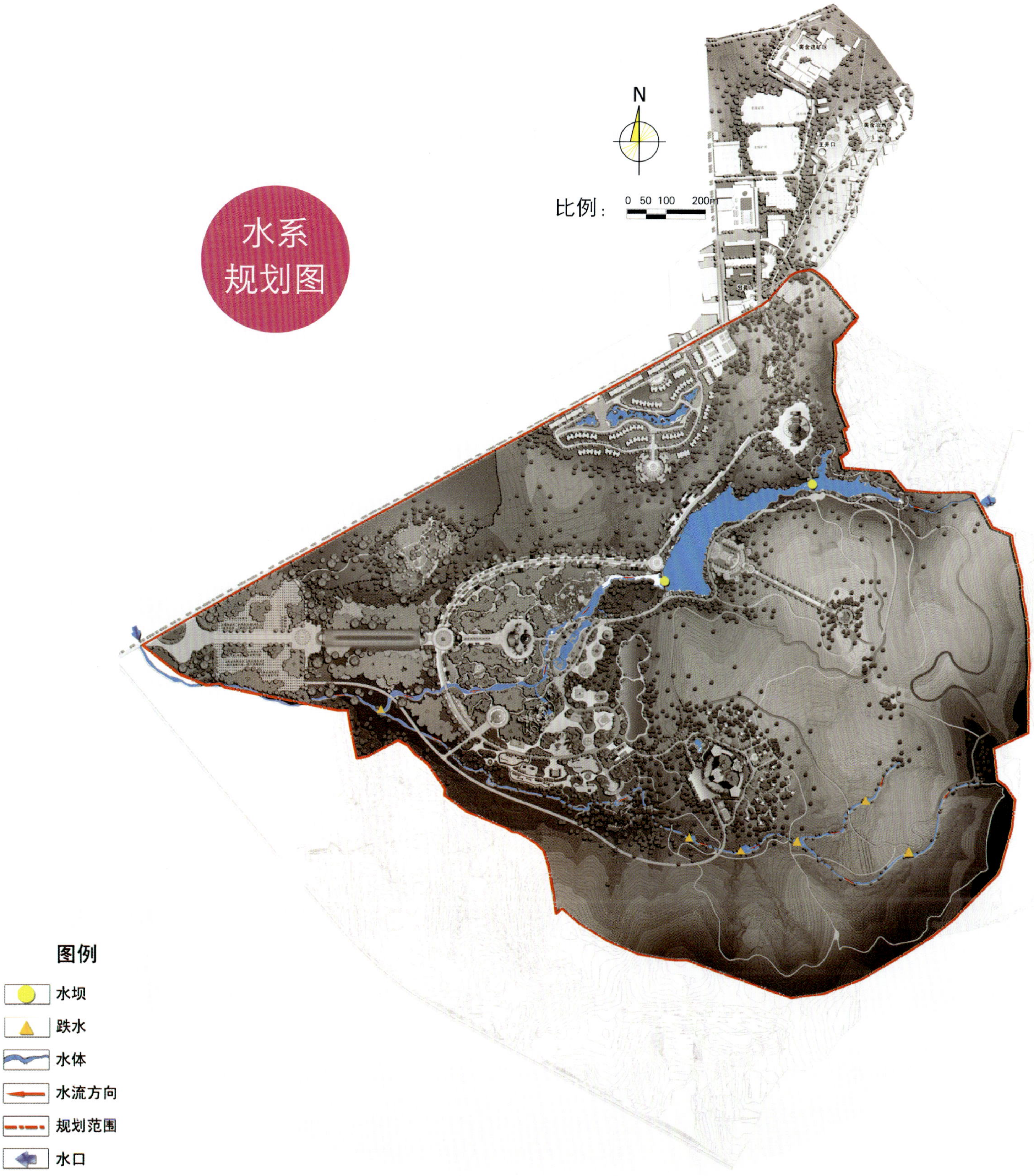

6 主体工程规划 /Planning of Major Projects

第一节 总体布局

将景区布局为“五大区域”、“十六大板块”。

五大区域包括：黄金主题区、现代游乐区、园林景观区、自然生态区、黄金地产开发区。

十六大板块包括：金牛广场、太阳神广场、黄金城、现代金矿街、黄金古镇、黄金博物馆、黄金生产观光区、三情园、黄金祈福台、淘金谷、现代游乐城、拓展中心、黄金休闲度假区、园林景观区、自然生态区、黄金地产开发区

Section Ⅰ Overall Layout

The scenery site is divided into "Five areas" and "Sixteen Forums".

Five Areas: Gold Theme Area, Modern Amusement Area, Garden Area, Natural Ecological Area, Gold Estate Development Area.

Sixteen Forums: Gold Bull Square, Sun God Square, Gold City, Modern Gold Mine Street, Gold Ancient Town, Gold Museum, Gold Producing and Viewing Area, Sanqing Park, Gold Blessing Platform, Gold Finding Valley, Modern Amusement City, Expanding Center, Gold Leisure Area, Garden Area, Natural Ecological Area, and Gold Estate Development Area.

第二节 功能及形象描述

五大区域：

一、黄金主题区：黄金文化主题公园的主体园区。所有景观建筑、景观雕塑和景观小品都紧紧围绕着“黄金”这一主题，是一个集黄金文化展示、黄金生产体验于一身的黄金文化大观园。

二、现代游乐区：黄金文化主题公园的补充园区。是一个会聚现代游乐设施、体验现代游乐情趣的现代游乐城。

三、园林景观区：黄金文化主题公园的人工园艺区。是一个集果园、花卉、人工园艺于一体的人工园艺景观区，是游人参与种植、采摘观赏等园艺体验的场所。

四、自然生态区：黄金文化主题公园的自然生态区。是滨海城市不可多得的具有江南风韵的天然氧吧，是游人回归自然、缓解都市生活疲劳感的绝佳去处。

五、黄金地产开发区：黄金文化主题公园的黄金住宅小区和黄金度假别墅区。

Section Ⅱ Function and Image

Five Area

Ⅰ Gold Theme Area: As the main area of Gold Cultural Theme Park, and with all the landscape constructions, landscape sculptures and landscape accessories reflecting the theme "gold", this area is a superb garden with the demonstration of gold culture and experience of gold producing.

Ⅱ Modern Amusement Area: With modern amusement facilities, it is a complementary of Gold Cultural Theme Park for visitors to experience modern interesting.

Ⅲ Landscape Area: The artificial garden area of the Gold Culture Theme Park consists of orchard, flowers and artificial gardening area where visitors can participate gardening activities like planting, picking fruits and so on.

Ⅳ Natural Ecologic Area: Being a rare natural resort with fresh air and southern charm, the natural ecologic area of Gold Culture Theme Park is the best place which makes people return to nature and release them from tense life.

Ⅴ Gold Estate Development Area: It includes residential area and villa area of Gold Cultural Theme Park.

总平面图

序号	用地类别	面积（m²）	比例（%）
1	总用地面积	1201000	100
2	总建筑面积	59621	——
3	建筑占地面积	21738	1.81
4	道路广场面积	128315	10.68
5	绿化面积	880310	76.45
6	水体面积	132754	11.06

建筑密度：1.81%
综合容积率：0.049
绿地率：76.45%
停车位：1210个

图例

- 规划建筑
- 规划道路
- 停车场
- 水体
- 挡土墙
- 规划范围

十六大板块

第一板块：金牛广场——黄金文化主题公园的门景系统和游客集散地，由生态导入景观、停车场、门景系统等构成。

第二板块：太阳神广场——园区管理和服务系统由金乌负日雕塑、园区商务系统（影像服务部、电子商务服务部、邮政银行服务部）、管理系统（导游服务部、安全保卫服务部、寄存服务部）以及黄金城入口等构成。

第三板块：黄金城——黄金饰品展示区，由园区标志性景观建筑黄金迷宫、黄金首饰商场、与黄金相关的工艺品商场、园区景观建筑工艺品商场等构成。

第四板块：现代金矿街——园区旅游商业街，由黄金文化长廊、各类旅游纪念品商场、土特产商场，以及黄金饮食街等构成。

第五板块：黄金古镇——园区金品加工作坊街，由古镇门景、黄金饰品加工作坊、黄金工艺品加工作坊、古镇风情街（酒肆、茶楼、咖啡屋、小吃坊、娱乐厅、博弈馆）以及景观帆船和水域风光等构成。

第六板块：黄金博物馆——由黄金博物馆单体建筑构成。此为黄金开采史、黄金冶炼史等黄金文化以及金洲矿业集团发展史展示区。

第七板块：黄金生产观光区——由采掘现场和选冶现场等组成。是游客进入亲临黄金生产现场的体验区。

第八板块：三情园——由婚庆广场、爱情园、亲情园、友情园、金翠湖、情侣谷等构成。

第九板块：黄金祈福台——由福星、禄星、寿星、黄金趣味园和金光大道构成。

第十板块：淘金谷——原始淘金体验区，由淘金谷门景、淘金工房、矿车、铁轨、淘金河、淘金水榭、景观水车、空中水渠、景观山体等构成。

第十一板块：现代游乐城——由黄金剧场、卡通城、欢乐迪斯科、过山车、摩天轮、飞天轮、极速之旅、旋转木马、遨游太空、弯月飞艇、疯狂飞碟、高空棱、跌倒的巨人、弹跳飞人等游乐设施，以及卡丁车训练场、山地车俱乐部构成。

第十二板块：拓展中心——由管理系统、服务系统和拓展训练场所构成。

第十三板块：黄金休闲度假区——由黄金大酒店、黄金会所、黄金度假别墅等构成。

第十四板块：园林景观区——由果园、花圃和人工园艺等构成。

第十五板块：自然生态区——由自然生态林、天然湿地构成。

第十六板块：黄金地产开发区——由小区民用建筑和度假别墅等构成。

Sixteen Forums

The First Forum: The Gold Bull Square—gate landscape of Gold Culture Theme Park and tourists distribution consisting of ecologic landscape, parking lots and gate landscape system.

The Second Forum: Sun God Square—The management and service system consist of the sculpture of "Jin Wu Fu Ri" (in Chinese, it means a gold crow carries the sun on its back), business system of the park (film service department, electronic business service department, post and bank service department), management system (guidance service department, security service department, and deposit department) and the entrance of the Gold City.

The Third Forum: Gold City—The exhibition area of golden ornaments consisting of landmark building of "gold maze", gold accessory store, crafts shopping centre related to gold and landscape building crafts shopping centre.

The Forth Forum: Modern Gold Mine Street—Tourist business street of the park consisting of gold culture gallery, Shandong Peninsula specialty, various kinds of tourist souvenir shops, Shandong folk culture show street and gold food and drink streets.

The Fifth Forum: Ancient Gold Town—Gold products process street in the park consisting of the ancient town door view, golden ornament process mill, gold craft process mill, folk street of ancient town (inns, tea houses, coffee houses, snack stalls, amusement house) and landscape ship and water views.

The Sixth Forum: Gold Museum—it is the exhibition area of gold culture such as gold mining history and gold refining history.

The Seventh Forum: Gold producing and sightseeing area—it is an experiencing area for tourists to go into the gold producing spot, consisting of excavation spot and smelting spot.

The Eighth Forum: Three Relations Park—it consists of wedding plaza, Loving Park, Kinship Park, Friendship Park, Jincui Lake and Lover's Valley.

The Ninth Forum : It consists of Golden Avenue, Fu, Lu stars, Star Sculpture and Gold Gout garden.

The Tenth Forum: Gold Valley—It consists of the original gold scene of gate view, gold workshop, mine cart, trail, gold river, gold watershed pavilion, landscape waterwheel, air water channel and landscape mountain.

The Eleventh Forum: Modern Amusement Center—It contains pleasure facilities such as the gold theater, cartoon town, happy disco, roller coaster, ferries wheel, journey speed, Merry-go-roundmerry-go-round, tour in a spaceship, crescent airship, crazy flying saucer, upper air ridge, fallen giant, leaping man, cart training ground, and mountain bike club.

The Twelfth Forum: Outward Development Centre—It consists of outward development management system, service system and outward development training equipment.

The Thirteenth Forum: Gold Leisure Vacation Area—It consists of Gold Hotel, Gold Club and Gold Resort Villas.

The Fourteenth Forum: Sight-seeing Garden—It consists of fruit garden and flower nursery area and maze, etc.

The Fifteenth Forum: Natural Ecologic Area—It consists of natural forest and everglade.

The Sixteenth Forum: Golden Real Estate Developing Area—It consists of folk house and resort villas, etc.

功能分析图

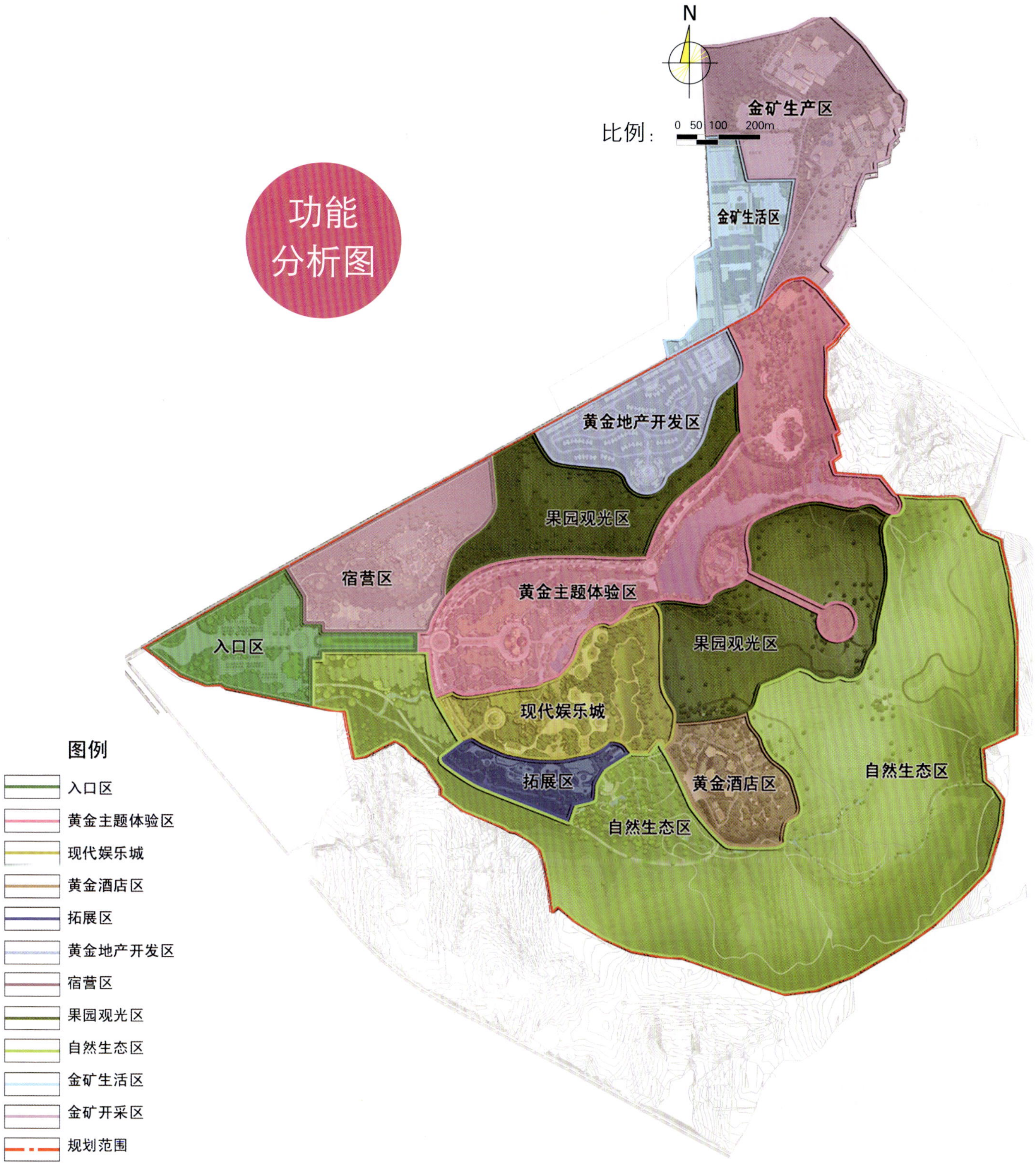
N
比例：
0 50 100 200m
金矿生产区
金矿生活区
黄金地产开发区
果园观光区
宿营区
黄金主题体验区
入口区
果园观光区
现代娱乐城
拓展区
黄金酒店区
自然生态区
自然生态区
图例
入口区
黄金主题体验区
现代娱乐城
黄金酒店区
拓展区
黄金地产开发区
宿营区
果园观光区
自然生态区
金矿生活区
金矿开采区
规划范围

7 基础设施规划 /Planning of Infrastructure

第一节 道路设施

一、出入口的选择：按照《乳山旅游发展总体规划（2005～2020年）》要求和用地特征，黄金文化主题公园的主要出入口设置在规划用地的西侧，与通河南村大道连接；次入口设置在规划用地的北侧，与现有矿区道路接轨。

二、道路骨架：依托通河南村大道，组织黄金文化主题公园内部交通，考虑车行的便捷性、步行的流畅性、游赏路线的合理性以及对景观的影响程度。在连接通河南村大道的道路中，设置一条进入黄金公园的主干道，引领游人进入园区；在生态停车场布置一条9米宽的主干道引车流进入酒店区和别墅区。在节点分流的地方由景观广场贯穿连接，用于协调组织用地的交通，使得整个道路网骨架清晰、流畅、合理。

三、消防通道：黄金文化主题公园内的道路网，形成“大环套小环”的车行交通系统，使各个功能区内部形成消防环道。黄金文化主题公园内的建筑间距以及与周围建筑间距均满足建筑消防规划，并满足疏散距离及宽度要求。园区内的主干道在平时作为人行道、电瓶车使用，一旦起火时可兼作消防通道使用，从而使得各个分区都有消防安全通道，各项消防设施均满足规划要求。

四、流线组织：黄金文化主题公园的道路系统在功能上为人车分流制。车行系统分为三级：即消防通道、果园观光道路和山地车游览道路，形成环行的车行交通系统，与广场、景观区相结合，形成连续的绿色空间。

五、道路绿化：同一道路的绿化宜保持统一的风格，不同路段的绿化形式可有所变化。

六、停车场：当前，自驾游已经成为一种潮流，考虑到停车、游线、管理等因素，在规划中采用集中停车和分散停车方式。规划在主入口停车场可以停放约1150辆；另一部分为地面临时停车和室内停车，分设在酒店和别墅区，可停放30辆。

七、竖向设计：在进行总体布局时，首先充分利用地形和环境，尽可能地顺应地形和环境，减少土石方工程量、挖填方量对用地环境的破坏，尽量在园区内平衡土石方量，力求土石方不外运。规划中保证主次道路线形的合理平顺，满足道路交通的技术要求，划分出的地块也相对完整，有利开发建设。主干道纵坡坡度不大于8%，次干道不大于9%，在土石方相对平衡情况下，建筑用地坡度调整在25°以内。

Section Ⅰ Road Facilities

Ⅰ Entrance–exit: Selection: in accordance with the requirement and land use property of General Planning on Rushan Tourism Development (2005~2020), the main entrance–exit of the theme park will be located in the west side of the planned land used to connect the South Tonghecun highway and the secondary entrance–exit located in the north side to connect the existing road in the mining area.

Ⅱ Road Framework: Backed by the South Tonghecun Highway, organizes the internal traffic preparation around the theme park and considers driving convenience, walking fluency, tourist track rationalization and the possible influence caused by such factors. A main road with quick access to the theme park shall be created to guide the tourist and a main road at the ecological parking area with easy access to the hotel and villa area created to guide the traffic stream. Just at the joint division, the landscape square shall connect all paths to coordinate the traffic until a clear, smooth and appropriate road framework.

Ⅲ Fire–fighting Access: The road network in the gold culture theme park shall enjoy a convenient traffic system as "small ring road surrounded by the big ring road outside" so as to form a fire–fighting access in each functional area. All space and distance between buildings or between buildings and surrounded counterparts shall not only meet the requirement of the fire–fighting planning, but also satisfy the need of evacuation. The main roads in the park for pedestrian and electro mobile use on ordinary occasions and shall be used as fire–fighting access in case of fire, so that each functional area has its own fire–fighting access to meet the planning requirement of fire fighting.

Ⅳ Streamlining Organization: The road network in the park shall be recognized based on functions as "walking–driving system". The driving lane shall cover 3 grades, i.e. fire–fighting access, orchard sightseeing path and upland driving touring lane. An annular traffic system, in combination with the square and landscape, shall generate a continual green space.

Ⅴ Roadside Greening: The greening style shall be kept identical on the same roadside, but may change on the different section of the same road.

Ⅵ Parking Lot: At present, self–driving travel is more and more popular. Considering the factors like parking, itinerary and management, two parking modes are adopted in this planning, i.e. centralized parking and dispersed parking. It's planned that the ecological parking area at the main entrance shall accommodate 1150 cars. Besides, there shall be ground parking area for interim use and indoor parking area mainly located in hotel or villa with accommodation 30 cars.

Ⅶ Vertical Design: When the general layout is under consideration, making full use of terrain and background shall take priority so as to adapt itself to the environment and decrease the damage to the environment. It's also required to make balance of earth–rock volume in the park with no earth–rock transporting outward. The planning shall ensure an appropriate and smooth line shape of the main or minor roads so as to suit the technical requirement of road traffic and guarantee the divided plot in relative integrity for further development. The longitudinal grade shall be no more than 8% for main roads and no more than 9% for minor roads. On the basis of the earth–rock balance, the longitudinal grade for construction land use shall be adjusted within 25°.

道路分类 Item	长度 Length	宽度 Width	面积 Area
主干道（消防道）Main road(fire–fighting access)	5411m	9m	48451m^2
山地车道 Upland driving lane	4708m	6m	32485m^2
园区路 Campus network	12181m	4m	481149m^2
合计 Total	——	——	562085m^2

交通
流线图

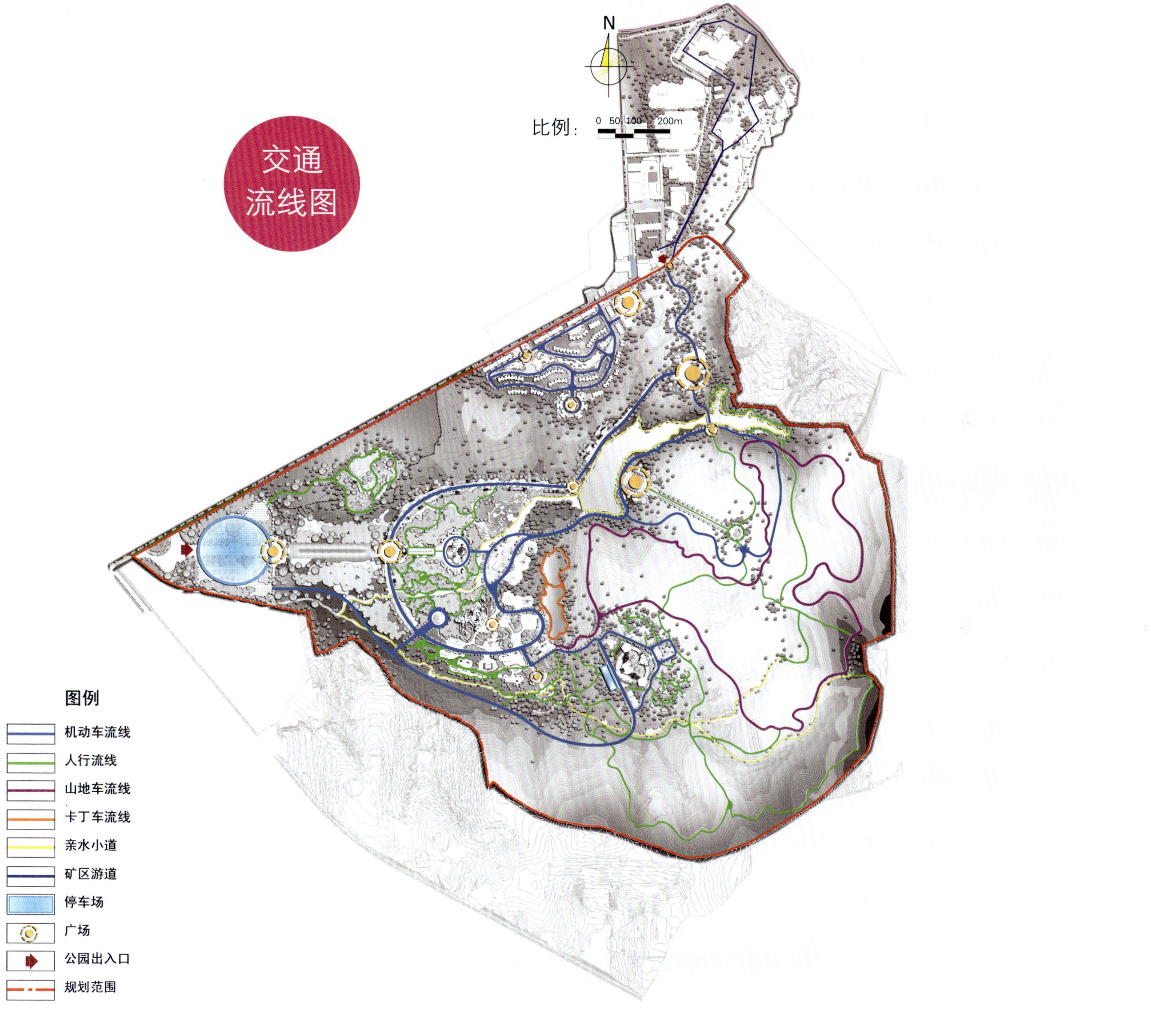
N
比例：
0 50 100 200m
图例
机动车流线
人行流线
山地车流线
卡丁车流线
亲水小道
矿区游道
停车场
广场
公园出入口
规划范围

第二节 给水设施

一、水源的选择：饮用水源选择的原则是先地表，后地下；先节水，后远距离调水，北部厂区、黄金公园内的各建筑采用自来水供水。非饮用水取自金翠湖，其取水量应不大于河流枯水期的可取水量。

金翠湖的保护

1. 金翠湖周围半径 100m 的水域内，严禁打捞、游泳和从事可能污染金翠湖的任何活动，并设置明显的标志。

2. 金翠湖水域内，不得排入生活污水，其沿岸防护范围内不得堆放废渣，不得设立堆站或装卸垃圾、废弃物，不得从事有可能污染该水域水质的活动。

3. 将金翠湖沿岸划为卫生防护带。

二、供水设施建设

输配水管道的管径应满足规划期给水规模和近期建设的要求，宜沿规划道路敷设，并应尽量缩短线路长度，减少跨越障碍次数。

三、用水量预测——分类加和法

分类加和法是分别对各类用水进行预测，求出各类用水后再加和，得到用水总量：

生活用水按 180L/(人·d)，公共建筑用水按 $100m^3$/(万 m^3·d)，绿化和道路用水 1 L/(m^2·d)，停车场地面冲洗水按 2L/(m^2·d)，未预见水量按上述之和的 20%计算，则黄金文化主题公园可预见水量为：5628568L/d；不可预见用水量为 1125705L/d；总用水量为 6754273L/d。

第三节 污水工程

一、污水量

污水量按照用水量进行预测，园区内设计污水量包括居民生活污水量和公共建筑生活污水量。综合生活污水量按综合生活用水定额的 80% 预测。

二、污水排放

1. 本园区内的污水由污水管道排出流入城市污水管道。排水管道应布置在排水区域内或地势较低或便于雨污水汇集的地带。排水管宜沿园区内道路敷设，并与道路中心平行。

2. 污水管沟系统根据园区的规划和建设情况统一布置，分期建设。污水管沟断面尺寸应按远期规划最大流量设定，并考虑公园远景发展的需要。管沟平面位置和高程应根据地形、土质、地下水位、道路情况、原有的和规划的地下设施以及施工条件等因素综合考虑确定。

三、雨水工程

雨水管沟系统规划要充分利用自然水渠，采用分散雨水口，充分利用金翠湖的调蓄作用。在地势平坦区埋设深度或出水口受限制时，可采用渠道（盖板沟）排除雨水。

Section Ⅱ Water Supply Facility

Ⅰ Selection of Water Source: The selection of water source shall conform to the principle of "surface water in priority over ground water; water saving in priority over long–distance water transfer; the plants in north and all buildings in the park with water supply from the pipe; the volume of non–drinking water from the Jincui Lake no more than the available volume during the low–flow season".

Jincui Lake Protection

1. Within an area of 100m radius circle water, any refloatation, swimming or any other activity which may contaminate the Jincui Lake shall be prohibited with recognizable marks in place.

2. No domestic sewage shall be allowed to drain into the Jincui Lake. Besides, no waste residue shall be allowed to pile up at the protection bank, no stock dump or waste loading and uploading. Any activity which is possible to contaminate the water shall be prohibited.

3. The sanitary protection zone shall be set up at the both banks of the Jincui Lake.

Ⅱ Construction of Water Supply Facility: The diameter of water supply pipe shall meet the requirement of water supply volume at the planning stage and the following construction, and shall be placed along the planned road with shortest length and fewest spanning.

Ⅲ Estimation of Water Consumption—Categorized Additive Approach

Categorized additive approach is applied to calculate the total water consumption by predicting the required water in various fields and adding up the estimated figures. Household water consumption shall be estimated as 180L per person per day, public construction water use 100L per 10 000SQM per day, greening and road water use 1L per SQM per day, flushing water use at the parking 2L per SQM per day, other unpredictable water use 20% of the total water consumption. Therefore, the water consumption for the theme park shall be as follows: the estimated figure 5 628 568L per day, the unpredictable water use 1 125 705L per day, total figure 6 754 273L per day.

Section Ⅲ Sewage Engineering

Ⅰ Sewage

Sewage quantity shall be forecast based on the water consumption. The planned sewage in the park includes household sewage and public sewage. The general sewage flow shall be estimated as 80% of the water consumption norm.

Ⅱ Sewage Discharge

1. All sewage in the park shall be drained by the sewage line into the designed public pipeline system. The sewage line shall be set at the discharge area, at the low–lying place or at the location where the rain water is easy to flow together, and shall be set along the road in the park to make sure it is parallel to the central line of the road.

2. The sewage discharge system shall be set in unification but constructed in phases. The section size of the sewage pipe shall be set according to the max flow rate of long–range planning and the further development of the theme park. The plane location and altitude of the discharge system shall be determined based on the terrain, geology, water table, road conditions, original underground facility, existing underground facility and construction conditions.

Ⅲ Rainwater Engineering

The rainwater pipe system shall make full use of the natural channel and the Incubi Lake's impound function by utilizing dispersive rainwater outlet. The trench cover shall be used to discharge the rainwater when the laying depth or rainwater outlet is affected at the level ground.

给水排水规划图

图例

- 规划红线
- 给水管道
- 雨水管道
- 污水管道
- 污水处理站
- 雨水排放口

第四节 电力设施

一、电源

10kV 电源由北部厂区 35kV 变电站进两路 10kV 市政供电。整个规划区根据用电分区负荷计算表和用电负荷量，在场地北部设 10kV 开闭所 1 座，再以放射式向各 10kV/0.4kV 变电所供电，开闭所采用双路供电，互为备用，自动切换。10kV 及以下线路全部采用埋地敷设方式。

二、用电负荷预测——综合用电水平法

根据单位消耗用电量来推算各分类用电。

标准及用电量 / 项目	标 准	用电量	备 注
生活区用户	10kW/ 户	11400 kW	1140 户
建筑	70W/m^2	4.17 kW	59621m^2
公共绿地及道路广场用地	10kW/hm^2	25.8 kW	2.58hm^2
用电总量	--	11429.97 kW	--

实际用电量以总用电量的 70% 计算，即为 8000.98kW。

第五节 电信设施

一、电话系统：电话线路由园区北部市政电信主干网分别引入规划区内，再分支到各建筑物。

二、有线电视系统：直接由乳山市有线电视网引入电视电缆到有线电视前端箱。

三、数据通信系统：由乳山市数据网采用光纤引入到数据管理房，之后用数据专用线敷设到各功能区及其他需要的地方。

四、闭路监控系统：在规划区内各主要节点设置监控点，监控信号集中引至园区内保安监控中心，各建筑物内由本楼内保安监控室负责监控，在需要设置监控点之处设监控设备，在园区保安监控中心与各建筑物保安监控室之间设置报警电话，报警信号及监控信号联络。

五、消防自动报警控制系统：在黄金博物馆内设消防自动报警中心，负责火灾报警及控制，在需要设置监控点之处设监控设备，各建筑物消防自动报警控制室设置报警电话，报警信号及控制信号联络。

Section IV Power Facility

I Power

The 10kV power supply shall be transformed into two 10kV line by 35kV power substation in the North plant. According to the power load estimation and load rate, a switching substation shall be set up in the north site and supply the power to all 10kV/0.4kV power substations by radial ways. Double line for power supply with backup for each other and free switch automatically shall be applied in the switching substation. All wire for 10kV power or below shall be buried in the ground.

II Estimation of Power Load—Horizontal Approach for General Power Consumption

Estimation on power consumption in different fields based on power consumption per unit

Standard&Power / Item	Standard	Power consumption	Note
Resident	10kW/house	11400kW	1140house
Construction	70W/m^2	4.17kW	59621m^2
Public green land and land use for road and square	10kW/hm^2	25.8 kW	2.58hm^2
Total power Consumption	--	11429.97kW	--

The real power consumption shall be calculated as 70% of the total, namely 8000.98kW.

Section V Telecommunication Facilities

I Telephone system: Telephone line will be introduced respectively into planning area by the main municipal telecom network in northern part of the park and embranched to buildings.

II Cable television System: Television cable will be introduced directly into headed box of cable television by cable television network in Rushan city.

Ⅲ Data Communication System: Data network in Rushan city introduces data communication system into data management house, and then uses leased line to lay it to functional zones and other places needed.

IV Closed Circuit Television System: Set up monitory point at primary nodes in planning area and introduce pilot signal into safety monitoring and control center in the park. Security guards in control & security monitor room take charge of the building. Install monitoring equipments in places needed and set up telephone alarm device, alarm signal and monitoring signal liaison between safety monitoring and control center of the park and building.

V Automatic Fire Alarm Control System: Automatic Fire Alarm Control System which is responsible for fire alarm and control will be set up in the Gold Museum. Monitoring equipments will be available in monitoring points and people will also have access to alarm reporting telephone, alarm signal and control signal liaison in automatic fire alarm control room.

供电规划图

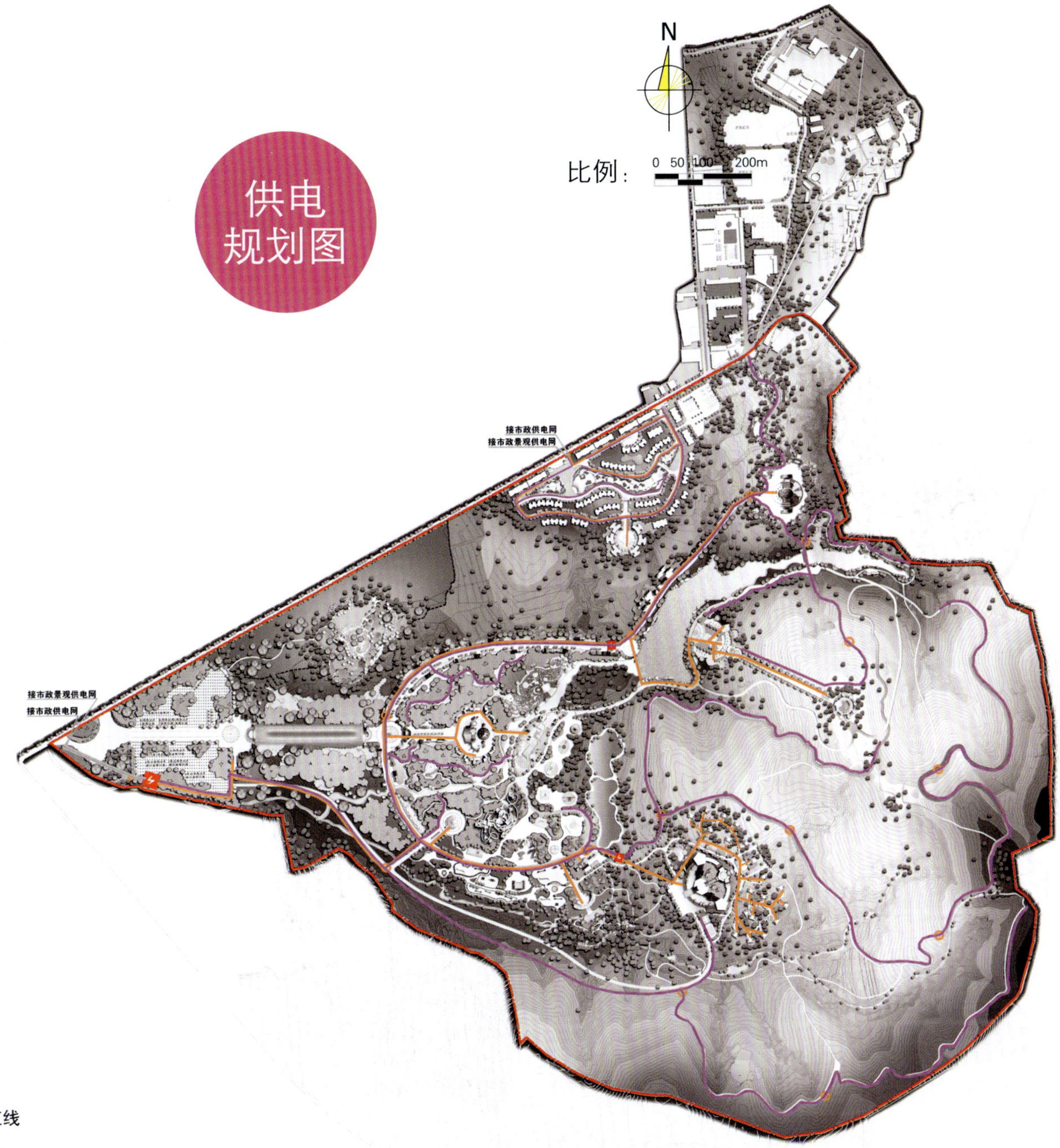

图例

- 规划红线
- 检查井
- 变电箱
- 市政供电网
- 照明供电线网
- 景观供电线网
- 市政景观公电线网

8 环境防灾规划 /Environmental Protection Planning from Disasters

第一节 绿地规划

一、景观系统

1. 金牛下初广场、太阳神广场、情人谷、金翠湖、婚庆广场、淘金谷、湿地景观、建筑小品、绿地、节点广场等形成一条贯穿整个园区的环形景观轴带。

2. 精心设计健身设施、休闲步道、景观小品等，使其与黄金主题建筑、自然景观等相互融合，交织成一幅“虽由人作，宛自天开”的画面。

3. 巧妙利用山地、坡地处理建筑的起伏和层次，形成丰富的天际线为景观增色。

二、园林景观配置原则

植物配置上要符合生态原则，营造四季有花，四季各有季相特色的生态景观，争取做到在不同时节都有观赏亮点。使人们时刻都感到生态园“可览”、“可游”、“可赏”，达到终年游客不断的目的。

三、建筑景观配置原则

根据整体规划，分期开发原则，黄金文化主题公园的建筑将按经营目标进行全面规划，然后按实际需要逐步配套，以保证园区建筑布局合理、格调一致，避免杂乱无章、互不关联，尽量减少前期投入和后期维修养护费用。

第二节 生态保护

黄金公园按照总体规划要求，要对规划用地内原有的自然资源环境进行保护和合理适度的开发；要对用地内的环境资源，包括河流水景、环境景观、树木植被、山头林带等进行分类保护，保护自然景观，防止黄金文化主题公园的自然生态环境人工化和城市化。

一、河流水景

黄金文化主题公园内各建筑和规划用地内的景观用水的排放，对生活污水及其他各种污水处理，处理达到《污水综合排放标准》的一级标准后方可排放，避免对金翠湖造成污染。

二、绿化植被

要维护规划区内的风景环境的自然本色，保护现有的滨河的植被和规划区内的良好树种，保护自然景观和山头绿地的生态景观。

Section Ⅰ Greenbelt Planning

Ⅰ Landscape System

1. "Jinniu Xia Chu"Plaza, Sun-god Plaza, Valley of Love, Great Lake, Wedding Plaza, Gold-Valley, wetland landscape, street furniture, greenbelt, node squares, etc.—all these have formed a circular landscape axis running through the whole park.

2. The exquisite design of fitness facilities and leisure trail with scenes goes harmoniously with gold theme architectures and natural landscapes, interweaving a landscape of "although people do, resembles natural".

3. Masterfully use highlands and sloping fields to handle rise and fall and layers of the building as well as form a colorful skyline so as to add hues to landscape.

Ⅱ Park Landscape Match Principles

Plant furnishing should conform to ecological principles, which is meant to create ecological landscapes in four seasons with respective characteristics in an effort to present highlight of aesthetic enjoyment in different seasons so as to make people feel that the eco-park is worth "enjoying", "visiting" and "appreciating" all the time and serve for the purpose of having tourists coming one after another in all round year.

Ⅲ Building Landscape Match Principles

According to the overall planning and principles of phasing development, the buildings in the Gold Cultural Theme Park will be planned comprehensively in line with business objectives and assembled gradually according to practical needs so as to ensure that the building layout in the park is reasonable, agreeable, in order and in connection with one another and minimize pre-phase investment as well as post-phase maintenance costs. Building landscape match principles are as follows:

Section Ⅱ Ecological Protection

In accordance with master plan requirements, the Golden Park should protect and develop reasonably the original natural resources and environments in planning land use as well as provide classified protection to environment resources, including water landscape, environmental landscape, vegetation cover, hills and forest belts, in order to prevent artificialization and urbanization of natural eco-environment in the Gold Cultural Theme Park.

Ⅰ Water Landscape

The discharge of building water and landscape water in planning land use in the Gold Cultural Theme Park as well as management of sewage in daily life and other sewages should reach the first-level Integrated Wastewater Discharge Standard so as to avoid pollution in Jincui Lake.

Ⅱ Green Vegetation

Maintain natural landscape environment in planning area, protect vegetation cover along Binhe River and fine tree species as well as protect natural landscape and ecological landscape in hilltops and greenbelts.

绿地
规划图
N
比例：
0 50 100 200m
图例
道路绿化
景观绿化
果园
自然生态绿地
规划范围

三、环境卫生

1. 废弃物处理：固体废弃物要从源头治理，要争取按分类收集垃圾、分类运送及处理。有的可以再生利用，变废为宝，有的可以燃烧，变成沼气或能源，有的需要运到废弃场掩埋。

2. 黄金文化主题公园规划以贯彻“全面规划、合理布局、美化环境”的指导思想，使其展现出清洁、优美、宜人的旅游环境。建立环卫设施，加强维护管理，配置相应设施，应配有保洁废物箱、水冲式公共厕所、垃圾中转站等清洁设施。

第三节 安全防护

一、消防措施

1. 消防建设应认真贯彻“预防为主、防消结合”的方针，以“高起点、高水平、高标准”的要求，建立完善的消防体系。

2. 消防给水、消防车道、消防通道等设置严格按国家现行规范、标准进行建设，确保消防安全。

3. 各项建设必须执行国家有关防火规范，健全消防设施和管理制度，保证消防通道畅通，提高预防和补救的综合能力。园区内的建筑严格按照总规要求，根据消防规划要求设置消火栓、灭火器等。

二、防洪措施

1. 加强流域内水土流失的治理，保持、恢复植被，重点治理薄弱地段。

2. 金翠湖要加强工程控制调度管理，提高防洪、抗洪能力。建立防洪报警系统和防洪抗洪指挥系统，加强防灾、减灾工作。

3. 加强金翠湖河道管理，禁止向河内倾倒垃圾和废弃物。

三、防地质灾害

1. 在规划区内道路选线应顺应地形，在满足交通景观和消防的前提下，结合地形，避免高切坡、深开挖，防止建设工程引发人为地质灾害，陡坡与堆积土地段应采取相应措施。

2. 在项目定点、修建过程中，应充分考虑滑坡、塌陷、泥石流等地质灾害的影响。加强排洪措施、植树种草及按规范放坡等综合工程治理，在建设场地的斜坡地段可采用设置挡墙和各种排水渠道等。

3. 在对项目进行规划设计的时候，必须严格遵守“先勘察，再设计，后施工”的原则。

Ⅲ Environmental Hygiene

1. Salvage and Scrap Disposal: solid waste should be handled from its source and efforts should be made to collect wastes separately as some of them can be recycled and turned into valuable materials, some of them can be burned into biogas or energy and some others can be transported and buried in junkyard.

2. Gold Cultural Theme Park planning implements the guidance of "comprehensive planning, reasonable layout and beautified environment" in order to present a clean, beautiful and agreeable tourist environment, establish environmental hygiene facilities, strengthen maintenance and management, assemble correspondent facilities as well as equip cleaning facilities such as cleaning waste-bin, water flush public lavatory, and garbage transfer stations.

Section Ⅲ Safety Precaution

Ⅰ Fire Protection

1. Fire protection construction should carry out the policies of "put prevention first and combine prevention with elimination" and the requirements of "high starting-point, high level and high standard", and establish sound fire protection system.

2. The installation of fire water supply, fire-fighting passage and fire fighting access strictly adheres to current national codes and standards in order to ensure fire safety.

3. The construction should implement the related national fire safety rules, improve fire control facilities, ensure smooth fire engine access and increase comprehensive precaution and remedy capabilities. The buildings in the park should strictly conform to general rules, and install fire hydrant and extinguisher in line with fire planning and requirements.

Ⅱ Flood Mitigation Measure

1. Intensify the treatment of water loss and soil erosion in river basin, maintain and recover vegetation cover with an emphasis on the governing of the weak sections.

2. Reinforce engineering control scheduling management, increase flood prevention and control capacity in Jincui lake; establish flood prevention alarm system and flood prevention and control command system and intensify disaster prevention and alleviation work.

3. Strengthen river management in Jincui Lake and prohibit dumping garbage and refuse into the river.

Ⅲ Geological Disaster Prevention

1. In the planning area, road location should conform to the topography, combine with the terrain environment under the premise of meeting transport landscape and fire prevention, avoid highly incised slope and deep excavation, prevent construction engineering from incurring man made geologic hazard, and adopt correspondent measures in slope and cumulose soil sections.

2. During the process of project location and construction, full consideration should be given to the impact of geological hazards like landslide, collapse and debris flow. Intensify comprehensive engineering treatment such as water release measures, tree and grass planting and sloping inn line with the codes and adopt barricade and various drainage channels.

3. While planning and designing the projects, strictly adhere to the principles of "first surveying, then designing and finally construction".

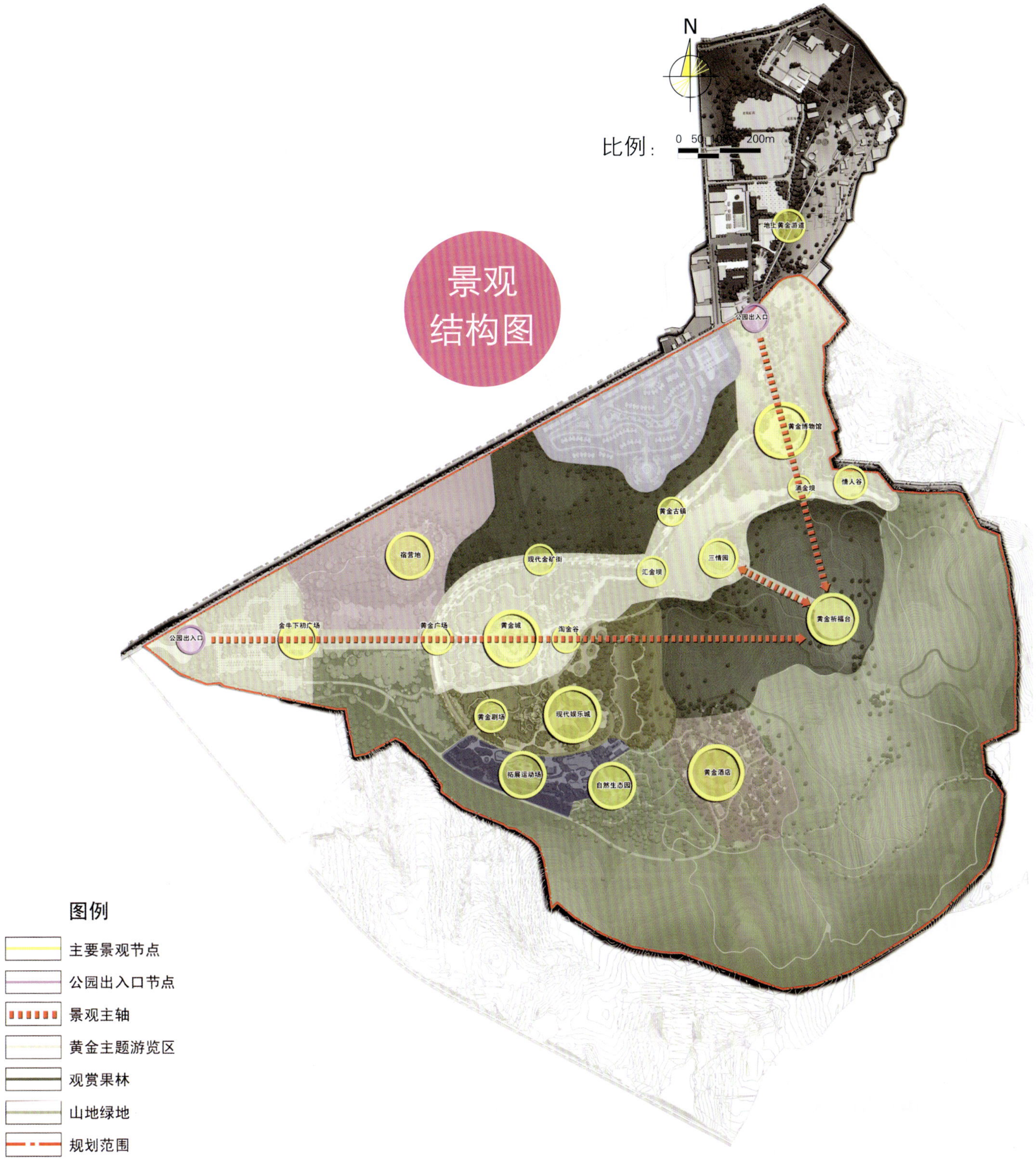
景观
结构图
N
比例：0 50 100 200m
地上黄金游道
公园出入口
黄金博物馆
洒金坝
情人谷
黄金古镇
三情园
汇金坝
现代金矿街
宿营地
黄金祈福台
公园出入口
金牛下初广场
黄金广场
黄金城
淘金谷
黄金剧场
现代娱乐城
拓展运动场
自然生态园
黄金酒店
图例
主要景观节点
公园出入口节点
景观主轴
黄金主题游览区
观赏果林
山地绿地
规划范围

植物配置图

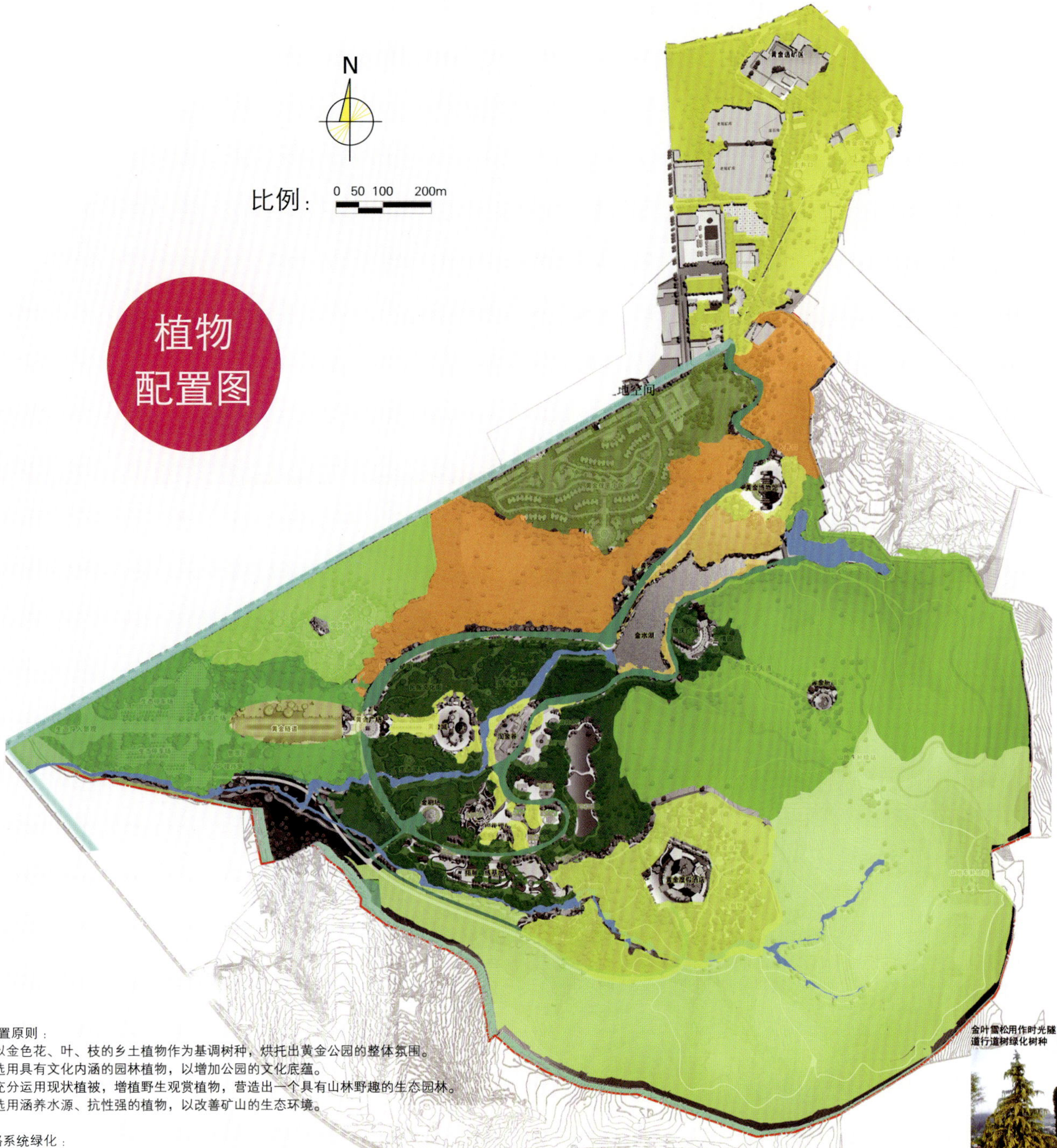

植物配置原则：

（1）以金色花、叶、枝的乡土植物作为基调树种，烘托出黄金公园的整体氛围。
（2）选用具有文化内涵的园林植物，以增加公园的文化底蕴。
（3）充分运用现状植被，增植野生观赏植物，营造出一个具有山林野趣的生态园林。
（4）选用涵养水源、抗性强的植物，以改善矿山的生态环境。

1. 道路系统绿化：

- 生态入口、生态停车场绿化：选用水杉、银杏、金叶红瑞木等植物，对称与自然式布置相结合的手法，构成一个生态环保、适用优美、大气磅礴的公园入口。
- 外围道路绿化：选用黑杨作为行道树种，较为紧密栽植，对公园起到围护作用。
- 时光隧道中央绿化：选用金叶雪松，池杉、海棠作为绿化树种，形成山峰的感觉。
- 公园内部道路绿化：选用糖槭、银杏、雪松、白蜡、伊朗金缕梅等作为行道树，营造出一个季相明显，独具特色的公园道路系统。

2. 公园主题参观区绿化：

- 公园中心区绿化：选用梅花、桂花、牡丹、蜡梅、杏、竹等具有浓厚历史文化 的园林植物，使其与黄金文化相融合。
- 建筑周边植物配置：选用金钱松、海棠、菊花、萱草等庭院植物，营造出简洁大方的建筑环境空间。
- 宿营地植物配置：选择小苹果、杏、雪松等树种，营造一个开敞，灵活的自由活动场所。
- 生态湿地绿化：选用莩蓬草、黄花鸢尾、千屈菜，睡莲、菖蒲等水生植物，营造一个大型生态湿地空间。

3. 公园自然生态系统绿化：

- 果园观光绿化：选用苹果、杏、石榴、枸橘、柿树等“金果”作为主体树种。
- 板栗原生植被区域充分利用现有的高大板栗，在林下增加棣棠、鸡爪槭等树种，营造一个四季山花烂漫的森林景观
- 松柞混交林原森植被区：在现有松、柞林的基础上，增植黄金槐、元宝槭、梓树、黄刺玫等树种，营造出周边“金山”的氛围。

金叶雪松用作时光隧道行道树绿化树种

黄金槐枝、叶金黄，速生树种用作公园山体大面积体块绿化树种

伊朗金缕梅用作公园次干道行道树

金钱松为世界五大庭院观赏树种之一用作建筑周边绿化树种

糖槭用作公园主路行道树绿化

9 服务设施及游线规划 /Service Facilities and Tourism Planning

第一节 基础服务系统

一、住宿服务：在园区设置特色客栈（自驾游宿营地）、星级宾馆和酒店别墅，既顾及一般游客的住宿需求，又满足中高档游客的入住要求。

二、饮食服务：在园区设置酒吧、风味小吃、大众餐馆（黄金饮食街）和高档酒楼（黄金大酒店），以满足不同层次游客的消费需要。

三、娱乐服务：为丰富游客旅游生活，园区配置娱乐厅和体育健身等场馆，可满足游客文体娱乐需要。

四、游购服务：为满足游客购物需要，园区建有黄金饰品购物一条街和其他商品购物场所，以方便游客为亲朋好友购买馈赠礼品。

五、拍摄服务：为使游客留下珍贵的旅游镜头，园区设有摄影服务部，专门为游客提供拍摄、冲印服务，拍摄器材出租服务，以及出售胶卷、电池以及摄影器材等服务。

Section Ⅰ Fundamental Service System

Ⅰ Lodging: Special inns (drive camp), stars hotels and villas are available in the park to meet the lodging needs of both ordinary and distinguished tourists.

Ⅱ Diet: Bars, flavor snack, public restaurants (the Prime Diet Street) and great hotel (Gold Grand Hotel) to satisfy visitors of all levels.

Ⅲ Entertainment: Clubs and gymnasiums are needed to enrich visitors' tour.

Ⅳ Shopping: A street selling gold ornaments and other commodes places are in the scheme for visitors to buy presents for their relatives and friends.

Ⅴ Photography: In order that visitors can leave precious travel photographs, photography services are set to provide visitors with photography, print, rent photography equipments, and sell films and photography equipments.

第二节 信息服务系统

一、功能牌示：建立规范完整的标识系统，诸如管理标志、方向标志、限制标志和解说标志等。在园区入口处设立大型全景立体导游图牌；在重要岔路口设立交通示意图或指示牌；在重要景观前设立说明牌。所有牌示设计应与园区协调一致，并充分考虑其美观、耐用等因素。所有牌示一律采用中英文互文形式。

二、导游中心：建立功能完善的导游中心，为游客提供旅游咨询、安排导游员（含英文导游员），提供导游书籍和导游图（含英文版本）。

三、公用电话亭：为方便游客通信联系，在园区接待中心设置公用电话，在园区主要景点和路段设置公用电话亭。

四、商务网络中心：为方便游客传输旅游照片，发送电子邮件，查阅有关电子资料，在景区设置专门的商务网络服务中心。

Section Ⅱ Information Service

Ⅰ Functional Sign: To establish regular and completed sign system, such as signs of management, direction, limit and explanation: to set grand guiding map board of the whole scenic spots at the entrance of the park, traffic diagrams or direction boards at the important forks and explanation boards before important sceneries. All expressions are in two languages of Chinese and English.

Ⅱ Guiding Center: To establish guiding center with perfect functions to answer tourists' inquiries, arrange guiders (including English speaking guiders), and provide tourists with guiding books and maps (including English editions).

Ⅲ Public Telephone Booths: To equip public telephone at the reception center and public telephone booths at the important scenic spots and locations in order that tourists can get convenient communication.

Ⅳ Business Web Center: To set special business web center in the park so as that tourists can transport tourism photos, send emails and check relevant electronic materials.

第三节 安全服务系统

一、导游服务：建立健全导游解说系统，向游客出售导游图或其他导游资料，设置导游标志，提供导游语音系统，为游客指明游览路线，以提高游览效果，避免游客迷失方向。导游资料必须充分体现指示性、规定性、说明性、解释性和宣传性，并以中英文互文形式出现。

二、警示服务：在园区不允许触摸、攀摘、攀爬以及可能出现危险的地段和有关高危设施处设置专门的警示标志，同时派专人巡逻，以免发生景观受损或游客人身伤亡事故。警示牌要醒目、美观；用文字表述的，要有中英文对照。

三、安全防护：为游客提供下井观光游览的安全帽、工作服和手电筒等安全防护用品，以确保游客下井安全；在地形险要的游道，以及游人集中，而边缘临空高差超过 1m，容易发生跌落等人身事故的场所，设置防护栏和其他安全保护设施；对较为惊险的活动派专业人员对游客进行专门保护。

四、医疗服务：在园区设置医疗机构，为游客提供简易医疗服务和急救处置服务。

五、配置防火设施：按国家标准在园区设置安全防火设施。

第四节 游览路线组织

根据金洲黄金文化主题公园内的景观布局和项目设置，将景区的游览线路按不同的主题大致分为以下三类：

一、黄金主题游线：

以体验黄金文化为主要内容的游览线路。游线安排如下：金牛广场—黄金隧道—太阳神广场—现代金矿街—黄金古镇—黄金博物馆—黄金生产观光区—涌金坝—三情园—黄金祈福园—汇金坝—淘金谷—现代游乐城—太阳神广场。

二、婚庆主题游线：

以婚礼、庆典参加者为主要受众设置的主题游览线路。游线安排如下：金牛广场—黄金隧道—太阳神广场—黄金城汇金坝—婚庆广场。

三、生态主题游线：

以满足游客回归自然、享受自然的需求，设置的主题游览线路。游线安排如下：金牛广场—黄金隧道—太阳神广场—拓展训练基地—山地车俱乐部—自然生态园—山地车游览道—情人谷。

Section Ⅲ Security Service System

Ⅰ Guiding Service: To set up and complete guiding and explanation system, sell guiding maps and other guiding materials to tourists, set guiding signs, provide guiding sonic system, direct tourism routes to visitors so as to keep them from lost and improve their travel effect. Guiding materials should embody fully indication, regulation, description, explanation and popularization, also in both languages of Chinese and English.

Ⅱ Warning Service: To put warning signs on the places where touching, picking and climbing are prohibited, high and dangerous facilities are set, and danger may happen. Meanwhile, patrolmen are needed to keep landscapes from destroyed and tourists from hurt. Sign boards should be eye-catching, beautiful and any character expression should be in both Chinese and English.

Ⅲ Security Service: To provide tourists with personal protective equipments such as safety helmets, work uniform and electric torches for sightseeing in the pits so as to keep their safe; set protection railings and other security facilities in the places where visitors are crowed.

Ⅳ Medical Care Service: To provide tourists with simple medical care service and first aid by setting medical care in the park.

Ⅴ Fire Prevention Facilities: To set fire control facilities in accordance with national norm.

Section Ⅳ Travel Routes

According to landscape layout and items setup, the scenery area can be classified generally into three travel routes.

Ⅰ Gold Theme Route

It is a route for experiencing gold culture. The route is: Golden Bull Plaza—Golden Tunnel—Sun God Plaza —Golden City—Modern Gold Mine Street—Old Golden Town—Golden Museum—Viewing Area of Gold Process—Gold Flatland—Three Relation Park—Gold Blessing Park—Gold Concentrated Flatland—Gold Finding Valley—Modern Amusement Center—Sun God Plaza.

Ⅱ Wedding Theme Route

It is a route for the participators of weddings and ceremonies. The route is: Golden Bull Plaza—Golden Tunnel—Sun God Plaza—Golden City—Gold Concentrated Flatland—Wedding Plaza.

Ⅲ Ecologic Theme Route

This route is set to meet tourists' needs of returning and enjoying nature. The route is: Golden Ox Square— Golden Tunnel—Expending Training Base—Mountain Bike Club—Natural Ecologic Garden—Pavilion for Mountain Bike—Valentine's Valley.

10 经济技术指标 /Economic and Technique Target

规划用地平衡表

序号	分类	面积（hm^2）	比例	备注
1	规划总用地面积	252.29	100%	
2	建筑用地	4.56	1.81%	
3	道路广场用地	12.83	5.08%	
4	绿化用地	203.21	80.54%	含公共绿地、果园、山头绿地、景观绿地
5	水体面积	13.27	5.26%	
6	其他用地	18.43	11.4%	含黄金住宅小区、游乐设施用地、等

▲建筑密度：1.81%

▲容积率：0.049

▲绿地率：80.54%

▲停车位：1180 个

▲规划用地总面积：252.29hm^2

注：厂区内用地及长 1768m 的地上黄金栈道面积未计入本规划用地平衡表中。

Balance of Planned Land Use

NO.	Classification	Area(hm^2)	Proportion	Note
1	Planned Area for Total Land Use	252.29	100%	
2	Total Construction Area	4.56	1.81%	
3	Land use of squares and Roads	12.83	5.08%	
4	Land use of forestation	203.21	80.54%	Including public green land, orchards, hills and landscape
5	Water area	13.27	5.26%	
6	Other land use	18.43	11.4%	Including golden residential area, land use of entertainment facilities and so on

▲ Construction density: 1.81%

▲ Volume rate: 0.049

▲ Greening rate: 80.54%

▲ Parking lots: 1 180

▲ Planed area of total land use: 252.29 hectares

Note: Area of land use in the plants and ground golden footway with 1 768 length is not included.

公共服务设施

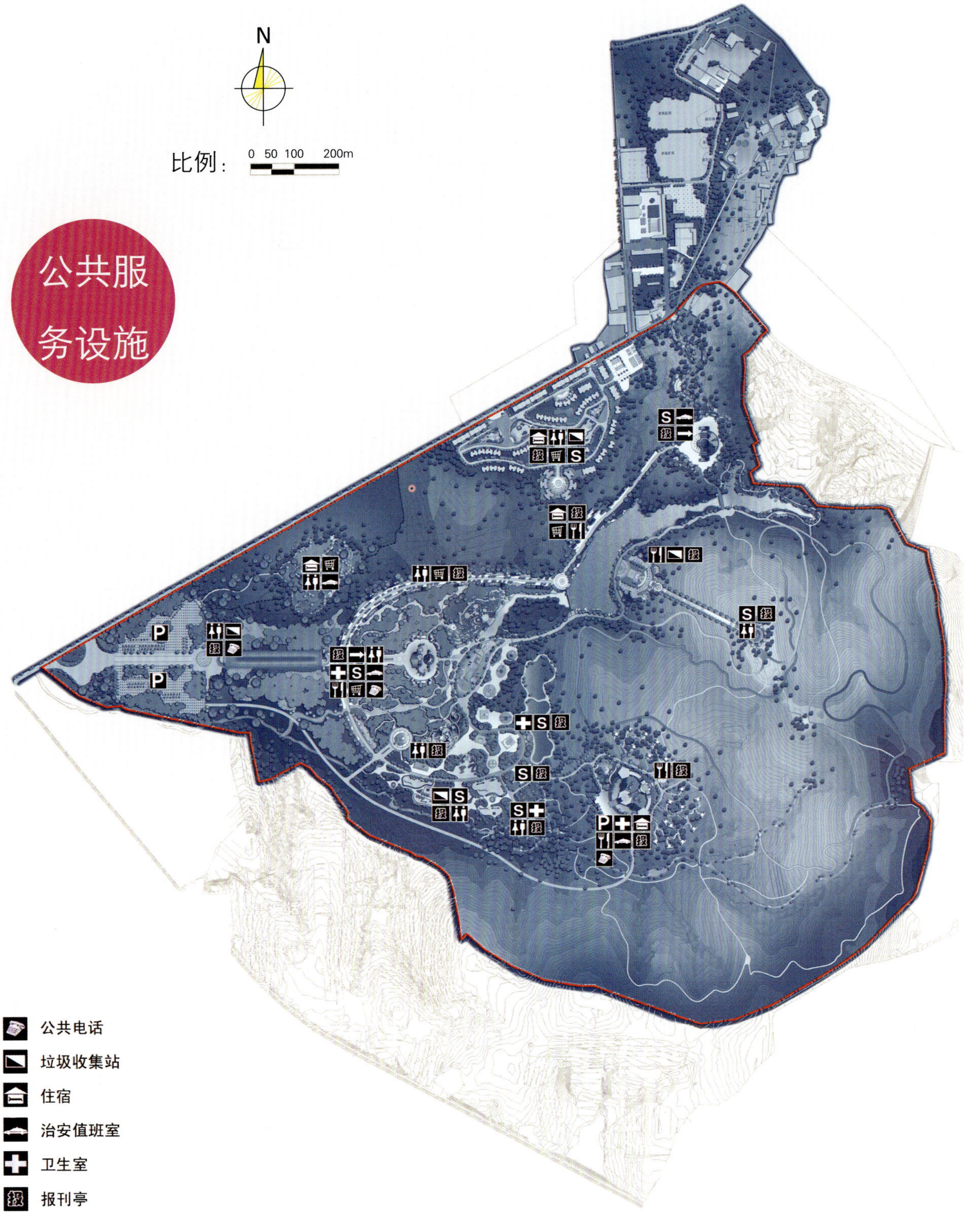

图例

- S 管理中心
- 餐饮
- 公共厕所
- 商店
- 导游图
- P 停车场
- 公共电话
- 垃圾收集站
- 住宿
- 治安值班室
- 卫生室
- 报刊亭

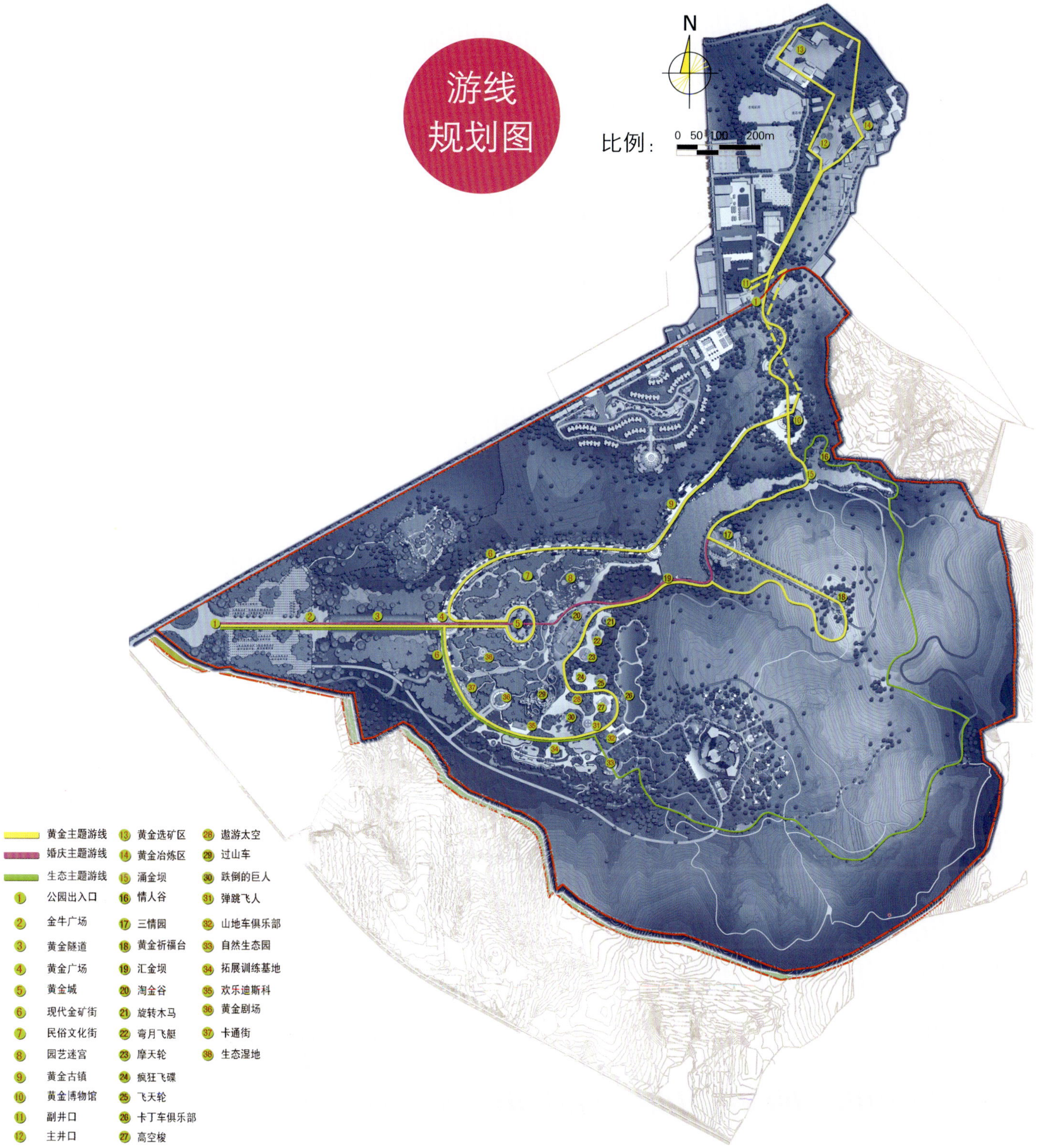
游线
规划图
N
比例：
0 50 100 200m
黄金主题游线
婚庆主题游线
生态主题游线
1 公园出入口
2 金牛广场
3 黄金隧道
4 黄金广场
5 黄金城
6 现代金矿街
7 民俗文化街
8 园艺迷宫
9 黄金古镇
10 黄金博物馆
11 副井口
12 主井口
13 黄金选矿区
14 黄金冶炼区
15 涌金坝
16 情人谷
17 三情园
18 黄金祈福台
19 汇金坝
20 淘金谷
21 旋转木马
22 弯月飞艇
23 摩天轮
24 疯狂飞碟
25 飞天轮
26 卡丁车俱乐部
27 高空梭
28 遨游太空
29 过山车
30 跌倒的巨人
31 弹跳飞人
32 山地车俱乐部
33 自然生态园
34 拓展训练基地
35 欢乐迪斯科
36 黄金剧场
37 卡通街
38 生态湿地

Shandong Gold Park

效果图件 Effective Pictures

黄金公园
全景鸟瞰

入口区鸟瞰图

生态导入景观：位于园区前沿，与通河南村大道相连。利用园区富余地块，配置相应的树丛花卉（树丛以黄色树叶为主，花卉以黄色花朵为主），同时，在路旁搁置和点缀些许大块金矿石和金矿生产设备，通过写意的、虚化的艺术表现手法将金矿与生态的元素符号有机地组合成一体，使其即具象征性，又具观赏性，从而成为既具有景观效果的园艺绿化带，又具有导入功能的黄金文化主题公园的导入景观，成为引领游人进入黄金文化主题公园，拉开黄金文化主题公园游览序幕的前奏曲。

Eco-lead landscape: It locats in the leading-edge of the park and connects to thoroughfare of Henan village. Make good use of the margin plot to plant correspondent groves and flowers (mainly groves with yellow leaves and yellow flowers). Meanwhile, arrange and embellish some blocks of gold ore and gold ore producing equipments, organically integrate gold ore with ecological elements through leisurely and virtual artistic devices so as to make it symbolic and appreciative and become not only greenbelts with landscape effects but also leading landscape with leading functions and in gold cultural theme park. It will also become the prelude that will lead tourists into and pull open the tour prologue of the gold cultural theme park.

①

金牛广场：黄金文化主题公园的门景系统和入园通道，由“金牛下初”雕塑和黄金舞台、黄金隧道、售票大厅、VIP接待处、公共厕所等组成，利用匠心独运的标志性雕塑，以及井架、破碎机、滑轮等金矿符号构筑的建筑景观，着力打造黄金文化主题公园独有的、与众不同的园区风貌，让游人一进入园区就眼前为之一亮，精神为之一振，获得一种独有的感观享受。

Golden Bull Plaza: Golden Bull Plaza is the landscape system and entrance of the Golden Cultural Theme park, consisting of the sculpture of "Jinniu Xiachu", Gold Stage, Gold Tunnel, Ticket Hall, VIP Reception Center and Public Lavatory. The plaza makes good use of gold ore symbols such as creative landmark sculptures, derrick, crusher, pulley, etc. to construct building landscapes in an effort to blaze a unique and different park landscapes in the golden cultural theme park, which will catch the eyeballs of tourists when they enter the park and give them a unique feeling and enjoyment by refreshing their energy.

售票厅：开放式发售门票服务厅。

Ticket hall: Opening service halls for selling tickets.

VIP 接待处：高端游客接待处。

VIP reception desk: High–end tourist reception desk.

①入口区局部鸟瞰图 ②表演广场 ③ VIP 接待室 ④售票厅

黄金舞台：位于园区出入口正中，以金矿破碎设备——球磨机剖面为主体造型，作为开园仪式及演艺活动的表演台。为吸引游客，增加游乐情趣，营造园区浓郁的游乐氛围，黄金文化主题公园每日必须定时举行开园仪式，在隆重、热烈、欢乐和喜庆的开园仪式中，将游客一拨一拨送入黄金文化主题公园。

The Gold Stage: The Gold Stage is located in the middle of the park entrance with gold detritus equipment—ball mill section as its main style. It is regarded as a performance stage for the park opening ceremonies and other related activities that can attract tourists and add pleasures to the park. The gold cultural theme park should hold opening ceremonies at regular time everyday. Tourists will be absorbed to the gold cultural theme park one wave after another during grand, ardent, joyous and jubilant opening ceremonies.

黄金舞台分开园仪式表演舞台和闭园仪式表演舞台，前面（金牛广场）为开园仪式表演舞台，后面（太阳神广场）为闭园仪式表演舞台。

The Gold Stage is divided into performing stage for opening ceremonies and performing stage for closing ceremonies. The performing stage for opening ceremonies is in front of the Golden Bull Plaza while the performing stage for closing ceremonies is at the back of the Sun God Plaza.

金牛雕塑：园区标志性雕塑。以“金牛下初”掌故为题，采用艺术演绎夸张的表现手法，整座雕塑由金牛、祥云和底座组合而成，白色的金属构件代表翻卷的祥云；通身金黄的金牛身长羽翅，表示其为上苍派遣到人间的神牛；金光闪闪的底座代表乳山黄金矿山。雕塑的寓意为：金牛乘祥云由天庭而下，秉玉皇大帝的旨意来到人间为乳山人民探矿寻宝，造福乳山人民；乳山的黄金宝藏是自然造化，更是上苍对勤劳智慧的乳山人民的垂爱。如此创意，既艺术地演绎了乳山黄金宝藏的由来，同时，也深刻地表达了乳山人民对金牛的深切怀念和无限崇敬。雕塑底座铭记“金牛下初”的传说故事概要。

Golden Bull Sculpture: The Golden Bull Sculpture is a landmark sculpture in the park with "Jinniu Xiachu" as the theme. The sculpture adopts artistic and exaggerate devices and is made up of golden bull, lucky clouds and golden pedestal. The white hardware represents rolling lucky clouds extending to the wings of the golden bull which is a heavenly bull dispatched by the heavenly god to the human world. The glittering pedestal represents the golden mine in Rushan Mountain while the implied meaning of sculpture is that the golden bull comes to the human world by riding the lucky clouds from the heaven, bearing an order from the jade emperor to help the people in Rushan Mountain explore mines and search treasures so as to bring benefit to the people there. Rushan Mountain is endowed with the natural gold treasure which shows dearly love of the heavenly god to the intelligent and diligent people in Rushan Mountain. Such creative work not only manifests the origin of the gold treasure in Rushan Mountain but also conveys profoundly the deep memories and endless respect of Rushan people towards the golden bull. Below the sculpture pedestal in engraved with the summary of folk tales of "Jinniu Xiachu".

①时光隧道及表演广场局部

黄金隧道：位于黄金舞台两侧，置身于人造山体之中，宽 36.9m，长约 226.8m。是进入黄金文化主题公园的入口。设计打破传统的大门入口形式，以模拟“时光隧道”和以金矿矿车演变而成特殊造型的“黄金火车”作为游客进入黄金主题公园的交通工具，着力追求超凡脱俗、出人意想的形式美感，刻意营造黄金文化主题公园特有的“神秘”色彩。游客由此乘坐小火车穿越黄金隧道进入园区。对于身处矿山的矿工来说或许不足为奇，但对于没有矿山体验的人来说就别有一番情趣。尤其是通过声光效果的处理营造一种遨游天体宇宙、穿越“时光隧道”的感觉；而穿出“时光隧道”，眼前又一片金碧辉煌，一下子又进入到金光灿烂的黄金世界，这种视觉上的强烈变化，体验上的强烈转换，必然会产生妙不可言的情趣，游人必然会为之欣喜若狂，惊奇万分。

隧道为回环式，一头进，一头出。为了方便进出，在隧道两侧还辟有步行通道，游客除了可乘小火车出入外，还可从隧道两侧的步行通道进出。

Gold Tunnel: located at both sides of the Gold Stage and in artificial mountains, the Gold Tunnel is 36.9 meters wide and around 226.8 meters long. It is the entrance of entering the gold cultural theme park, which has broken away from traditional entrance of the gate. "The Gold Train" which simulates time tunnel and evolves into a special shape from gold mine car is taken as transport tool for tourists going into the theme park. The train is in pursuit of an aesthetic pleasure of extraordinary rather than ordinary and without expectation so as to create intentionally the exclusive "mysterious" color in the gold cultural theme park. Those tourists who go into the park by riding the train are not a surprising phenomenon for mine workers. However, for those who have never travelled through the mine, such kind of experience is very exciting and fascinating, especially when they get the feeling of travelling in the celestial universe and running through "time tunnel" by using the lighting and visual effects. When they come out of "time tunnel", a resplendent and magnificent view will suddenly appear in front of them and they will once again enter the splendid golden world. Such a strong change in visual and intense transfer in experience will surely shock and please the visitors.

The tunnel is in loopback mode with one end in and the other end out. On both sides of the tunnel, there are walking passages for the convenience of tourists going in and out of the tunnel apart from taking the small train. (See renderings)

①时光隧道出口效果图 ②时光隧道透视图 ③时光隧道透视图

①

黄金城：由园区标志性建筑“黄金迷宫”、黄金首饰商场、与金字相关的工艺品商场、园区景观建筑工艺品商场等构成。整体建筑主要由黄金饰品变形夸张创意而成，主要用于展示黄金饰品和与金相关联的工艺品，以及微型的园区景观建筑工艺品。

Gold City: Gold City consists of the landmark building of "gold maze", gold accessory store, crafts shopping centre related to gold and landscape building crafts shopping centre. The whole building is mainly created and exaggerated by transforming the gold ornaments, which is mainly used to exhibit gold accessories and handiworks made by gold as well as miniature landscape architectural works in the park.

②

黄金城入口：按“柳暗花明又一村”的设计理念，在太阳神广场与黄金城之间筑景观门洞，人为地将太阳神广场和黄金城分隔开来，以增加园区的层次感，着力营造一种别有洞天的观感效果。

Gold City Entrance: Designed by the idea of "seeing the light at the end of the tunnel", a landscape doorway was built between the Sun–god plaza and the gold city, separating the two artificially in order to increase the layering effects of the park in an effort to create a totally different world with a feeling of deep impressions.

①黄金公园局部鸟瞰图 ②太阳神广场效果图

①

①太阳神广场局部效果图

太阳神广场：由“金乌负日”雕塑、“太阳神”演绎而成的服务系统、管理系统建筑组成。“太阳神”演绎而成的景观建筑作为园区的服务系统和管理系统。广场铺装为金光灿烂的太阳图案，与“金乌负日”雕塑及广场周边景观建筑的风格和色彩相互映衬，互为呼应。

The Sun God Plaza: The Sun God Plaza consists of service system and management system evolved from the sculpture of "Jin wu fu ri", meaning that a gold crow carries the sun on its back, and "the Sun God". The landscape buildings evolved from "the Sun God "is viewed as service system and management system in the park. The plaza pavement is decorated with the pattern of the splendid sun which goes well with the sculpture of "Jin Wu Fu Ri"and the style and color of the surrounding landscape buildings in the plaza.

金乌负日雕塑：黄金文化主题公园主要景观雕塑。以“金乌负日”神话传说故事为题，演绎“天上有太阳，地下有黄金，黄金就是金光灿烂的太阳的化身”的黄金文化理念，赋予黄金象征光明，如同太阳一样灿烂辉煌，如同太阳一样神圣和令人崇拜的深刻寓意。

The Sculpture of "Jin Wu Fu Ri" : "Jin Wu Fu Ri"is the main sculpture in the gold cultural theme park. Titled in the legendary of "Jin Wu Fu Ri", it plays with gold cultural idea of "the sun in the heaven with gold on the earth, incarnated in the form of the splendid sun", which gives the gold the symbolic brightness as if it were as spectacular as the sun and as if it enjoyed the profound meaning as the sacred and admirable sun.

①

黄金首饰商场：纯金制品展示区。建筑制式由各种黄金饰品和金矿符号变形夸张创意而成，主要用于展示金项链、金耳环、金戒指、金手镯等纯金制品，是黄金制品的购物天堂。

Gold Accessories Shopping Centre: Gold Accessories Shopping Centre is the display zone of our gold products. The architecture is created and exaggerated by transforming various kinds of gold accessories and gold mine symbols. The purpose is mainly for the showcase of gold necklaces, gold earrings, gold rings, gold bracelets and other pure gold products. It is the shopping paradise of gold-made products.

①黄金城局部效果图 ②景区电瓶车 ③景观小品 ④景观小品

①

黄金迷宫：园区最耀眼的建筑，是黄金文化主题公园的标志性建筑。位于黄金城末端，建筑面积约3652m²。通过奇思妙想的创意设计，赋予黄金迷宫梦幻般的浪漫和神秘情调，既极具观赏性，又不乏趣味性。游人既可在它面前浮想联翩，神游天外；更可进入其中，一直转上顶部，俯瞰周边景观。

Gold Maze: It is both the most eye-catching and landmark architecture in the gold cultural theme park. Located at the end of the gold city, the area of the building is about 3 652m². The novel and creative design has endowed the gold maze with dream-like romantic and mysterious sentiment, which is not only highly apperceive but also dramatic. Tourists can either think big and wildly as if they were wandering in a wonderland or enter the maze till its top and overview its surrounding landscapes.

①黄金迷宫效果图 ②黄金迷宫局部效果图 ③黄金迷宫夜景效果图

现代金矿街：以黄金生产设备符号结合现代建筑手段构建而成，给人以强烈的现代金矿特色和气息。由黄金文化长廊、胶东半岛特产、各类旅游纪念品商场、山东民俗文化展示街、黄金饮食街等构成。

Modern Gold Mine Street: It is constructed by gold producing equipment symbol combined with modern architecture, which gives people a strong modern gold mine characteristics and atmosphere. The street consists of gold culture gallery, Shandong Peninsula specialty, various kinds of tourist souvenir shops, Shandong folk culture show street and gold food and drink streets.

①黄金城金饰街效果图 ②现代金矿街透视图 ③现代金矿街局部效果图 ④现代金矿街局部效果图

山东民俗文化街：位于黄金广场的北部，主要展示、推介山东的民俗文化，如：鲁西南、鲁西北地区的剪纸、刺绣，潍坊、平度、高密、阳谷张秋镇、聊城、惠民清河镇的民间张贴木版年画，高密聂家庄、临沂褚庄、苍山小郭等地的泥玩具，郯城樊埝、鄄城刘家旋木玩具，曲阜的楷雕，菏泽的面塑，即墨、胶南、福山等地的雕果模等。

Shandong Folk Culture Street: Located at north of the gold plaza, Shandong folk culture street mainly displays and introduces Shandong folk culture, namely, paper cut and embroidery in both south and north of Luxi; folk paste wood engraving pictures in Weifang, Pingdu, Gaomi, Zhangqiu town in Yanggu county, Liaocheng and Qinghe town in Huimin county; mud toys in Niejia zhuang, Linyi zhu zhuang, Cangshan xiao guo in Gaomi city; wood toys in Fannian of Tancheng county and Liujia xuan in Juancheng county; sculptures in Qufu; dough modelling in Heze city; sculpted fruit modellings in Jimo, Jiaonan and Fushan in Shandong provinces.

①～⑧民俗文化街局部效果图

①

园艺迷宫：毗邻金翠湖、情侣谷，是一道极具景观效果、观赏性极强的人工园艺景观。设计注重线条和图案的组合，注重观赏性和娱乐性的有机结合。远观是一处颇具观赏性的园艺景观，游走其中如步入妙趣横生的绿色迷宫之中，给游人带来无限的乐趣。

Garden Maze: Abutting on the Jincui Lake and Lover's Valley, the garden is a landscape for sightseeing and appreciation. The design focuses on the line and pattern, composing organically sightseeing and recreation. It is a landscape seeing from faraway. When entering green maze, one can feel it is so interesting and funny.

①园艺迷宫鸟瞰图 ②迷宫大门效果图 ③迷宫局部效果图 ④迷宫局部效果图

①

②

黄金古镇：古风、古韵是黄金古镇的魂，粗犷的寨门、古朴的街巷把黄金古镇打造成一个风韵独具的原生古镇，着力营造一种返璞归真的怀旧情调。

Ancient Gold Town: Ancient wind and rhyme is the spirit of the ancient gold city. Rough village gates, ancient and simple streets and allies forge the ancient gold town into a raw ancient town with unique charms.

古镇门景：以马灯、铁锹等原始符号，以及简陋、残破的门墙和历经沧桑的百年老树创意而成，意在着力营造黄金古镇的古老、久远的氛围。

The Ancient Town Door View: Created by using the original symbol like hurricane lamp and shovel, broken gate and tempered one-hundred-year old trees, the ancient town door view is meant to create an age-old and time-honored atmosphere in the ancient gold town.

①黄金古镇效果图 ②黄金古镇小品效果图 ③黄金古镇入口效果图 ④黄金古镇局部效果图 ⑤黄金古镇局部效果图 ⑥黄金古镇帆船效果图

①

黄金博物馆：这是一座独具魅力的景观建筑。整个建筑以金矿金砖成型设备为造型，大气磅礴、富丽堂皇、匠心独运、出人意料。

Gold Museum: It is a landscape architecture with a unique charm. The whole building is modeled with forming equipment of gold mine and gold brick, majestic and incredible.

黄金博物馆主要通过文字、照片、音像、实物、雕塑、模型等方式，集中展示黄金历史、黄金制品、黄金矿产、黄金生产加工，以及黄金文化等内容，是集黄金知识、黄金科普、黄金文化于一体的博览场馆。同时，也兼具游客进入黄金采掘观光区的功用。游客参观完黄金博物馆后，即由此更换工装进入黄金采掘观光区进行黄金采掘体验。

The Gold Museum displays concentrate such contents as the history of gold, gold products, gold mine, gold production and process as well as gold culture though written languages, pictures, audio and video, real object, sculpture and model. The museum integrates the knowledge of gold, popularization of gold science and gold culture together. At the same time, the museum also possesses the function of enabling tourists to visit the gold excavation sight seeing area. When tourists finish the visit of the gold museum, they can change into work suit and experience gold excavation in the gold excavation sightseeing area.

①黄金博物馆效果图 ②黄金博物馆鸟瞰图 ③黄金采掘体验区入口效果图

黄金生产观光区：由采掘现场和选冶现场等组成。是游客进入亲临黄金生产现场的体验区。

地下黄金滑梯：进入黄金采掘体验区的运载工具，仿奥地利盐矿滑梯构建而成，总长约448m。以此进入方式，可增加游乐情趣。

黄金栈道：是供游人观看黄金选矿和黄金冶炼等生产工艺流程的空中通道，全长1768m，是架设在金矿选冶生产区域的一道亮丽风景线。

Gold Production Sightseeing Area: It consists of excavation spot and smelting spot. It is an experiencing area for tourists to go into the spot in person.

Underground Gold Slide: It is a carrier for entering the gold excavation experience area and it is constructed by simulating Australian salt mine slide with an overall length 448 meters. By this way, tourists can increase their pleasure and excitement.

Gold Plank Path: It is an air passage for tourists to watch production process flow of the gold mine selection and refinery with an overall length 1,768 meters.

①黄金采掘体验区隧道 ②黄金采掘体验区隧道 ③黄金选冶观光区栈道局部

⑥

④黄金选冶观光区栈道局部 ⑤黄金选冶观光区栈道局部 ⑥黄金采掘体验区出口效果图

①

①爱情港湾鸟瞰图 ②爱情港湾局部效果图 ③爱情港湾局部效果图 ④涌金坝效果图

黄金祈福台：由金光大道、福星、禄星、寿星雕塑和黄金趣味园构成，是游客祈盼幸福、祈求利禄、祈望长寿的祈福场所，是黄金文化主题公园最为神圣、最为庄严肃穆的地方。游客可以在这里虔诚地许愿求拜，祈求自己及家人福星高照，好运多福，一生顺遂；祈求自己及家人功成名就，仕途通达，财运亨通；祈求自己及家人一生平安，健康长寿……人一生所期所盼和所求不可谓不多，但归结起来大抵不过如此。黄金文化主题公园借“黄金”而建，自然蕴含了尊贵、权力、财富和幸福、吉祥、恒久等意味，迎合了人们追求权力和财富、幸福和吉祥，向往尊贵和恒久的心理，而“黄金祈福台”的构建也正满足了人们的这种精神追求和心理需要。

Gold Blessing Terrace: As the most sacred and solemn place of the Gold Cultural Theme Park, it consists of the Gold Avenue, the sculptures of Fu Star, Lu Star and Shou Star and the Gold Pleasure Park where visitors can pray lucky, wealth and long-living. Constructed on the "gold", the Gold Cultural Theme Park, beyond words, possesses the meaning of honor, power, wealth, happiness, and long-lasting, which caters people's mind of pursuing power, wealth, happiness and desiring honor and eternity, while the construction of the "Gold Blessing Terrain" is just meet this spiritual pursuit and psychological need of people.

①黄金祈福园鸟瞰 ②黄金祈福园局部图

金榜题名：黄金文化趣味园的第四个场景，由孔子雕塑，书写着论语的竹简和状元榜等组成。以此寓意人生勤奋努力、刻苦学习、求取功名的奋斗阶段。

Names of Successful Candidates: It is the fourth scene of the Gold Cultural Pleasure Park, consists of Confucius sculpture, the bamboo slip with the analects and Hall of Fame. It implies the stage of hard work and the struggle for fame.

金印紫绶：黄金文化趣味园的第五个场景，建于黄金文化趣味园的最高处，游客沿着台阶来到一个方形镜子前，地面上“平步青云”四个红色的字折射在镜子里；镜子的周围是紫色的飘带，以此作为位高权重的象征，寓意人生的成功阶段。

Gold Seal Surrounded by Purple Ribbon: This is the fifth scene in the detour founded at the highest location. Visitors may see in a square mirror 4 red characters "平步青云 (rapidly go up in the world)" reflected by the real counterpart on the ground when stepping up. In Chinese culture, the square mirror with purple ribbons around symbolizes high position and strong power, or the white moments of life.

金盆洗手：黄金文化趣味园的第六个场景，由一个大金盆、六个小金盆组成。大金盆里的水流入小金盆供游客洗手，表示洗去人世污垢，在以后人生道路上就可少出差错，少走“弯”路。以此作为人生的反思、自省阶段。

Washing Hands in Gold Basin: This sixth scene in the Gold Cultural Pleasure Park consists of one big gold basin and six small ones. The waster in the small gold basin flows from the big one for only use of tourist's hands washing. This act, as a higher stage of self-examination and introspection, implies "after washing away dust or dirt in this mortal life, one may enjoy fewer errors and less detours".

①金榜题名效果图 ②金印紫绶效果图 ③金盆洗手效果

求得真金：黄金文化趣味园的第七个场景，中心设计“六轮回”菩萨雕塑（人道、畜生道、天道、阿修罗道、恶鬼道、地狱道），四周布置哈哈镜，借以投射人生百态，暗喻人生的大彻大悟。

Pursuing Genuine Gold: The seventh scene in the gold culture park: with design at the center of "Six reincarnation" Buddha sculptures (human, animals, Heaven, asura, evil spirits Road, hell). It is surrounded by distorting mirrors to project all expressions of the world, and imply discerning and apprehending of life.

金石为开：黄金文化趣味园的第二个场景，由飞镖靶牌和老虎雕塑组成。通过飞镖投射，靶牌打开的形式，寓意正式进入人生的起步阶段。

Sincerity Can Make Metal and Stone Crack: This is the second scene of the Gold Cultural Pleasure Park. It consists of darts target and tiger sculpture. Through the darts, target card, it refers to the formal stage of entering the start of one's life.

一寸光阴一寸金：黄金文化趣味园的第三个场景，由布置在场地四周的沙漏组成，沙子洒落的过程象征时光流逝的过程，以此比喻时光易逝，人生短暂，不应虚度美好的时光。

Time is Money: The third scene in the Gold Cultural Pleasure Park is made up of sand clock placed around the site. The process of sand trickling down is just like the process of time lost, which signifies flying time and transient life deserves to be highly cherished.

①求得真金效果图 ②金石为开效果图 ③一寸光阴一寸金效果图

①

淘金谷：由门景、淘金工房、矿车、铁轨、淘金河、淘金水榭、景观水车、空中水渠、景观山体等原始淘金场景构成，是游人体验古老的原始淘金生活、感受沙里淘金、披沙拣金乐趣的极佳场所。

Gold Valley: Gold valley consists of the original gold scene of gate view, gold workshop, mine cart, trail, gold river, gold watershed pavilion, landscape waterwheel, air water channel and landscape mountain. It is a wonderful place for tourists to experience the ancient and original gold-washing life, feel gold-washing in the sand, and obtain the gold by washing it from the sand.

①淘金谷鸟瞰图 ②淘金谷大门效果图 ③淘金谷局部效果图 ④淘金谷局部效果图

②

③

④

①

③

现代游乐城：由黄金剧场、卡通城、欢乐迪斯科以及过山车、摩天轮、飞天轮、极速之旅、旋转木马、遨游太空、弯月飞艇、疯狂飞碟、高空梭、跌倒的巨人、弹跳飞人和卡丁车训练场、山地车俱乐部等游乐设施构成。

Modern Amusement Cente: It constitutes of pleasure facilities such as the gold theater, cartoon town, happy disco, roller coaster, ferries wheel, journey speed, merry-go-round, tour in a spaceship, crescent airship, crazy flying saucer, upper air ridge, fallen giant, leaping man, cart training ground, and mountain bike club.

黄金剧场：与现代游乐城相配套的四维电影城，是园区内高科技的影视放映场馆。场馆建筑的构想来源于“飞碟”（UFO——天空中的不明飞行物）、金矿磨矿设备和动画变形蜘蛛的外观形状，通过夸张变形创意设计而成。

The Gold Theater: It is a four dimension movie city matched with modern pleasure city. It is a high-tech movie projection pavilion. The structure of the pavilion comes from "flying saucer" (UFO), the gold mine grinding facilities and the shape of cartoon spider, and it is designed by exaggerated transformation.

游乐设施：过山车、摩天轮、飞天轮、极速之旅、旋转木马、遨游太空、弯月飞艇、疯狂飞碟、高空梭、跌倒的巨人、弹跳飞人等。

Pleasure Facilities: Roller coaster, ferries wheel, journey speed, merry-go-round, tour in a spaceship, crescent airship, crazy flying saucer, upper air ridge, fallen giant, leaping man, cart training ground, and mountain bike club.

①现代游乐城鸟瞰图 ②黄金剧场鸟瞰图 ③游乐设施

①

②

③

卡丁车训练场：卡丁车训练场根据卡丁车的结构形式设计而成，以两个大车轮作为卡丁车训练场入口大门主要造型。训练场包括管理系统、服务系统和训练系统，训练系统由培训教练区和比赛跑道构成，是游客进行卡丁运动体验的场所，更可作为乳山市卡丁车训练和比赛的基地。

Kart Training Ground: It is designed on the basis of the structure of the Kart, having two large wheels as the main entrance of the training ground. Training ground consists of management system, service system and training system, and training system including training area and competing lanes, where is the place for Kart campaign experience, it can also serve as a training and competition base for Kart of Rushan city.

④

①跌倒的巨人效果图 ②跌倒的巨人效果图 ③卡丁车跑道鸟瞰图 ④卡丁车俱乐部入口效果图

①

②

卡通城：专为少年儿童设计的游乐园，由适合少年儿童观赏、极富童趣的卡通形象构筑而成，是少年儿童的乐园。游人（携带儿童出游的游客）可将儿童交由园区管理人员负责照顾，自己可轻装上阵游玩其他景区。

Tonicity: A pleasure garden exclusively designed for children, constituted by cartoon images fit for children. Tourists who bring their children with them can hand their children to the administrators in the garden, and they themselves can therefore visit other places with relaxation.

①卡通城鸟瞰图 ②卡通城局部效果图 ③公厕效果图 ④卡通城垃圾筒效果图 ⑤卡通城垃圾筒效果图

KING HOTAL

①黄金大酒店鸟瞰图 ②黄金大酒店正面透视图 ③黄金大酒店正侧面图

黄金休闲度假区：由黄金大酒店、黄金度假别墅、黄金会所构成。

The Gold Leisure Resort: The Grand Gold Hotel, the Gold Resort Villa and the Gold Club.

黄金大酒店：位于规划用地南面。酒店外形取材于金矿井架构筑形制和金矿机器设备构件，通过夸张变异创意而成。造型独特新颖，完全打破了常规酒店宾馆的建筑模式，紧扣金矿朴拙粗犷的气韵和品质，从而形成黄金文化主题公园自己独有的建筑风格。整个建筑呈半围全圆弧状，新颖别致，风格独具。酒店前地面铺装以太阳图案和金矿转道图案构成，既与黄金文化紧扣，又充分展示金矿风貌特点。酒店附属设施有停车场、室内游泳池、室外游泳池、羽毛球场、网球场、娱乐场、酒店别墅等，为进入黄金文化主题公园的游客提供高品位休息和度假的场所。

Gold Hotel: Located at the south of the planned land, the figure of the Hotel is the creative exaggerated variation of the Gold derrick building shape and Gold mining equipment components. This unique and novel architectural style completely broke the conventional hotel construction method, but closely meets the simplification, innocent and rough quality of the gold mine so as to form the unique architectural style of the Golden Cultural Theme Park. The shape of the whole building is a ring without two semi-rounds connected on the top, novel and unique. The ground in front of the hotel is paved with the sun pattern and the pattern of gold diversion so as to closely link to gold culture and demonstrate thoroughly the characteristics of the gold mine. The hotel possesses ancillary facilities like parking, indoor pool, outdoor swimming pool, badminton courts, tennis courts, casino, and hotel villa for visitors to enjoy and take vacation.

黄金会所：位于黄金大酒店后面，建筑造型以四部矿车创意组合而成，是配套黄金大酒店的高端消费区。

Gold Club: Being the matched high-end consumption area of the Gold Hotel, the Gold Club is designed innovatively on the basis of four mine cars and located behind the Gold Hotel.

黄金度假别墅：建筑设计以金黄色为主色调，以金矿机器设备的零部件为主要点缀，建筑风格上追求特色化和别致性，建筑造型上追求几何化和简洁化。所有别墅一律为单体建筑。

Gold Resort Villa: The architectural design takes golden yellow as its main color and the parts in the gold mine machine equipments as the main decoration. The villa pursues differences and uniqueness in its architectural style as well as geometrization and simplicity in its structure. All villas are single unit building.

①黄金会所效果图 ②～④酒店别墅效果图

①

①酒店游泳池鸟瞰图 ②酒店游泳池正面效果图 ③酒店游泳池船效果图

公用设施及景观小品：由园区公共厕所、指示牌、休闲亭、休憩椅、灯饰、电话亭、垃圾箱、艺术景观小品等构成。

Infrastructure and Scenery Accessories: They are made of public washroom, guiding board, pavilion for resting and leisure, benches for rest, lamp, telephone booth, and litter bin, art sketched.

Buddhism Park of Multi–Country—Sino Tibetan World

多国佛教园——华藏世界

Theme Park
主题公园

Buddhism Park of Multi-Country

项目规划 Item Planning

1 项目建设地概况 Outline of Land Use

第一节 清镇市概况

清镇位于贵州省中部，乌江上游鸭池河南侧，东西长42.6km，南北宽56km。是贵州省经济、文化、旅游和交通较发达的地区之一，市内有汉、苗、布依、侗、彝、仡佬、水、回、壮、满、白等30多个民族，少数民族人口12.16万人。占总人口的24.34%，以苗族居多。

1992年经国务院批准撤县设市，1996年划归贵阳市管辖，1999年被省委、省政府列为全省首批建设的20个经济强县（市、区）之一；同时，经过清镇人民的共同努力，清镇市先后获得过“全国双拥模范城市”、“全国科普示范市”、“全省科技工作先进市”，“省级文明卫生城市”、“全省社会治安综合治理模范市”等称号。清镇是贵州重要的工业基地。境内有中央、省、市属大中型企业近20家（其中贵州水晶有机化工集团公司为“全国化工500强企业”），集化工、能源、磨料、磨具、纺织、冶金、采矿于一体。市境内有国家级风景名胜区红枫湖、省级风景名胜区百花湖和市级风景名胜区东风湖，现在城区建设正朝着红枫湖、百花湖及贵阳方向辐射延伸。

Section Ⅰ An Outline of Qingzhen

Qingzhen, located in the north of Wujiang River and the south of Jiachihe River, is in the middle of Guizhou Province. Generally, it is one of the developed territories in economy, culture, tourism and transport in Guizhou Province. There are more than thirty nationalities including Han, Miao, Buyi, Dong, Gelao, Shui, Hui, Zhuang, Man and Bai etc., with the minority 1.216 million, about 24.34% of the total population.

Qingzhen was promoted from a county to a prefecture approved by the State Council in 1992. And it has been controlled by Guiyang City since 1996. In 1999, Qingzhen was determined by the Provincial Party Committees and the government to be the first "one of the twenty Advanced Economic Counties or Prefectures" to develop in the whole province. Meanwhile, with the effort of Qingzhen people, Qingzhen has enjoyed the fames of "National Model City for Two Supports", "National Science Demostration City", "Techology Advanced City of the Province", "Provincial Civilized & Hyginical City", "Comprehensive Administration of Social Security of the Province" and so on. Qingzhen is the important industry base of Guizhou. Almost 20 national, provincial and prefectural enterprisements in the large or middle size (among them, Guizhou Crystal Organic Chemical (Group) Co.,Ltd. is the "Top National Chemical 500 Corportion") are in this territory, including chemical industry, enery, abrasives, testiles, metallurgy and mining etc. And also the national scenic spot Hongfenghu Lake, the provincial scenic spot Baihuahu Lake and the prefectural scenic spot Dongfenghu Lake. At present, the city is extending toward Hongfenghu Lake, Baihuahu Lake and Guiyang City.

第二节 清镇市人文自然环境资源概述

一、自然景观环境

清镇市地处苗岭山脉北侧。地形总趋势是东北、西南高，平均海拔高程为 1290m。

清镇市地貌在内外压力共同作用下，形成了丰富多样的地貌类型。山地、丘陵分别占全市总面积的 31% 和 59%，坝子占 5.5%。

二、水利资源

清镇市位于苗岭山地北坡，西北部界河为乌江上游三叉河及鸭池河。境内猫跳河及其支流、暗流河等均发源于安顺、平坝县境内，向北流入本市，注入乌江。全年水资源总量 8.93 亿 m^3，水力资源可开发 37.7 万千瓦。

清镇市境内建有红枫湖、百花湖和东风湖三大水库。其中红枫湖、百花湖是贵阳市和贵州省早期开发的重要旅游资源，现为贵阳市两大饮用水源地，旅游业开发受到限制。

Section Ⅱ Culture, Environment and Resources

Ⅰ Natural Environment

Qingzhen City lies on the north of Miao Mountain. The average level elevation is 1.29km.

Endogenetic and exogenetic force contribute to various of terrain in Qingzhen prefecture of which 31% mountain land, 59% hills and 5.5% flatland.

Ⅱ Water Resource

Qingzhen lies on the north side of the Miaoling highland, extending to Sanchahe River and Yachihe River which are the upper part of Wujiang River in the northwest. Miaotiaohe River and its branches and Anliuhe River, running north to Qingzhen city and finally to Wujiang river, provide Qingzhen with 0.893 billion cubic meters of water resources and 377.7MW power.

Three main reservoirs were contributed in Qingzhen Prefecture respectively in Hongfenghu Lake, Baihuahu Lake and Donfenghu Lake. Among them, Hongfenghu Lake and Baihuahu Lake are the important tourist resources that were developed in Guizhou in the early time. As the main water resource for people's life of Guiyang today, they are confined in developing tourism.

区位分析图

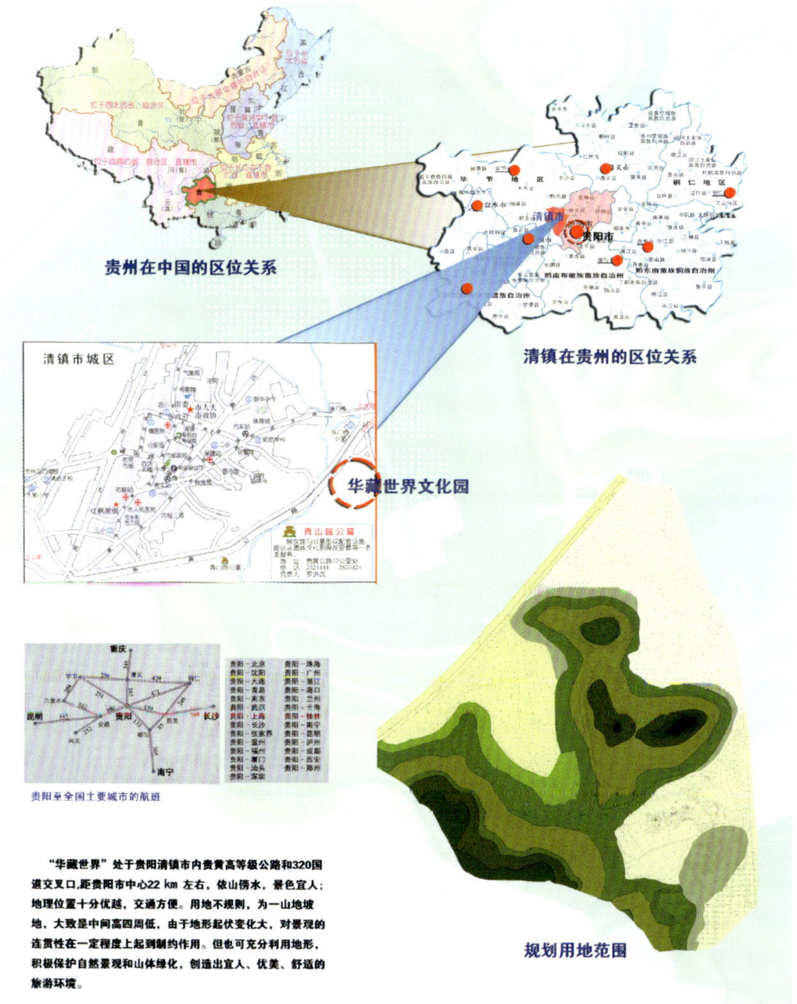

高程分析图

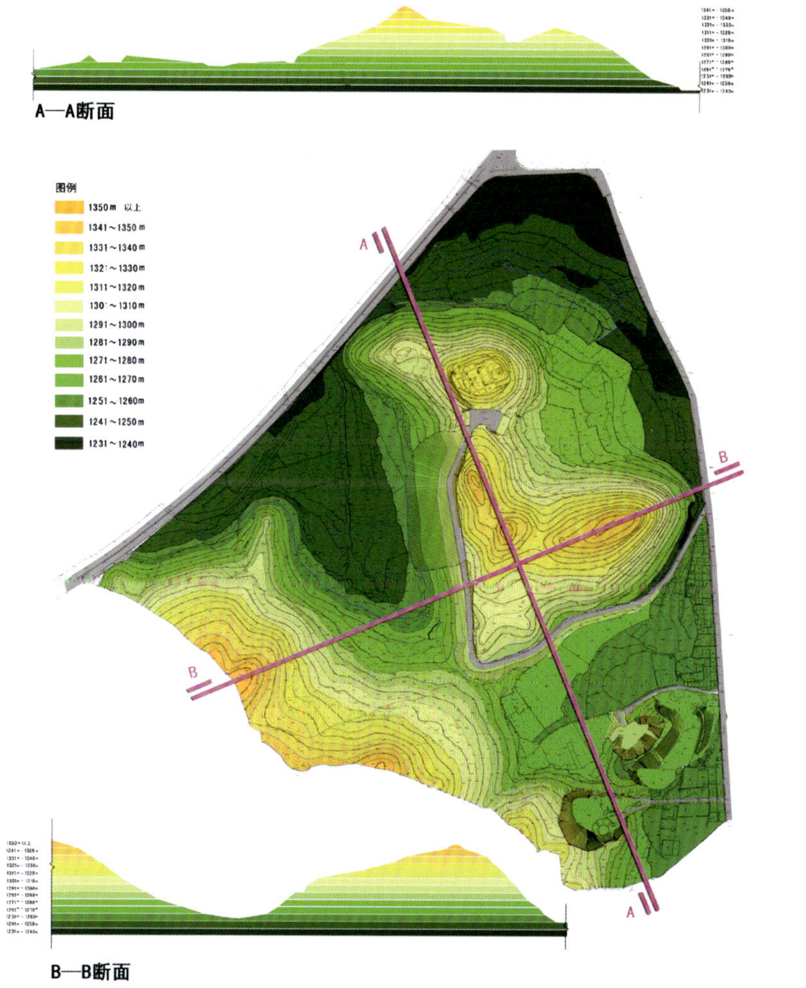

三、气候

清镇市属于亚热带季风湿润气候，其特点是气候温和，雨量充沛，雨热同季，多年平均雨量 1147.8mm，多年平均气温 22.6℃。

四、植被生态

植被属于亚热带常绿阔叶林和常绿落叶混交林。由于人为活动的种种破坏，多演替为次生的植物群落。其中以灌木草坡为主。全市现有森林面积 16785hm^2，森林覆盖率 11.5%。

五、宗教发展

灵永寺位于百花湖景区中十屯村中，始建于明初，民国 19 年（公元 1930 年）重建，占地 200m^2，有正殿和左、右厢房，现仅存正殿和梁架、房瓦以及三块残碑，据碑文记载，明建文帝曾隐居在此。

东山巢凤寺，始建于明朝初年（公元 1370 年左右），距今已有 600 多年的历史，期间几经毁坏和重修，2002 年，巢凤寺遗址被列为清镇市 21 处县级文物保护单位之一。

Ⅲ Climate

Temperate climate and plenty of rains and enough sunshine are characteristic of the weather in Qinzhen. The average rainfall is 1147.8mm and the average temperature is 22.6℃ .

Ⅳ Vegetation

The subtropical zones have large areas of evergreen broad–leaf forest and evergreen–deciduous broad–leaved forest, which mostly have become secondary plants (most of them are shrub and grass slopes) due to the destruction of men. At present, the forest area of Qingzhen is 16785 hectares, with the forest coverage rate 11.5%.

Ⅴ Religion

Lingyong Temple, built at the beginning of Min Dynasty and rebuilt in 1930, is in Shitun Country of Baihuahu Lake scenic spot. It occupies 200m^2, and was divided into main hall, left and right houses. Today, only remain the main hall, the frame, tiles and three broken steles. According to the inscription, the Emperor Jianwen in Min Dynasty had live here in seclusion.

Caofeng Temple of Dongshan, built at the beginning of Min Dynasty (around 1370 AD), has been about 600 years, and were wrecked and rebuilt several times during these years.

坡度分析图

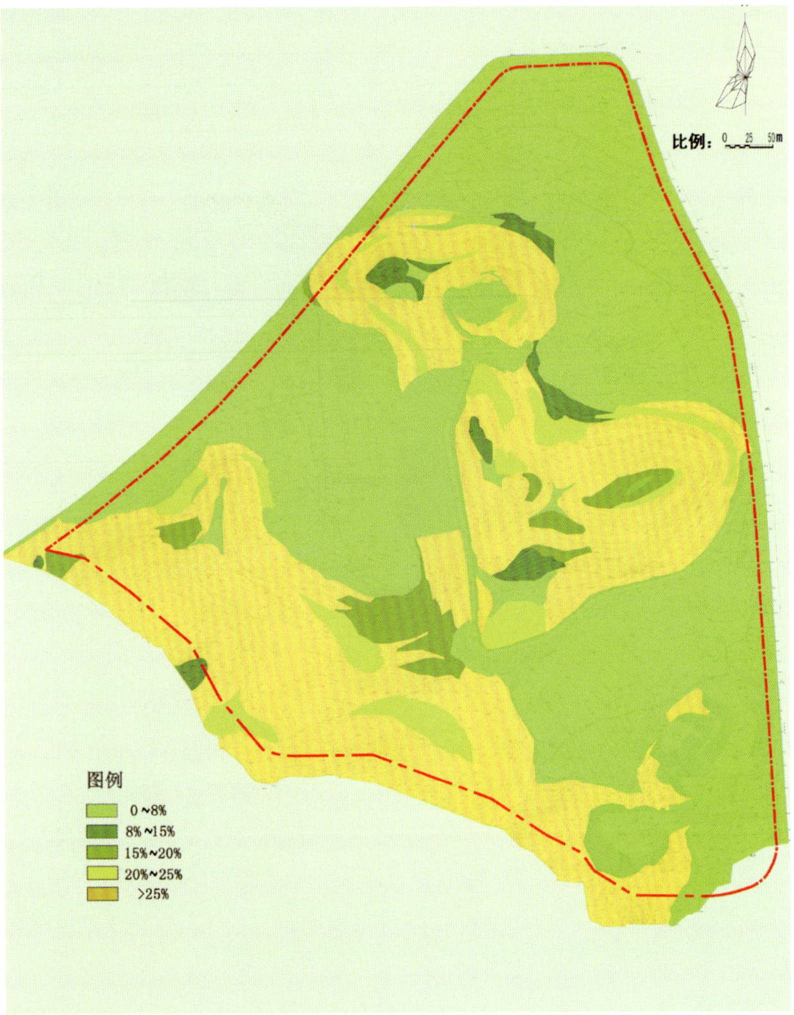

现状分析图

第三节 旅游经济环境分析

一、旅游资源的主要类型

清镇市旅游资源十分丰富，涵盖了自然资源类、水域风光类、生物景观类、人文资源类、消闲求知健身类等5类28种。

二、清镇旅游发展情况

2005年，全市接待国内旅游人数57.8万人次，实现旅游总收入5820.73万元。

Section Ⅲ Economic and Environment Analysis of Tourism

Ⅰ Types of Tourism Resources

In 2005, the number of tourists in Qingzhen were 0.578 million, totally achieved 58.2637 million yuan tourism revenue.

Ⅱ Situation of Tourism

In 2005, the number of tourists in Qingzhen were 0.578 million, and achieved 58.2637 million yuan tourism revenue.

2 项目文化旅游资源评价 Evaluation of Culture and Tourism Resources

第一节 项目规划起因

人类社会物质文明的高速发展，对于人类社会的前进无疑起到了极大的推动作用，但是，毋庸置疑，人类社会物质文明的高速发展，以及这种发展的失重和不均衡，同样也带来了一系列社会问题，造成了国与国之间、地区与地区之间，以及人与人之间、人与自然之间关系的紧张，带来了不同国家、不同民族、不同宗教的仇视和对立，不同国家、民族和宗教之间的纷争和冲突愈演愈烈，人类智慧和文明已经受到严重威胁，人类社会已经开始呈现“无序”和“失范”状态。就世界范围而言，尽管和平与发展是时代前进的主旋律，但是，社会动荡不安，暴力冲突不断，恐怖袭击频繁，贫富差距拉大，环境污染加剧，生态失衡严重，已经给人类社会蒙上了一层阴影，成了阻碍和制约人类社会进步的障碍。就当今中国而言，全面建设小康社会，构建和谐社会是全中国人民的共同心愿，但是，由于受西方文化的影响，也由于社会变革带来的悸动和不安，人与人之间的关系，人与社会的关系亦日趋显得紧张起来，贫富悬殊，物欲横流，道德沦丧，世风日下，心燥气浮，精神空虚，无所寄托的社会现象已经成了纷扰社会的渊薮，而由此造成的各种罪恶已经严重地威胁着社会的安宁与祥和。面对如此动荡不安的世界，面对仇视与对立、纷扰与冲突日益加剧的局面，面对物欲横流，道德沦丧，世风日下，心燥气浮，精神空虚，无所寄托的罪恶渊薮，祈盼和平，构建和谐、净化心灵、重修精神家园的呼吁已经成为当今世界之必须，历史发展之必然。

着力打造佛教胜地“多国佛教园——华藏世界”，就是志在启迪人们心智，借佛教之力制止邪恶，抑制人的贪恶之念，激发人的良善之心，为净化心灵、重修精神家园、构建和谐社会、谋求天下和美尽绵薄之力。

万恶因贪起，诸邪缘欲生。当今世界纷争不断，冲突迭起，拥有五千年灿烂文明的中华民族也开始面临礼崩乐坏、世风日下的威胁，其根源皆在一个“贪”字和“欲”字上。纵观人类社会，无论是国家交恶，还是民族纷争，无论是鸡鸣狗盗，还是劫杀奸淫，无不都是因贪而起，缘欲而生。要遏制罪恶，预防人类社会的“无序”和“失范”，就必须遏制人的贪欲恶念，

Section Ⅰ Background of the Planning

The rapid development of material civilization has surely greatly accelerated social development. However, the unbalanced development of material civilization and the development itself have also caused a lot of social problems: the national and regional crisis, the religious contradiction, and the tense relationship between men and nature, which were threatening seriously human wisdom and civilization, and the society have begun to present "disorder" and "uncontrolled". Although peace and development are the mainstream in the world, the existence of turbulent society, continual violent conflict, frequent terrorist attacks, growing wealth gap, and serious environmental pollution and unbalanced ecology cast a gloom over human society, and hinder the development of human society. In China, to build a well-off and harmonious society is the common aspiration of the Chinese. Nevertheless, due to the influence of Western culture and the turbulence and intranquility caused by social revolution, the tense relations between people and between people and society keep growing. The social phenomenon of wealth gap, acquisitive mind, and moral bankruptcy has been formed and become the evil to threaten social peace and harmony. Under this circumstance, people all over the world are asking for rebuilding peace and harmony and purifying their mind, which are the indispensability of the world and the necessity of historic development.

"Sino Tibetan World", created at this very time, was aimed to irradiate men's mind, curb evil and covetous thoughts, and irradiate men's mind with kindness with the power of Buddhism, so as to contribute their share to rebuild human mind, and harmonious and peaceful society.

Covetous thought is the source of all evil. China, with five thousand years of splendid culture, is facing up moral bankruptcy and declining courtesy on the background of continual conflict throughout the world, which, essentially, is the result of "rapacity". In the history, the evil matters from national conflict to individual delinquency are caused by covetous thoughts. Therefore, we have to curb people's covetous thoughts, promote their moral and mind quality, and purify their soul if we want to check evil and prevent "disorder" and "uncontrolled" of human society. While Buddhism, with the aim of pursuing harmony and fitting final and the task of undertaking historic mission and social responsibility, cherish the belief of saving and sympathizing everyone and the essence of doctrine of "purity, security and peace", which are just the best way for human to curb evil, save soul, spread kindness and build peaceful society.

提升人的道德品质和精神境界，净化人的心灵。而以追求圆满和谐为宗旨，以承担历史使命和社会责任为己任的佛教，其普度众生的宏图大愿，悲天悯人的慈悲心行，“心净则国土净，心安则众生安，心平则是天下平”的教义精髓，恰恰正是人类社会抑制邪恶、张扬良善，拯救灵魂、抑制恶行，整饬无序世界、构建和谐社会的良方。

作为世界三大宗教之一的佛教，传入中国已经两千多年，是世界三大宗教在中国传播最为广泛、影响最为深远的外来宗教。在两千多年的传播和演绎过程中，它通过与中国原有思想文化和信仰形态的相互碰撞、相互交融以及相互渗透和相互滋养，已经与儒家和道家文化共同成为中华传统文化的重要组成部分。作为世界三大宗教之一，佛教无疑具有其博大精深的宗教文化内涵；而作为与儒家和道家文化一道共筑中华传统文化大厦的外来宗教，佛教必然与中国传统文化、与中国本土文化的代表儒家文化和道家文化存有一种相交相融的共通性和天然自存的契合性。向世人展示和张扬佛教博大精深的文化内涵，展示和张扬佛教与中国传统文化相交相融的共通性和天然自存的契合性，也正是着力打造多国佛教园——华藏世界的又一用意。

“华藏世界”中文名字出自于大乘佛教的经典《华严经》，《华严经》被称之为大乘佛教的经中之王，也是大乘佛教经典当中最优美最华丽的经典。在这个经典当中主要是描述了佛教对光明、博大、美妙、快乐和安宁的理想世界的向往。因此从这个意义上说，“华藏世界”是精辟地阐述了佛教对于宇宙、对于世界的中心观点，我们希望借这个名字来加强这个项目的内在精神内涵和人文底蕴，同时也借这个名字表达民众对于共同地创造我们这个和谐的社会的一个基本想法。

Buddhism, as one of the three main religions in the world, has been to China for more than two thousand years. And it is the most widely spread and far-reaching impact of the three foreign major religions in China. Within the two thousand years of spread and interpretation, Buddhism has collided and blended with the original culture and faith of China to become, along with the Confucian and Taoist, the major pillars of the structure of traditional Chinese culture. As one of the three major religions of the world, Buddhism, doubtlessly, is of profound and expanding religious cultural connotation; while as a foreign religion constituting to traditional Chinese culture with the Confucian and Taoist, Buddhism must have something common with the Confucian and Taoist which are the sign of traditional and local Chinese culture. To demonstrate and publicize the intersection and natural existing commonality between Buddhism and traditional Chinese culture, which was the purpose to create "Sino Tibetan World".

The name "Sino Tibetan World" is from the classic "Flower Adornment Sutra" which was known as the king of Buddhist sutra and the most elegant and brilliant sutra of the classic ones of Mahayana Buddhism. The major idea that "Flower Adornment Sutra" elaborates is that Buddha conceives an ideal world full of brightness, permanence, beauty, happiness and peace. Therefore, in this sense, "Sino Tibetan World" brilliantly expounds the core point of view of Buddhism for the universe. We would like to use this name to strengthen the inner spiritual connotation and cultural heritage of this project, and also to express the public basic idea of creating a harmonious society of our own.

第二节 多国佛教园——华藏世界的旅游资源

自汉代佛教由印度传入我国后，在唐代，由于玄奘的西行取经，使我国佛教文化得到了完善。而后每朝每代都对佛教有不同程度宣扬，更使得佛教文化在我国根深蒂固。如今的中国佛教，其发源地虽然不是中国，但在百姓心中影响和传播超过了佛教发源地印度对当地百姓的影响。甚至可以说超过了本土宗教的影响。这种影响不仅在精神文化、物质文化，更是浸透在了民众的骨血中，在我国的历史文化遗迹上是不可缺少的一部分，充分反映了中国文化历史发展的个性特点。

多国佛教园——华藏世界是以“各国佛教文化的发展”和“中国佛教文化的发展”为思路而构筑的佛教旅游胜地，让人们在很短的时间里就能了解东南亚各国佛教文化发展史和中国佛教文化精华。从旅游开发的角度来看，佛教文化具有历史文化旅游资源和现实旅游资源的双重特征，既历史又现实，只要加以整合，人们便可在很短时间内了解各国佛教文化发展的大概和中国佛教文化的精华，帮助人们去思考，去遐想，帮助自己去其躁哗，恢复宁静心态。

Section II Tourism Resources of "Sino Tibetan World"

Buddhism has been introduced into China from India since the Han Dynasty. In Tang Dynasty, Buddhism of our country was improved thanks to Xuan Zang's pilgrimage to the West, and it was spread in different extent in every dynasty then that Buddhism has had its root in China firmly. Although China is not the birthplace of Buddhism, the impact and transmission in China are more profound and extending than in India where the Buddhism was born, and even more influential than local religion in China. This influence not only reflects on spiritual and material culture of Chinese people, but also penetrates into people's soul.

"Sino Tibetan World" is the Buddhism resort created on the thought of "the development of Buddhism in every country" and "the development of Buddhism in China" in order that people can know about the developing history of Buddhism in Southeast Asia and the cultural essence of Chinese Buddhism. From the perspective of developing tourism, Buddhism culture is of the characteristic of both tourist resources of historic culture and present culture, so that people can know generally about the development of Buddhism culture in every country within a short time. The essence of Chinese Buddhism is to help people to think and image all peace in life so as to leave them a tranquil mind.

佛教文化资源价值

1. 历史文化价值

佛教文化自存在至现在已有几千年历史，其文化影响涉及全世界各个角落，尤其是东南亚各个国家，更是将其推崇和发展为本国国教。佛教对中国文化发生过很大影响和作用，在中国历史上留下了灿烂辉煌的佛教文化遗产，许多佛教建筑已成为我国各地风景轮廓线突出标志。佛教还为中国文化带来了新的意境，新的文体，新的命意遣词方法以及天文、医药等科学技术的传习。

2. 产品价值

佛教文化具有良好的产品价值优势，一是佛教文化包含着物质、精神、交往等 3 个文化层面，既具有陈列展示、提供游、购、娱的文化内容，又具有参与生产、体验生活、感受历史的文化内容，使之可在精神文化得到更深的体会。

3. 市场价值

以集成的方式打造佛教文化旅游产品，一是国内外给我们提供了丰富的各类型佛教文化旅游的成功模式和经验；二是伴随城乡人民文化生活的日益提高，人们将不断地追求精神上的和谐和宁静。

第三节 多国佛教园——华藏世界文化旅游产品的 SWOT 分析

一、优势

1. 项目建设得到了政府的大力支持

清镇地方政府积极扶持旅游事业，政策优势明显，开发条件好。

2. 具有独特的文化生态

在多元文化丛林中，佛教文化具有典型意义，是独特的文化遗产，宗教人士的精神家园。特别是信仰者，多国佛教文化对他们有着极大的吸引力。

3. 区位优势突出，客源市场广阔

规划用地位于清镇市贵黄高等级公路和 320 国道交叉口，交通便利，景色宜人；也是通往国家风景区黄果树、龙宫的必经之路；区位甚佳，距贵阳市 22km。清镇市又是贵州城市水平相对较高区域之一，交通便捷，可进入性好，居民出游率较高，旅游消费增长较快。

4. 清镇市较好的旅游环境氛围条件

清镇山清水秀，景色宜人，冬无严寒，夏无酷暑，境内繁花似锦，有红枫湖、百花湖、鸭池河、猫跳河、苗族、布依族风情和自然人文景观，是贵州省著名的自然生态文化旅游区。这为加快其旅游业的发展步伐创造了极为有利的条件。

Value of Buddhism Culture

1. Value of Historic Culture

Buddhism culture has been existing for thousand years when, which had broad impact throughout the world, especially in all countries of Southeast Asia where Buddhism has been developed as national religion. Buddhism has also played an important role in Chinese culture, and left splendid Buddhism cultural heritage in Chinese history that many Buddhism architectures has become outstanding signs of historic sites of China. Besides, Buddhism has injected new imagery, new literary style, new ways of words using, and application of scientific technology such as astronomy and medicine into Chinese culture.

2. Product Value

Buddhism culture, covering three cultural aspects of material, spirit and communication, is of providing exhibition, travelling, shopping and entertainment, and also of participating producing, experiencing life and reliving history. Moreover, it creates deep feeling on mental culture for people.

3. Market Value

There are two reasons that we created this tourism product of Buddhism culture with the way of integration: first, we have plenty of all types of successful modes and experiences of Buddhism culture tourism; secondly, with the promoting of people's life, they will pursue spiritual harmonies and tranquil continually.

Section Ⅲ SWOT Analysis of Cultural Tourism Products of "Sino Tibetan World"

Ⅰ Strengths

1. Government Support

The local government of Qingzhen supports tourism industry positively that the policy advantage is obvious.

2. Specific Culture

Buddhism culture, as the unique cultural heritage and the spiritual support for religious persons, is of typical meaning in the diversity of culture. "Sino Tibetan World" has great attraction with the Buddhism culture to the believers.

3. Broad Market

The planning place, with convenient traffic and pleased scenery, is at the cross of Guihuang highway and 320 state roads and the only way to Huangguoshu scenic spot and Longgong palace. Moreover, Qingzhen prefecture is one of the areas with comparatively high urbanization level.

4. Pleased Tourism Environment

With pleased landscapes and comfortable climates, Qingzhen prefecture is a famous beauty spot with natural ecological resources, where Hongfeng Lake, Baihua Lake, Yachihe River and Maotiaohe River are running by, which, along with colorful flowers and plants, ethnic flavor like Miao and Buyi and natural humanistic landscapes, contribute greatly to speed up tourism development.

二、劣势

1. 国内市场竞争激烈

我国佛教文化开发进入激烈竞争阶段，若想在激烈的竞争中独树一帜，崭露头角，需要更多的投入；特别是周边省份大力度的佛教开发对本区域形成压力。不仅省内以宗教为主题的旅游产品开发项目有很多，就是国内外以宗教为主题的景点不胜枚举，这对产品的唯一性提出了挑战。

2. 贵州有关人才相对缺乏

多国佛教园——华藏世界中集中式文化建筑是一种投资大、风险高的新型旅游产品和旅游市场，对于管理企业的人力资源的水平要求较高，既要求决策者有战略眼光，管理者要掌握经营管理的各方面知识，而且还要有市场运作能力和产品创新能力。

3. 对省外游客具有一定的依赖性

贵州属于全国经济发展的欠发达地区，地方对佛教文化的消费有一定局限。因此，多国佛教园——华藏世界必须具有吸引省外游客的能力。

三、机遇

1. 旅游业蓬勃发展，市场广阔

非观光旅游产品已经逐渐成为主流，旅游者以双向甚至多向式参与旅游，休闲、保健、文化参与型旅游产品已逐渐成为时尚，多国佛教园——华藏世界正好具有这样的特征。

2. 贵州旅游产业结构调整的需要

贵州缺少佛教文化性集中地，特别缺乏体现东南亚各国佛教文化的旅游产品，多国佛教园——华藏世界正好满足这一要求，填补了这一空缺。

四、挑战

1. 佛教产品在省内面临的挑战

2. 省外存在一定的竞争压力

3. “酒香不怕巷子深”的时代已经过去

Ⅱ Weakness

1. Fierce Competition in Domestic Market

The development of Buddhism culture in our country has come to the stage of fierce competition that we have to invest more if we want to win a position in this situation, especially facing up the pressure from great development of Buddhism of the surrounding provinces. This is a big challenge of uniqueness of products since the projects of tourism development with the theme of religion are numerous, and so do the worldwide.

2. Relative Deficiency of Relevant Talents

The concentrated cultural architectures in "Sino Tibetan World" are a new kind of tourism product and market with high investment and venture, so that the managers should be qualified enough with strategic view and massive knowledge about manage, as well as the ability of marketing and productive innovation.

3. Relying on Tourists from Foreign Provinces

Guizhou province is in the relatively undeveloped area of China that the consumption of local people is confined. In this case, "Sino Tibetan World" must be attractive enough to draw tourists from other provinces.

Ⅲ Opportunities

1. Prosperous Tourism

Tourism products of no seesighting has been become mainstream gradually that tourists can participate these tourism products like entertainment, keeping fit, and culturalizing, which has become a kind of fashion. "Sino Tibetan World" is just of these properties.

2. Tourism Industry Structure

Buddhism in Guizhou province is not concentrated, and the tourism products reflecting Buddhism culture of Southeast Asia is especially deficient, while "Sino Tibetan World" can just fill this vacancy.

Ⅳ Threats

1. Challenges in our province.

2. Competition pressure from both our province and the others.

3. The time that "Good wine needs no bush" has passed.

3 项目性质定位 Property Orientation

第一节 规划对象的性质与主题定位

多国佛教园——华藏世界以浓缩中国佛教精华核心，集合东南亚各国佛教精华于一城，将印度、韩国、日本、柬埔寨、泰国、缅甸、尼泊尔、斯里兰卡、中国9个国家的佛教特色集中在文化园区中体现，集中再现佛教文化和佛教精神，并将中国佛教重点突出，使人们认识到佛教由发源国传到中国后，如何逐渐演变成具有中国特色的宗教大家。从而形成贵州境内独具特色、绝无仅有的新型旅游景观，构造清镇旅游强势产业，打造贵州旅游精品。

Section Ⅰ Orientation of Teature and Theme

Sino Tibetan World" concentrates Chinese Buddhism essence, as well as the one of Southeast Asia, demonstrating Buddhism features of nine countries: India, Korea, Japan, Kampuchea, Thailand, Burma, Nepal, Sri Lanka and China, and presenting Buddhism culture and spirit, and also highlighting Chinese Buddhism to make people know how Buddhism gradually become the religion with Chinese style after its arrival from India. In this way, "Sino Tibetan World" will form an unique new scenic spot in Guizhou to create a strong tourism industry in Qingzhen prefecture.

多国佛教园——华藏世界以清镇市政府发展休闲度假旅游和商业配套旅游产业为主的全新旅游发展观为指导，在充分展示佛教文化、佛教精神的同时，尤其注重旅游观光和商业服务的完善结合，城内除构建令人耳目一新、让人流连忘返的各国佛教文化区和中国佛教区，还精心配套了道场区，多国塔林区、慈善公益事业区、禅修区、佛学文化区、溶洞观赏区等一系列与之相关联的游览和商务设施，以满足游人旅游、观光和休闲、娱乐、消费的整体需求，不仅体现了佛家文化净化心灵，更丰富游人休闲旅游项目，弥补清镇现有旅游景区商务服务单薄甚至不配套的缺陷，让游人有耳目一新的观赏景观，又可以购买琳琅满目、可纪念的佛家特产、工艺品。

多国佛教园——华藏世界肩负打造清镇新型旅游产业、振兴清镇旅游产业之重任，通过自身品牌的塑造，将与区内景区形成合力，即通过“华藏世界”连接区内各旅游景点，串珍珠似的把红枫湖公园、国家自然风景保护区红枫湖、省级风景名胜区百花湖和市级风景名胜区东风湖等省内知名度较高的景区串联起来，形成市内旅游产业链，丰富旅游项目，提升旅游品位，激活、兴旺整体旅游资源，促进旅游产业可持续发展。

多国佛教园——华藏世界是以佛教文化为主题兼容各国佛教文化和民俗文化，将环境景观作为其载体，而创作出的一种佛家情境的大游憩空间。总体风格是：神圣、庄严、古朴。

把佛教“辉煌的精神文化”用艺术手法提炼出来，结合现代旅游“形式”所需要素，综合运用陈列展示、参观学习、表演欣赏和主体参与四种设计思路，经过三至五年或更长时期的努力，创建一个以佛教文化特色体验为主的集人文景观、休闲、文化、娱乐为一体、传统与现代交相辉映的文化都城，为贵州本地及海内外游客提供一个了解佛教历史变迁的文化基地。这一定位的文化内涵表现为：

1. 感受各国佛教文化大乘
2. 体验佛教神秘文化背景
3. 享受宁静情境休闲时空

With the guidance of new concept of developing leisure vacation travel and commercial tourism by the government of Qingzhen prefecture, the refreshing and impressive Buddsim cultural areas of many countries are built and the relative tourist and commercial facilities are delicately embellished such as the Rite Area, Pagoda Forests Area, Social Charity Activities area, Buddhism Culture Area, and Cave Viewing Area and so on so as to meet the overall needs of tourists' travel, sightseeing, leisure, entertainment and consumption with fell elaboration of Buddhism culture and spirit, and perfect combination of tour and commercial facilities. This not only embodies the purification of Buddhism culture to soul, but also enriches programs of leisure travel for tourists, Moreover, it makes up with the unsound commercial services so that tourists can enjoy the wonderful landscapes and buy Buddhism accessories and other crafts which are valuable to be memorized.

"Sino Tibetan World" takes the responsibilities of creating new type of tourism industry and strengthening tourism industry of Qingzhen. Shaping its brand, it will uniform with the other scenic spots of this region to become stronger. In other words, all the scenic spots of this region including Hongfenghu Park, Hongfenghu national preserve, Baihuahu Lake Provincial Scenic Spot, and Dongfenghu Prefectural Scenery Site etc. will be linked by "Sino Tibetan World" like linking pearls to form a tourism chain of Qingzhen, so as to enrich tourism program, enhance tourism taste, active and prosper tourism resources in the whole and achieve sustainable development of tourism industry.

"Sino Tibetan World" was created as a type of great recreational space in Buddhism atmosphere with the theme of Buddhism culture along with the ones and folks of other countries, and the carrier of environmental landscapes. The overall style is: holy, serious, and primary.

A cultural city, traditional and modern, major in experiencing Buddhism culture specialties along with humanistic landscape, leisure, culture and entertainment will be constructed by purifying "the magnificent spiritual culture" of Buddhism with artistic means, combining with the factors of modern tourism "ceremonies", applying comprehensively the four designing concepts of demonstration, sightseeing, performance and participation, and using three to five years or more times.

1. Feel the high leveal Buddhism culture of every country
2. Experience the mistrial cultural background of Buddhism
3. Enjoy the peaceful space and recreational time

第二节 旅游项目设计与开发原则

一、源于生活，高于生活

本规划兼顾真实性和艺术性，使项目品位不落俗套，采取了源于生活、高于生活的创作理念。以一定气势、体量和规模，创造佛教崇高、神圣、神秘的感应气氛，使游人产生由衷的敬仰之情。

二、源于历史、高于历史

在时间定位上，从人到物，从古到今，保持佛教的原始风味，形成与现代生活的巨大反差。而且，这种原始风味既包括各国佛教文化，又包含佛教传入我国后进行的演化，最后形成有中国特色的佛家文化，使游客感受到千百年来佛教的精神演化和文化演化，以各种展览、表演等活动为需求要素，向旅游者提供多种活动服务的旅游项目，体现佛教的宽容、慈悲。

Section Ⅱ Design and Developing Principle of the Project

Ⅰ Based on Life but above Life

This planning is both real and artistic and adopts the creative concept of basing on life but above life, so that the taste of it can be beyond blame. With certain magnificence, volume and scale, a lofty, holy and mysterious atmosphere can be created so that the tourists will cherish it with sincerely respect and reverence.

Ⅱ Based on History but above History

On the orientation of time, Buddhism will be kept in its original flavor from human to material and from ancient to present, which forms huge difference between primary and modern Buddhism. Moreover, this original flavor is presented through Buddhism culture of many countries and the one with Chinese characteristics after its arrival and evolution in China so that visitors can feel spiritual and cultural evolution of Buddhism after thousands of years. In addition, tourists are provided with tourism programs of various activities services such as exhibitions, performances and so on in order to embody tolerance and mercy.

三、文化符号

根据游客的消费心理，以尚存在或曾经存在的名人、名（史）事、名著、名物、名景、名情等文化符号为蓝本，以巧妙的手法再创造出来的一种构思新颖的人造景观，体现出现代人对佛教的理解。

四、人文、宗教、景观和谐统一

保护资源环境，统一规划布局，尊重地域景观，在《总体规划》的指导下，合理布局多国佛教园——华藏世界的配套功能设施，充分考虑人文、宗教、景观的和谐统一，科学预测景区发展的各种需求。

五、整体效果与个体风格并存

多国佛教园——华藏世界的规划设计，应坚持整体与个体的协调统一，既要注重整体构图和空间效果，又要关照个体的独特性；既要保证各建筑整体风格的协调统一、相互关联，又构成有机整体，产生遥相呼应、相互辉映的空间效果，要保持各建筑个体的独特性，使之各具特色，各有韵味，自成体系。

六、因地制宜、有效利用、降低投资

因地制宜，合理有效利用地形和环境，尽可能顺应地形，减少建设中的土石方工程量及对用地环境的破坏，填挖尽量就近平衡，同期平衡，降低工程综合投资，力求达到综合开发的最佳效益。

Ⅲ Cultural Signs

To create with a delicate way a novel artificial landscape in accordance with people's consumption psychology and the cultural signals of the ever or existing famous people, events, works, matters, landscapes and emotion ect. in order to show people's understanding of Buddhism in the modern time.

Ⅳ Harmonious and Uniformed Culture, Religion and Landscape

We should preserve resources, make proper layout and respect regional landscape. Meanwhile, with the guidance of the "overall planning", we should consider fully the harmony of culture, religion and landscape of "Sino Tibetan World", and predict scientifically all needs of the development of this scenic spot to set relative functional facilities reasonably.

Ⅴ Overall Effect and Individual Style

The design and layout of "Sino Tibetan World" should insist on the unifying of entireness and individual, which means we should either emphasize on overall structure and special effect to form organic entireness or pay attention to specific of individuals to demonstrate their characteristics respectively.

Ⅵ Positioning on Local Conditions

We should make layout according to local conditions, make good use of environment and terrain so as to decrease earth-rock work and damage of environment. Fill and cubage of excavation should be balanced in certain times and nearby places in order to lower synthetic investment of the project and achieve the best benefit of the synthetic development.

4 项目规划目标 Planning Object

一、通过佛教熏陶和感染，平和人们的心态，净化人们的心灵，培养人们的嘉言懿行，为清镇市营造一方和谐安宁的社会小环境。

二、拓展清镇旅游产业，构建旅游观光、休闲度假、会展商务和宗教熏陶一体化的旅游新格局，提升清镇市旅游品位，开创清镇市旅游观光的全新局面。

三、以多国佛教园——华藏世界为中心，连接清镇各旅游景点，金线穿珍珠般地将红枫湖、白花湖等旅游景点串联起来，使之融为一个有机的旅游整体，整合旅游资源，从而形成区域旅游产业链，构建区域旅游强势，促进和推动清镇市旅游产业高效有序和可持续发展。

四、立足清镇，放眼全省乃至全国，依托贵（阳）黄（果树）高速公路的便利条件，通过精心打造多国佛教园——华藏世界，提升清镇整体旅游品位，将清镇建成贵州旅游观光、休闲度假、会展商务和宗教熏陶的集散地和截流地，汇集、截流各方游人、香客，并逐步与黄果树、龙宫等西线名胜旅游景区形成互动，结成旅游伙伴，力求形成贵州黄金旅游线。

五、通过对这个项目的规划让我们做到四个和谐：第一个是旅游与人文的和谐；第二个是建筑与环境的和谐；第三个是佛教文化与本土民俗文化的和谐；第四个是经济效益与社会效益的和谐。

Ⅰ To create a small harmonious and tranquil social environment of Qingzhen by tempering people's mind, purifying their soul and fostering their good manners with the influence and impact of Buddhism culture.

Ⅱ To improve tourism taste and bring about a new situation of tour in Qingzhen by expanding tourism industry of Qingzhen and constructing new pattern of tourism of sightseeing, recreation, business exhibition and religious influence.

Ⅲ To link the scenic spots including Hongfenghu Lake, Baihuahu Lake and so on with the center of "Sino Tibetan World", so that they will combine with each other to form an organic tourism unity, integrate tourism resources and form regional tourism chain, and eventually accelerate tourism industry of Qingzhen to develop effectively and sustainably.

Ⅳ To enhance the whole tourism taste of Qingzhen by exquisitely constructing the "Sino Tibetan World", so as to build Qingzhen Prefecture into a distribution place with tourism resources of sightseeing, recreation, business exhibition and religion atmosphere relying on the convenient conditions of Guihuang (from Guiyang to Huangguoshu) Highway, so that it can attract visitors and pilgrims, and gradually interact with the west scenic spots like Huangguoshu and Longgong of Guizhou, and form prime tourism line of Guizhou.

Ⅴ To achieve harmonies in four aspects by making planning of this project: the harmony between tourism and culture, architecture and environment, Buddhism culture and local customs and economic effect and social effect.

5 项目主体工程规划 Planning of Major Projects

第一节 总体布局

第一板块：佛教园区：由主出入口、东南亚各国佛教园、中国佛教园、素斋园、服务场馆等组成。
第二板块：道场区：由方丈园、道场和上客堂组成。
第三板块：多国塔林区。
第四板块：慈善公益事业区：老年公寓、孤儿园、安养园、极乐园组成。
第五板块：禅修区：由禅修斋、居士区组成。
第六板块：佛学文化区：由佛学院、武僧馆组成。
第七板块：溶洞观赏区。

Section I Overall Layout

The First Part: Buddhism area: consisting of the main entrance and exit, Buddhism park of every country of South Asia and China, vegetarian food park, service center etc.

The Second Part: Rite area: consisting of abbots' gardens, rites, and distinguished guests' house.

The Third Part: Pagoda forests area.

The Forth Part: The Social Charity Activities Part including Apartment for the Old, Orphanage, Amusement Park and Sanitarium.

The fifth Part: Buddhism Cultivating Area consisting of Buddhism Cultivating House and Laymen Area.

The Sixth Part: Buddhism Culture Area: consisting of Buddhism Culture Institute and Monk House.

The Seventh Part: Cave Viewing Area.

总平面图

第二节 功能分区

一、佛教园区

1. 出入口

▲ 主大门：采用一个融整个亚洲佛教建筑经典为一体的建筑形制，运用典型的印度建筑风格，并采用大量的“华严壁”。

▲ 广场：客源集散地及文体活动场地。“华严壁”置于广场大门两侧，华严壁面采用佛教图案用浮雕手法展示。

2. 东南亚各国佛教园

印度佛教园、尼泊尔佛教园、斯里兰卡佛教园、日本佛教园、韩国佛教园、泰国佛教园、缅甸佛教园、柬埔寨佛教园

3. 中国佛教园

单体大佛像、毗卢殿、药师殿、弥陀殿、观音殿、地藏殿、文殊殿、普贤殿、四大天王

4. 服务场馆

禅憩园、俗家食府、素斋馆、禅茶苑、放生池、医务室

二、道场区

弘法道场、登高塔

三、多国塔林区

佛塔群、功德碑林

四、慈善公益事业区

老年公寓、孤儿园、安养园、极乐园

五、禅修区

禅修斋、居士寮

六、佛学文化区

佛学院、武僧馆

七、溶洞观赏区

开展溶洞观光活动

Section Ⅱ Functional Zoning

Ⅰ Buddhism Area

1. Entrance and Exit

Gate: It adopts a style which was mingled with whole architectural essence of Asia, the typical India architectural style and great many of "Huayan walls".

Square: It is the place for visitors and public activities. "Huayan Stones", with the patterns of Buddha in sculptures, are set on the two sides of the square gate.

2. Buddhism Parks of Every Country of Southwest Asia

Buddhism park of India、Sri Lanka Buddhism Park, Japanese Buddhism Park,Thailand Buddhism Park、Burma Buddhism Park, and Cambodian Buddhist Park

3. Chinese Buddhism Park

Single Buddha Statue, Mile Palace, Guanyin Palace, Manjusr, Vabhadra Palace, and Four Gods.

4. Service Area

Rest Park, Restaurant, Vegetarian Restaurant, Tea House, Releasing Pool, and Clinic

Ⅱ Rites Area

Rites of Preaching and Observation Pagoda

Ⅲ Multinational Pagodas Area

Pagoda Group and Forest of Steles of Virtue

Ⅳ The Social Charity Activities Part

Apartment for the Old, Orphanage, Sanitarium, and Amusement Park

Ⅴ Buddhism Cultivating Area

Buddhism Cultivating House and Laymen Area

Ⅵ Buddhism Culture Area

Buddhism Institute and Monk House

Ⅶ Cave Viewing Area

To branch out activities of cave viewing.

第三节 规划期限

项目分为二期工程，建设周期为三年：

一期工程：建设工期为 2007 年 1 月—2008 年 8 月，用地 328.274 亩，总建筑面积 3.12 万 m^2，建成佛教园、方丈园和道场多国塔林等项目。

二期工程：建设工期为 2008 年 8 月—2009 年 12 月，用地 558.276 亩，总建筑面积 4.02 万 m^2，建成养老园、孤儿园、佛学园、住宿区等子项目。

功能分析图

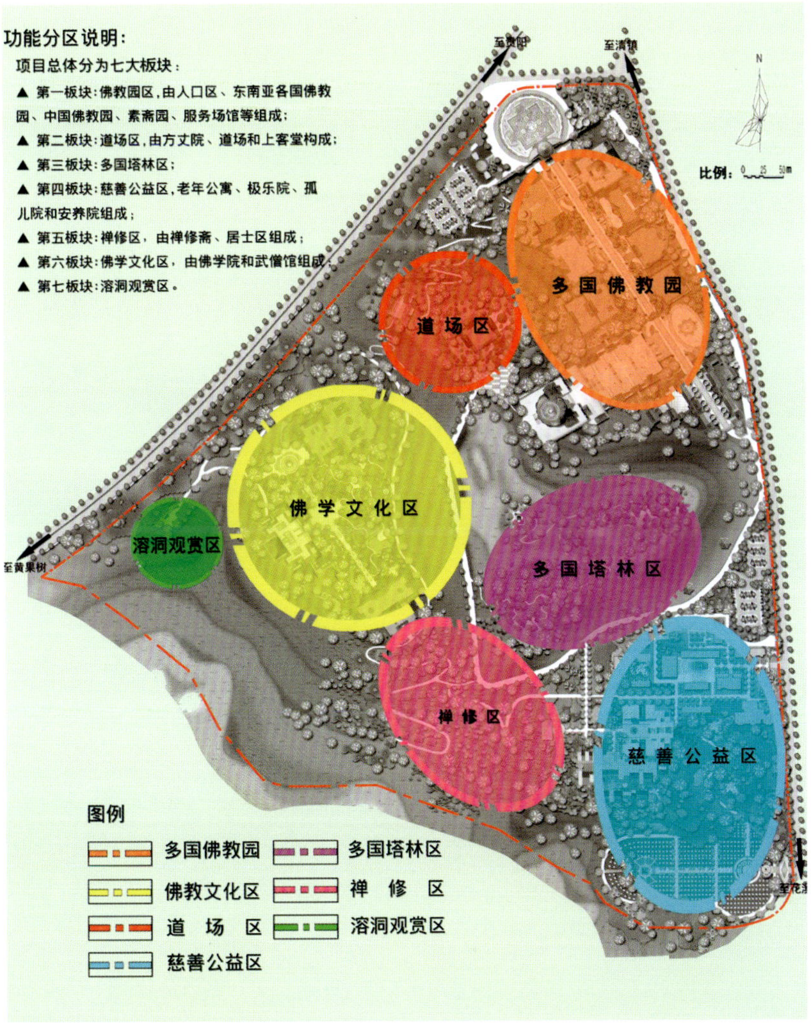

Section III Planning Deadline

We plan to finish this project in two phases, within three years:

In the first phase: we plan to construct Buddhism Park, Buddhist Abbot Park and Rite Square and so on from January 2007 to August 2008, using 328.274 mu lands, and the total built place is 31.2 thousand square meters.

In the second phase: we plan to construct the sub-projects of Old Park, Orphanage, Buddhist Park and lodging area from August 2008 to December 2009, using 558.276 mu lands, and the total built place is 40.2 thousand square meter.

6 环境保护与绿地规划 Environmental Protection and Greenland Planning

第一节 环境现状评述

一、自然环境

多国佛教园——华藏世界用地不规则，为一高山坡地，大致是中间高四周低，由于地形起伏变化大，对景观的连贯性在一定程度上起到制约作用。但也可充分利用地形，积极保护自然景观和山体绿化，创造出宜人、优美、舒适的旅游环境。

二、环境质量评价

环境质量特点如下：

天气：大气污染源分布密度小，排放强度小，空气质量总体较好；

噪声：声环境质量良好；

生态：荒山水土流失严重，岩石裸露程度高。

Section Ⅰ Current Situation of Environment

Ⅰ Natural Environment

The terrain where "Sino Tibetan World" located is on the slopes with a general shape of high center and low around. The continual landscape is restrained as the terrain changes like waves. However, we can make good use of this terrain and protect natural landscapes and green mountains to create a pleasing, beautiful and comfortable tourism environment.

Ⅱ Environmental Quality

Features of environmental quality:

Weather: the density of pollution sources is small, so that air is in good conditions generally;

Noise: little noisy;

Echology: barren hills and serious soil erosion make large extending exposed rock.

第二节 景观环境建设

绿化是形成宁静优美环境的主要因素，林存景在，林茂景美。规划用地内植物资源丰富，雨水热量充足，为景观绿化奠定了良好的基础。在保护好现有山体绿化植被的前提下，丰富植物多样性，配置更具地方特色的植物，创造良好的景观环境。巧妙利用山地、坡地处理建筑的起伏和层次，片状景观增色。

一、环境保护

1. 风景资源保护：多国佛教园——华藏世界处于贵黄高等级公路和320国道交叉口（位置），按照《总体规划》要求，要对规划用地内的自然资源环境进行保护和合理适度的开发；要对自然资源，包括山体景观、林木植被等进行分类保护，防止遭受人为破坏和污染。

2. 生态环境保护：多国佛教园——华藏世界要维护风景环境的自然本色，禁止开采沙石和树木砍伐。加强景观林的建设，培育乡土树种，保护自然景观和生态环境。

3. 污水处理：控制多国佛教园——华藏世界内的污水排放，对生活污水统一回抽经管网至污水处理站处理，实现对多国佛教文化园污水“零排放”，保护水域环境。

4. 清洁卫生设施：多国佛教园——华藏世界坚决贯彻“全面规划、合理布局、美化环境”的指导思想，着力营造清洁、优美、宜人的游览观光环境。按照《总体规划》要求，建立环卫机构，加强维护管理，配置保洁废物箱、水冲式公共厕所等相应的清洁卫生设施。

Section Ⅱ Construction of Landscape Environment

Afforesting is the major factor of creating peaceful and beautiful environment, so where there is forest, there is beautiful landscape. The planning place is of good conditions for greening with rich plant sources and raining and heat. On the premise of keeping the good current vegetation, we will enrich the variety of plants and add more plants with local features to create a nice landscape environment. Meanwhile, we should design the buildings in accordance with the rise and fall of the hills and slopes.

Ⅰ Protection of Environment

1. Protection of Landscape Sources: "Sino Tibetan World" lies on the cross of Guihuang Highway and 320 State Road, so, according to the "Overall Planning", the natural environment of the planning place should be reserved and developed properly; the natural sources including the hills and woods should be kept from destroyed or polluted.

2. Protection of Ecologic Environment: We should keep the natural landscape as it should be in "Sino Tibetan World", prohibit any quarrying of stones and cutting of trees, and strengthen the plantation of forest for sightseeing and the protection of natural landscape and ecologic environment.

3. Sewage Treatment: wastewater should be restrained and recycled to the polluted water disposal station to deal with, so that we can prevent wastewater from flowing into rivers and keep the water environment clean in "Sino Tibetan World".

4. Sanitation Facilities: With the guidance of "comprehensive planning, reasonable layout and beautiful environment", we will create a clean, nice and pleasing landscape for visitors. According to the requirement of "Overall Planning", we will set environment & sanitation institutions, strengthen protection and equip with litter cases and public toilets.

二、植物配置

绿化是形成优美环境的主要因素，林存景在，林茂景美。多国佛教园——华藏世界按照《总体规划》，在保护好现有山体绿化植被的前提下，应注意植物高矮顺序与游人视线的关系，植物色彩美与季节的关系和颜色相配的协调性。提倡开发运用乡土树种，既能丰富植物多样性，又可配置更具地方特色的植物，创造良好的景观环境。

II Afforesting Disposition

Afforesting is the major factor of creating peaceful and beautiful environment, so where there is forest, there is beautiful landscape. On the premise of keep current vegetation of mountains and, according to the "Overall Planning", we should consider the relations between orders of plants' height and visitors' vision, and the good matches of plants colors and seasons. We should advocate planting trees which are adaptive to local environment so that the diversity of plants can be enriched and eventually the nice landscape can be created.

绿地规划图

植物配置图

7 基础设施规划 Planning of Infrastructure

第一节 道路交通

一、道路骨架：依托贵（阳）清（镇）公路，组织多国佛教园——华藏世界的外部道路交通。多国佛教园——华藏世界的内部道路交通采用人车分流制，内部道路由三部分组成。第一部分为佛教园中间 11m 宽的景观步行主干道。第二部分为车行道，车行道位于多国佛教园右边是利用场地内原有车行道路加以扩展延伸改建而成，路面宽为 7m，以上道路均可作为消防通道使用，由于受到场地条件的约束使得消防通道无法形成环状，则在消防车道尽端设置了生态消防回车场。第三部分为通向各个景点的支路，支路宽为 1.5m ~ 2.5m。

二、停车场地：多国佛教园——华藏世界的停车场地主要是室外生态停车场。主要生态停车场设置在主入口广场区部位；而在服务区、道场区、塔林区、佛学院区和禅修区等处均设有临时停车点，以满足停车的需要。

三、停车位：多国佛教文化园停车位约有 350 个。

第二节 水电工程

一、给排水工程

多国佛教园——华藏世界给水排水工程包括生活给水、消防给水、污水和雨水处理三个部分

1. 旅游服务给水

(1) 多国佛教园——华藏世界的给水有两处源头：一是直接从市政给水管道引入，二是从地下水中提取，分别建高位水池蓄积输送，以满足多国佛教文化园的用水要求。

(2) 多国佛教园——华藏世界的给水量：公共建筑用水按 100m^3/(万 m^2·d)，绿化和道路用水 1L/(m^2·d)，未预见水量按上述之和的 20%计算，则"华藏世界"用水量为 875.08m^3/d。

(3) 多国佛教园——华藏世界给水管网系统采用给水母管、支状相结合的形式，给水主干管沿佛教园道路埋地铺设，以支管接入各功能区，整个多国佛教文化园给水管网的管径采用 DN50~DN300。

2. 消防给水系统

(1) 根据多国佛教园——华藏世界的规划旅游规模以及《建筑设计防火规范》GB50045—95，多国佛教园——华藏世界消防系统平面分区与功能分区一致，在各区同一时间内火灾次数为一次，分设消防水池、消防泵房及消防给水管网系统。

(2) 在接待中心、酒店以及其他公共场所另设置室内消火栓系统。同时，根据《建筑灭火器配置设计规范》在要求部位相应配套设置灭火器，并在不通车地段设置灭火器。消防用水按每 300m^3 设置消防水池一座。消火栓沿主干道每隔 120m 设置，消火栓的最小管径为 DNl00~200，以保证消防用水安全可靠。

Section Ⅰ Road and Traffic

Ⅰ Road Structure: The outside traffic net of "Sino Tibetan World" can be constructed with the major road of GuiQing Highway. The traffic nets in the "Sino Tibetan World" consist of three parts: the first part is the main road of 11 meters width for walking in the middle of Buddhism Park; the second part is the driveway which is on right of the Multi-State Buddhism Park and built by extending the original driveway to 7 meters width; (these two parts of ways can be used as fire control passageways which can't be in the cycle shape due to the features of the planning place.) the third part is the branches stretch to each scenic spot with the width 1.5m to 2.5m.

Ⅱ Park: We plan to build ecological parking lots in "Sino Tibetan World" with the main lot in the square of the main entrance and many other temporary ones respectively in Service Area, Rite Area, Pagodas Area, Buddhism Area and so on to meet the needs of parking.

Ⅲ Park Lots: The positions are around 350 in "Sino Tibetan World".

Section Ⅱ Hydropower Project

Ⅰ Water Supply & Drainage System

It includes three parts: supply water for life and fire fighting, and disposition of wastewater and rainwater.

1. Water Supply for Tourism Service

(1) Water supply can be achieved by two ways, one is introducing water directly from the municipal pipes of water supply, the other one is extracting the underground water; pools which offer water to the "Sino Tibetan World", should be constructed to accumulate water from the two ways respectively.

(2) Water capacity of "Sino Tibetan World": We estimate that we need 100 cubic meter water for every ten thousand square meter of public constructions, 1kg per square meter for watering plants and roads, and the others are 20% of the sum of these two, totally it is 875.08 cubic meter per day.

(3) We adopt main feedwater pipes and branch pipes to form the feedwater pipes system. The main feedwater pipes are set and buried along the avenue of Buddhism Park with branches linking to each functional area, the diameter of these pipes are from DN50 to DN300.

2. Feedwater for Fire-fighting

(1) According to planning area scale of "Sino Tibetan World" and "Code of Design on Building Fire Protection and Prevention" (GB50045—95), the planar partition of fire-fighting system should suit its functional partition with fire pool, fire pump room and fire water supply network system established among it. The amount of water use used for fire control in every fire,which would last 2 hours,should be 72 m^2 when the water consumption is 10 liters/second.

(2) Indoor hydrant system should be set in reception center, hotel and other public places. At the same time, according to "Code for design of extinguisher disposition in buildings", specified position should place fire extinguishers, so do those location where close to traffic. The main road must have fire hydrant with pipe diameter DN200 every 120 meters and fair pool every 300 cubic meters, in order to ensure the safe and reliable of fire water.

3、排水工程

多国佛教园——华藏世界采用雨、污分流排水系统。

(1) 污水处理：

污水量按多国佛教园——华藏世界用水量的 80%计，为 700.064m³/d。景区内污水管分别收集所排污水，餐厅污水需经隔油池预处理，粪便污水需经一级（化粪池）处理后，就近排入特设的两个污水泵站，经污水处理后一部分可作绿化灌溉用水，另一部分回抽或自流，接入市政排水管网系统（详见给水排水规划图）。

(2) 雨水处理：

建筑雨水经各建筑雨水管收集后分别就近汇入雨水管道中，用于农田和林木灌溉，其余雨水经统一收集后，经排水管回抽至公路路边的排水系统，排入市政雨水管网系统中（详见给水排水规划图）。

3. Drainage Engineering

The drainage system of this project adopts separate rain water and sewage systems.

(1) Sewage treatment

Sewage quantity is calculated as 700.064m³/d in line with 80% of total water consumption. Within the area, the sewage pipe collects sewage separately. Restaurant wastewater needs to be pretreated by oil separator, and fecal sewage can be treated by employing the following either methods after primary (septic tank)

(2) Rainwater treatment

The rainwater will import into storm sewer conduit separately after collected by building storm-water pipe for the use of farmland and forest irrigation. The rest of the rainwater will be pumped back to drainage system or discharge into Liujiang River tributary after collection.

交通流线图

给排水规划图

二、电力工程

1. 供配电：按照《总体规划》要求，多国佛教园——华藏世界供电依托清镇市 110kV 变电所，通过 10kV 输电线路输入景区箱式变压器，再用电缆向各个功能区输送。消防负荷等级为二级，其他为三级。

2. 负荷预测：

采用单位面积负荷指针法，其中住宅用地、公建用地等按建筑面积负荷指针进行预测。

按照有关规范规定，结合清镇实际用电水平，规划用电量为：

项目 \ 标准及用电量	标 准	用电量	备 注
华藏园建筑	10W/m^2	714KW	71400m^2
公共绿地及道路广场用地	10KW/hm^2	406.6KW	40.66hm^2
总用电总量		1120.6KW	

3. 电网规划：

供电电压等级为 10kV，多国佛教园——华藏世界内需采用双回 10 kV 电力线路供电。由箱式变压器向各建筑物供电，均采用环网方式供电，平时开环运行，适当位置设置联络开关，便于在其中一回线路故障或检修时能互倒互带，确保供电的安全及可靠性。

4. 配电变压器：多国佛教园——华藏世界内的配电变压器采用箱式变电站或结合建筑物设置户内式变电站，形成良好的景观生活环境。多国佛教园——华藏世界内设有两处地上变电所，配电变压器采用低耗节能型变压器，低压供电半径应＜ 200 m 。

5. 防雷接地：多国佛教园——华藏世界工程按三类防雷考虑。

三、电信工程

1. 电话系统：直接由清镇市电话网引入电话电缆到电话接线箱。在场地内设多部公共电话，在接待中心区、文娱休闲区等处预留适当数量的电话接口。

2. 有线电视系统：直接由清镇市有线电视网引入电视电缆到有线电视前端箱。

3. 数据通信系统：由清镇市数据网采用光纤引入到数据管理房，之后用数据专用线敷设到各功能区及其他需要的地方。

Ⅱ Electrical Power Engineering

1. Power Supply and Distribution:

According to requirements in General Planning, power supply in "Sino Tibetan World" relies on 110kVA transformer substations of Qingzhen City, and imports electricity to package transformer through 10kV transmission lines and then deliver electricity to all function areas by using electric cable.

2. Load Forecasting：

Load forecasting means to adopt unit area load indexing, of which, residential land, public construction sites, etc. will conduct forecasting in accordance with load indexing of construction area.

Project \ Standar Consumption	Standard	Power Consumption	Note
Area for Tibetan Park	10W/m^2	714KW	71400m^2
Public Green land, Roads and Square Land Use	10KW/hm^2	406.6KW	40.66hm^2
Total Consumption		1120.6KW	

3. Power Grid Planning:

Power voltage class is 10kV. In the planning area, it is necessary to adopt 10kV double-circuit lines for power supply. To supply electricity to buildings through package transformer will employ looped network for power supply. Generally, the looped network will be open for operation and set interconnection switch at appropriate place so that it can work with each other while one-circuit line is broken down or overhauling and ensure safety and reliability of power supply.

4. Distribution Transformer: In the planning area adopts cubicle-type substation or combines with buildings to set up indoor substation and for a good landscape environment. Distribution transformer utilizes low consumption and energy conserving transformer and the radius of low voltage power supply should be<200m.

5. Lightningproof Grounding: The third lightningproof is considered in "Tibetan World".

Ⅲ Electronics and Communications Engineering Technology

1. According to Building Layout in the Planning Area: A telecommunication cable Cross Connections Closet will be set up in the entrance for the purpose of installing communication equipment.

2. Cable Television System: Municipal government data network employs optic cable and introduces cable television system to central data room, and then use leased line to lay the system.

3. The road in the Planning Area: It will be set up with telecommunication cable duct, and all the telecommunication line should be included in cable duct laying for the purpose of local cable as well as non-voice service cable and optic cable application.

第三节 消防规划

火灾预防系统：多国佛教园——华藏世界内的建筑多为低层建筑，属于二类防火建筑，必须执行国家有关防火规范，健全消防设施和管理制度，保证消防通道畅通，提高预防和补救的综合能力。按《高层民用建筑设计防火规范》的要求，多国佛教园——华藏世界必须按照消防规划要求设置消防通道，便于消防车通过，配置消火栓，不通消防车的地方可用灭火器、防火门、防火卷帘、自动喷淋等设施。有关消防器材必须选用专业厂家产品。同时，还必须充分考虑消防疏散需要，严格按照《高层民用建筑设计防火规范》对疏散距离和人数的要求，在建筑内设置必要的安全疏散口。

消防报警系统：多国佛教园——华藏世界内设置集中报警系统，在各公共设施处设置感烟探测器，在室内车库设置感温探测器，并设手动报警按钮及对讲电话插孔；消火栓直接启动消火栓泵，并将信号送至景区消防控制中心。重要场所如变配电房、水泵房等处设消防直通电话。

电力电信

Section Ⅲ Firefighting Planning

Fire Prevention System: Buildings in "Sino Tibetan World" are low-rise buildings. As the second class of building fire protection, we must implement the National Building Fire System, improve fire control facilities and management system, keep clear of the fire control passageways and enhance the synthetic ability of prevention. According to the requirement of "Norm of Fire Prevention of Design of Multi-Buildings", we must set fire control passageways, in order that the fire control cars can come through safely and smoothly.

Fair Alarm System: The concentrated fair alarm system should be set in "Sino Tibetan World", as well as the detector of fire smoke in public places, detector of fire temperature indoors, and manual fire alarm button and jack for two-way telephone. Also the water hydrant should be linked directly to the pumper and send signals to the Fire Control Center of this area. To set telephone directly linking with the fire control center in the important places such as power distribution house and pump house.

8 旅游服务配套设施规划 Planning of Relative Tourism Service

第一节 旅游服务设施规划

一、旅游住宿

游客住宿主要依托于清镇市和贵阳市的酒店及度假别墅。

清镇市现已拥有各种星级宾馆，这些酒店设施不仅能满足普通中小型会议及一般旅游团队的需求，也能满足大型会议和特殊旅游团队的需求。

多国佛教园——华藏世界中的“居士寮”、“禅修斋”可以为不同的游客提供住宿，让游客体验山中修行。多国佛教园——华藏世界本身不提供较大的住宿功能。

二、旅游餐饮“多国佛教园——华藏世界”与素食园

餐饮是其产品的一个重要卖点，是营造人气、创造客源、引起重复消费和带动其他项目的要素，力图使其成为本地（清镇市）餐饮的又一大品牌，也成为本地市民喜爱的餐饮场所。素食园的餐饮开发为这样几块：

(1) 素食餐饮：提供个性化的服务，设计一系列的素食斋菜食谱。素食斋菜作为我国菜系中的一种，不仅有益身心健康，而且也可以看到佛教徒不杀生的慈悲之心，林园高士的淡泊之志。

(2) 特色餐饮：放在素食一条街，充分展示清镇市的特色餐饮系列，如各种风味小吃等，注重开发风味特色食品和生态型食品（如粗、野、杂菜）。目前，这些菜系已有了一定的发展，可以采用引资或合作的形式吸引已经开发成型并有意在本地发展的餐馆加盟，实行承包经营或共同经营。

三、旅游产品

多国佛教园——华藏世界的旅游特点设计要立足于特色性、宗教化。多国佛教园——华藏世界内旅游特点有以下几块：

(1) 节日：宗教诸佛节旦庆典
(2) 游客直接参与宗教中的各种活动
(3) 宗教中的各种祭祀表演
(4) 整合园区的游览
(5) 香客对佛学的学习
(6) 溶洞的观赏

Section I Planning of Tourism Service Facilities

I Tourism Lodging

Visitors can lodge in the hotels of Guiyang City and Qingzhen City and the holiday villas.

There are many hotels of all grades of stars, which can meet the needs of small or medium meeting, as well as large scales of meetings and special travel troupes.

Visitors can also get lodging in "Buddhist Dwelling" and "Buddhism Meditation Dwelling" of "Sino Tibetan World" to experience Buddhism Meditation in the mountain. Grand hotels are not available in the park.

II Cater Service —"Sino Tibetan World" and the Vegetarian Park

Cater is an important commodity, which is an indispensable factor of winning popularity, attracting visitors, inducing repeating consumption and driving other projects. So we should strive to make Vegetarian Park to be another brand of local catering and the residents' favorite place of catering. Following are the two aspects of developing the Vegetarian Park:

(1) Vegetarian Food: personalized service should be supplied, as well as a system of vegetarian food, which, as one type of cuisine of our country, is good for people's health and mind.

(2) Special Food: we emphasize on develop special food and ecological food like coarse cereals and wild herbs. At present, these cuisines have been developed in a certain level, so that we can adopt attracting capital or cooperating with those restaurants which have developed maturely and intend to expand their business in this region. Either contract management or joint management can be considered.

Ⅲ Tourism Product

When design the "Sino Tibetan World", we must stress specialty and religion, which is mainly demonstrated from the following aspects:

(1) Festivals: celebration for every Buddha
(2) Visitors' direct participation of every religious activity
(3) All religious fete ceremony
(4) Sightseeing around the whole park
(5) Pilgrim's learning of Buddhism
(6) Viewing caves

四、旅游商品

佛教旅游物品销售，放在各个佛教园区旅游物品销售处。

购物系列：以佛教历史、吉祥寓意为背景的各种特色产品为主。在多国佛教园区销售含有各国佛教特色的佛教纪念品。

此外，应大量开发地方土特产品和工艺产品，特色工艺品、民间工艺品、中草药材、常见旅游用品、工艺品、纪念品等，特别是轻便易带，既有纪念意义又有实用价值的产品；还应创造一系列仿古食品和菜肴，以满足游人的怀古之心理需求。

纪念系列：为海内外旅游者及香客提供一处举行留念活动之地，如刻石、植树、赠送礼品等。

Ⅳ Tourism Product

Buddhism commodities are put on the sale departments of tourism articles of every Buddhism gallery.

Shopping: The articles are mainly reflecting Buddhism history and auspicious implies and sold in many Buddhism Parks,

In addition, native products and crafts should be greatly developed, as well as special crafts, folk crafts, traditional Chinese herbal medicine, and common tourism products, especially the articles memorable, practical and easy to bring with. Also some ancient imitation of food and cuisines should be created to meet visitors' needs of memorizing ancient.

Memorizing: Some activities should be held for foreign visitors like carving stones, planting trees or some other presents.

第二节 旅游信息系统规划

一、游客服务中心：

解释多国佛教园——华藏世界的历史、文化、背景、功能的指示牌。包括管理标志、方向标志、限制标志和解说标志等。指示牌除用中文书写外，一般同时应用英文。

提供导游书籍和导游图，同样应有外文版本

接待和问询处

商店和纪念品销售处

照相设备，主要是胶卷销售和冲印以及数码冲印

出租业务

公用电话

网络中心

医疗急救服务

二、导游中心：

提供旅游咨询、印刷品解说，联系导游人员现场讲解。

三、指示系统：

在入口设立大型全景立体导游图牌；重要的岔路口，立交通示意图指示牌；所有重要的吸引物前，立说明牌。牌示设计应突出佛教风格，从色彩、造型规格等方面应与环境相协调，考虑到经济、耐用等因素。

另外，按照旅游区（点）国家标准，建立规范的标识系统。

Section Ⅱ Systemic Planning of Tourism Information

Ⅰ Service Center

Signal boards are used to explain history, culture, background and function of "Sino Tibetan World" including signs of management, direction, prohibition and elaboration in both Chinese and English.

Books and maps of guidance should be available, also in both Chinese and English

Reception and inquiry center

Sale department of commodities and souvenir

equipment for taking photo

renting

public telephone

internet center

first aid center

Ⅱ Guidance Center

The explanation system consists of guidance center, signs board and explanation of "Sino Tibetan World".

Ⅲ Direction System

Completed sign board system should be built. A large size dimensional tourist map of overall scenery should stand at the entrance, traffic signs at the important crosses, and explanation boards at important attractions. The design of all sign boards should highlight Buddhism style, and be coordinated with environment from color, shape, and size and so on, also considering the economical and durable factors.

In addition, normative sign system should be built according to national standard of scenic spot.

第三节旅游安全规划

1. 提供或设置导游标志，如导游图、警示牌、宣传牌等，为游客指明游览路线，提高游览效果，避免迷失方向。标牌应体现指示性、规定性、说明性、解释性、宣传性等。

2. 各种游人集中且容易发生跌落等人身事故的场所，地形险要的游道，要设置安全防护设施。

3. 各种游人活动较频繁，且边缘临空高差超过 1m 的场所，要设置护栏设施。

4. 各种装饰性、示意性和安全防护性的构造物，严禁采用锐角、利刺等形式。

5. 设置景区医疗机构，既为疗养需要，也同时满足游客安全防范需要。

6. 按国家标准设置防火设备。

Section Ⅲ Planning of Tourist Security

1. Guiding marks like guiding maps, warning boards, and billboard ect. should be supplied or set to improve tourism effect, and avoid being lost. All the sign boards should embody direction, regulations, illustration, interpretation, publicity.

2. Security and protection facilities should be set to keep visitors from fall in the dangerous places or path.

3. Protection facilities should be set in the places where visitors concentrate and of which the edge is more one meter height from a flat.

4. Prohibit any sharp edge or pointedness in the decorative, schematic and security constructions.

5. Medicine institutions should be set both for sanitarium and visitors security.

6. Set fireproofed equipment should be set according to national norm.

9 旅游市场开发策略 Developing Strategy of Tourism Market

第一节 客源市场定位

根据多国佛教园——华藏世界的交通区位、资源条件，充分考虑旅游市场的发展态势，确定在未来 5 年的期限内，旅游开发的客源市场定位为：以本市客源市场为主导和基础，以省内其他地区、邻省及省外经济发达地区国内旅游市场为依托，以开发专项特色旅游市场为主要目标。

一、国内旅游市场

根据区内旅游资源的特点、规划导向与旅游市场大环境分析，着重考虑到区位交通特点，本区的客源市场大致可分为三级：

(1) 一级市场：核心市场。目标贵阳市区及大市区内的各市、县。

目前不仅在清镇市、贵阳市存在着许多佛教类型旅游景点，就是在全国范围内佛教类型的旅游景点也数不胜数，但多国佛教园——华藏世界是打破传统佛教景区单一的特点，汇集了东南亚各国佛教和中国佛教的特点于一身的佛教景区，在全国尚属首例。

旅游资源在全省范围内占有一定的优势，在一日游程范围之内，市民出游方便，因而本地游客将是风景区最稳定的基础客源市场。客源包括短程休闲、度假、会议等类型。

Section I Orientation of Market

Considering fully the development situation of tourism market and the traffic position and resources conditions of "Sino Tibetan World", we make the orientation of tourist market in the following five years, that is, with native visitors as the major and basic target market and the dependence of other scenic spots of our province or the near provinces or the ones of developed area of our country to achieve the goal of developing special tourist market.

I Domestic Tourism Market

We divide the tourist market into three classes considering the features of tourism resources, planning guidance and the overall situation of tourism market of this area, and especially, the traffic features :

(1) First class market core market: Guiyang city and every cities and counties of grand prefectures.

At present, scenic spots of Buddhism type are numerous not only in Qingzhen city and Guiyang city, but also around China. However, "Sino Tibetan World", concentrating the Buddhism features of every country of Southeast Asia, is the unique in China.

The native tourists will be the stable customers as they can finish their tour in "Sino Tibetan World" within one day, which is convenient for travel, relaxation, vacation and meeting.

(2) 二级市场：基本市场。目标西南地区各城市。

这些城市距清镇市在 200km 左右，适合开展周末二日游。但目前本市缺乏知名度，如果能进行有效的团队促销，将会收到显著成效。

省外客源市场分成两类：一类是来自清镇、贵阳度假、参观游览黄果树和龙宫、探亲访友的近程邻省的旅游者，由于我省几条出省高速公路的修建，自助进入我省旅游的游客将愈来愈多；另一类是团队游客，一般以途经贵阳、清镇的过境游客居多。

(3) 三级市场：机会市场。目标省内其他地区及其他国内游客。

对此类客源市场的开发，主要在于做足线路组合的文章。如何让到贵阳其他景区的游客中转游览本市，是市场开发中需考虑的重要问题。

三级市场，多国佛教园——华藏世界可以作为我省的特色专项产品，挂在省内重要旅游线上，吸引东南沿海经济发达地区的旅游者，可以依靠贵阳市的重要活动或自己策划活动吸引某些方面的专业人士。

(2) The second class market: the basic market: every city of Southwest region.

We can develop weekend travel (two days) for these people who are living around 200 kilometers from Qingzhen. Nevertheless, Qinzhen has not been famous by now that it need some effective strategy of promotion to achieve outstanding effect.

Customers of other provinces can be classified two types. One type is the visitors who are living not far away from Guizhou, coming to Qingzhen and Guiyang to have vacation, tour Huangguoshu and Longgong and visit relatives and friends. With several highways towards foreign provinces, visitors driving to our province for travel by themselves are more and more. The other type is the group travelers, generally, they are travelling by the way of Guiyang and Qingzhen.

(3) The third class market: the opportunity market, tourists of other areas of our province and other provinces.

To develop this type of market, we need make good use of resource integrity, that is, we should consider how to make visitors take way of Qingzhen when they want to visit other scenic spots.

The cultural park of "Sino Tibetan World", as the special and important product, can be hung along the important travel route in our province to attract the visitors from developed regions of Southeast; also we can rely on the important activities of Guiyang City or scheme by ourselves some activities to draw attention of the professions of the relative fields.

二、国际旅游市场

服务于对中国佛教历史文化有兴趣的外国游客。

(1) 一级市场：港澳台、日本、韩国、东南亚的文化旅游及商务旅游客源市场。

(2) 二级市场 ：欧美市场， 主要是商务旅游和各种专项文化旅游市场。

(3) 机会市场：国外组团的专项旅游客源市场。

II International Tourism Market

It is for the foreign tourists who are interested in Chinese Buddhism history.

(1) The first class market: cultural travel and business travel of Hongkong, Macao, Tianwan, Japan, Korea and Southeast Asia.

(2) The second class market: the market of Europe and America, major in business travel and special cultural travel.

(3) Opportunity market: foreign group travelers.

第二节 市场开发策略

从游客不断发展的求新、求奇、求异的需要和不断变化的市场态势出发，不断追踪研究、策划和推出多国佛教园——华藏世界的更新、替代、后续产品。多国佛教园——华藏世界的寿命比较起以自然景观或历史古迹类的旅游区更短，而且旅游者的口味变化很大，因此要保证多国佛教园——华藏世界的不断发展，持续吸引旅游者，就必须在公园建设方面不断投入，不断更新公园内的娱乐项目和产品类型，甚至在必要的时候主题也必须增容或更新。

Section II Strategy of Developing Market

To meet the need of innovation, strangeness and difference of the tourists and the changing market situation, we have to keep studying to scheme and make out new products to renew and replace the old ones of "Sino Tibetan World". Compared to scenic spots of natural landscape and historic sites, "Sino Tibetan World" is short in lifespan as the visitors will keep changing in their appreciation taste, so that we need to invest more in the construction of park, and keep updating the entertainment programs and product types, or even add or renew the theme if necessary to make sure the sustainable development and attraction of "Sino Tibetan World".

一、基本对策　I Capital Strategy

根据对现代旅游消费趋势和我省客源市场的分析，其市场开发目标的分阶段策略是：

Analyzing modern tourism consumption and tourist market of our province, we make the strategy of different stages of developing market：

开业第一年，以清镇市民和贵阳市市民（尤其是本区市民）作为主要客源市场，倡导周末和节假日观光休闲，通过开展各种活动（如节庆）和开发有特色的餐饮、活动，成为清镇市、贵阳市的家庭聚会、团体活动的好去处，成为清镇市民和贵阳市民的近郊游憩场所。

At the first year of opening, the citizens of Qingzhen and Guiyang will be treated as the main customers. Through various activities like festival celebrations and creating distinctive food and drink, Qingzhen will be the ideal place for citizens of Qingzhen and Guiyang city to spend weekend and holidays.

开业第二年，以独特的产品形式，将多国佛教园——华藏世界建设成佛教文化内涵深厚的拥有众多可参与主题的佛教胜地，吸引省内其他地区来贵阳游玩的游客。同时，对来贵州旅游的省外游客及喜好佛教文化旅游的海外游客构成一定吸引力。由于佛教文化的性质所决定，景观只能是相对固定的，只能以各种活动创新作为发展的主流。

At the second year of opening, "Sino Tibetan World" will be constructed as the holiest site of Buddhism with deep cultural connotation and plenty of themes to attract the visitors from other region, including the tourists from other province or overseas. Landscape of "Sino Tibetan World" will be relatively unchangeable confined by the features of Buddhism culture, so we can only change it mainly on various creative activities.

第三年起，多国佛教园——华藏世界在完善服务设施及丰富产品内容，逐步增加活动的文化含量，使其成为我省有突出特色的产品。世界之窗在经营实践中摸索出“新增项目＋活动策划＋艺术表演＋节日庆典＝市场”的规律，要求每年有新项目推出，以培育新的旅游消费热点和新的利润增长点。大型高水平文艺表演和各景点的表演是多国佛教园——华藏世界最具生机与活力的部分。因此，“创一流服务，建旅游精品”是多国佛教园——华藏世界不懈追求的目标。

From the third year on, we have to improve service facilities, enrich product content and increase cultural connotation of activities gradually to make "Sino Tibetan World" become an outstanding tourism product of our province. With rich experience of running business, "The Window of the World" groped the rule "new program+activity scheme +artistic performance+festival celebration=winning market", which asked them to make out new program every year in order to create new consumption and increasing point of benefit. High-quality and large art performances and the performance of every scenic spot are the most alive and active parts of "Sino Tibetan World". Therefore, "First-rate Service, First-rate Tourism product" is the goal that "Sino Tibetan World" pursues unremittingly.

景观布置示意图

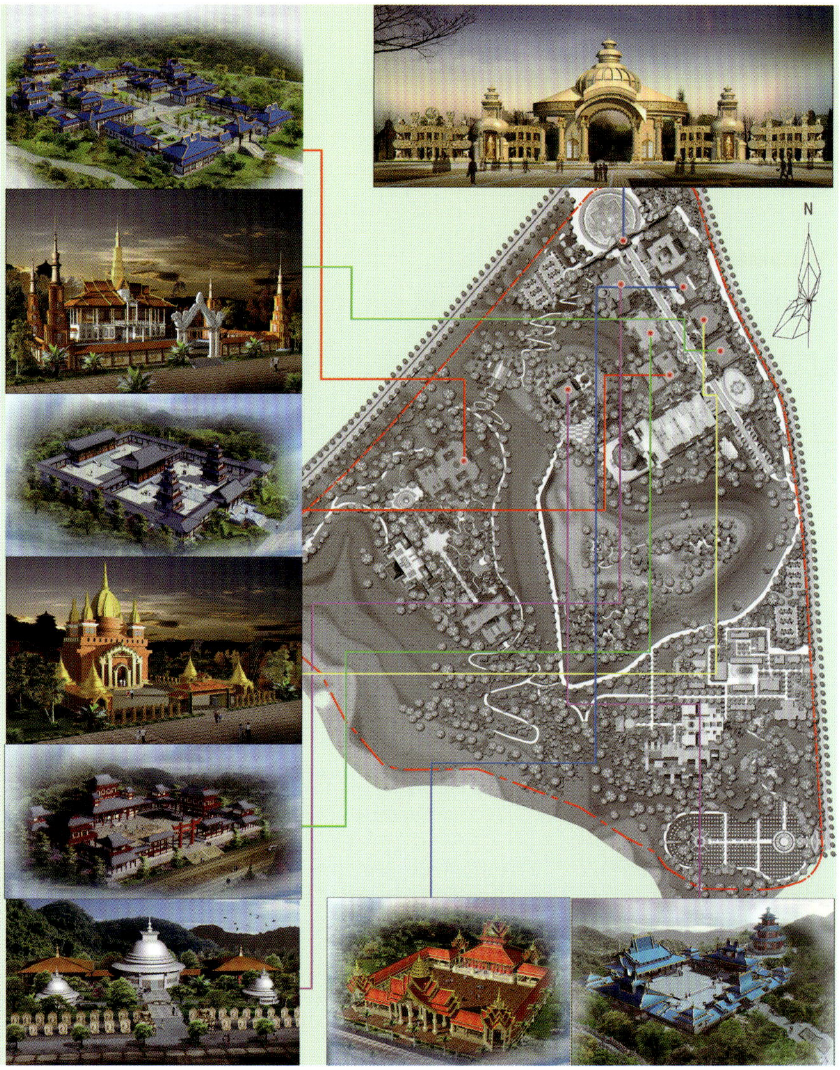

二、宣传促销策略 II Strategy of Advertisement and Promotion

根据市场需求和产品生命周期理论，任何一个产品都经历介绍期、成长期、成熟期和衰退期四个阶段，规划要逐一分析旅游产品生命周期的状况，根据市场发展趋势不断创新，使旅游产品体系保持整体的吸引力和竞争力。因此，采取各种方式强化市场营销，就成为多国佛教文化园的制胜法宝。

According to market demands and life cycle theories of product, any product should experience the four stages of introduction, growth, maturity, and recession. So we have to analyze the situations of each stage of life cycle, so as to making innovations unremittingly in accordance with market development to keep tourism products attractive and competitive entirely. Therefore, taking various methods to strengthen marketing has become the invaluable asset of "Sino Tibetan World".

1. 采用现代信息技术，开展网上信息发布和旅游促销，是当今最便利的营销手段。

1. The most convenient method is to adopt modern informative technology to publish information and tourism promotion in web.

2. 创作拍摄以佛教文化为题材或背景的影视作品，借助电影、电视和文学作品等形式，提高景区的知名度和美誉度，也同样是不可缺少的。

2. It is indispensable to shoot films with the subject or background of Buddhism culture, so that the fame and reputation will be improved with the works of films, television and literatures.

3. 收集整理出版佛教传说及故事集，组织学术交流会，对于发掘潜力，创作新的产品，也具有同样重要的意义。

3. To collect, compile and publish the Buddhism legends and stories collections and hold academic communication meets are meaningful to tap potential and create new products.

4. 参加收视率较高的电视专栏节目或与媒体共同开辟专栏，策划佛教祭祀大活动；争取记者团、旅行社客户团来佛教园参观考察，与他们合作推介本地旅游线路，向旅游中间商和客源地区的旅游行业管理部门、同行业旅游企业和资源互补性旅游目的地邮寄宣传资料；参加海内外相关的旅游展销会，旅游商品展销会。

4. To join the television programs with high ratings or cooperate with media to open new columns and scheme great Buddhism sacrifices; to make effort to accomplish the view and investigation of journalists and customers of travel agency, in order to cooperate with them to introduce and publicize the local tourism lines and send the tourist agents, management department of tour of the target market, tourism enterprises and tourist attractions with mutual complementary resources promotional materials; to take part in the domestic and oversea fairs of tourism and tourist products.

景观布置示意图

5. 邀请有宣传效应的活动（如比赛会、研讨会等）来佛教园内举办，印制宣传画册和散发各种旅游宣传品，在旅游目标市场地区，进行佛教旅游产品、户外佛教等活动的模拟展览；举办有关景区的摄影展等，也同样重要。

6. 向游客赠送、向各相关单位寄发旅游景点景区明信片，制作风光 VCD、DVD，在游客服务中心和宾馆饭店的闭路电视上播放，免费赠送有关单位放映，制作和改进电视广告并在本市电视台上选择恰当时机进行播放，也在本市或外地电视、报刊上做广告或制作广告牌进行宣传，都将获得良好的效果。

5. It is also important to hold the activities with publicity effect like contests, academic meets and so on in the Buddhist parks, make corporate catalogs and distribute all kinds of promotional products of tourism, and hold analog exhibitions of Buddhist tourism products and outside Buddhist activities in the areas of target market, as well as picture shows about relative scenic spots.

6. To give the visitors or send the relative department cards printing the scenic spots, to record VCD and DVD of sceneries for the tourist service centers and hotels to play, and advertise in local and foreign television channels or newspapers, which will work eventually.

三、价格政策　Ⅲ Price Policy

多国佛教园——华藏世界建成后，将本着以社会效益、环境生态效益为主，兼顾经济效益的原则，根据不同时期的需要，严肃、认真、科学地制定合理的价格政策，近期拟将推出的主要做法如下：

After the construction of "Sino Tibetan World", we will seriously, strictly and scientifically make a reasonable price policy according to the principle of emphasizing on social and ecological & environmental benefits and considering economic returns, and the demands of different period. Following is the price policy we plan to try:

1. 严格按照国家、省、市物价部门要求及相关政策法规，参照市场实际情况进行科学定价、变动价格和实行优惠价格。对团队和散客、节假日和平日分别定价。

1. Strictly follow the policies and regular of national, provincial and prefectural price department, and consider the practical situation of market to price scientifically, change and lower the price. The price should be different between group and individual, and on festivals and ordinary days.

2. 为吸引更多消费者，采取低价门票的策略，用价格策略招揽游客，扩大市场份额。

2. The tickets should be cheap to attract more consumers, so as to expand market.

3. 向游园客人微利销售小册子、导游图等有关宣传本园的资料、纪念品。

3. Sell the promotional materials and souvenirs like pamphlets and guiding maps to visitors at low prices.

4. 优惠提供快餐、饮品、贵州特产及佛教商品。

4. Supply visitors with fast-food, drinking, specialties of Guizhou and Buddhist commodities at low prices.

景观布置示意图

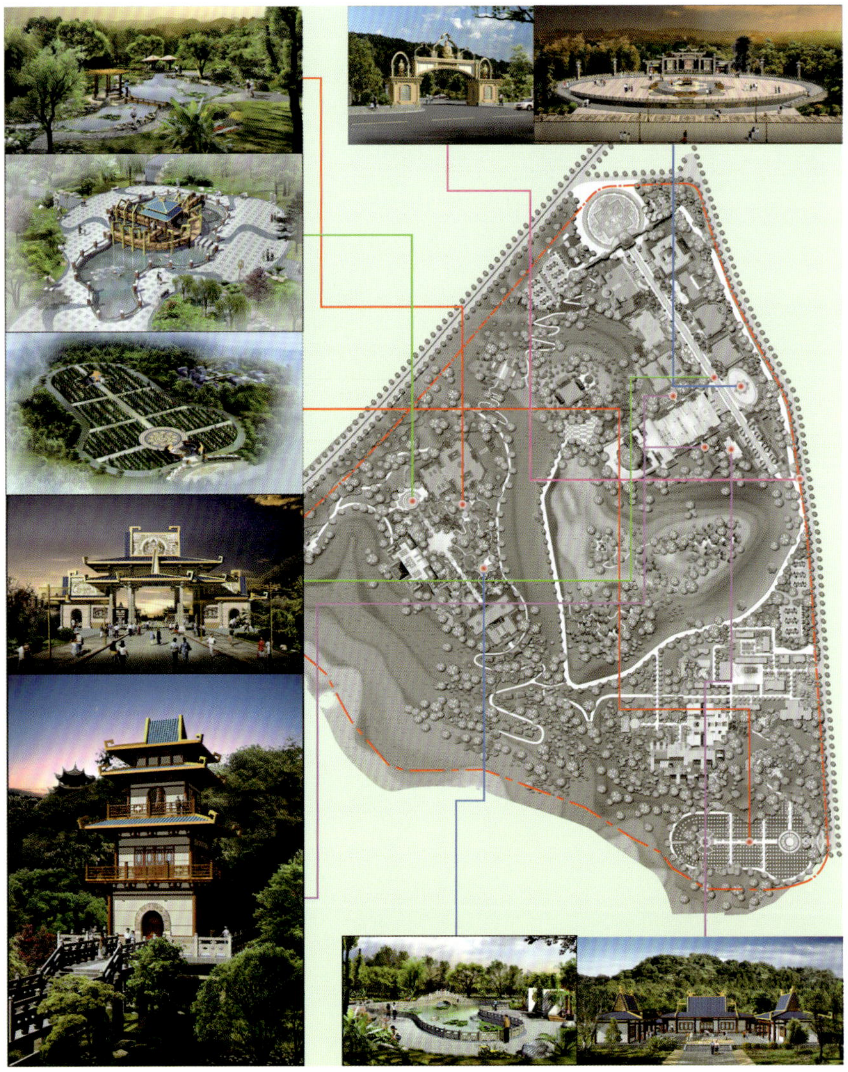

10 项目规划主要技术指标 Major Economic and Technique Target

第一节 用地平衡 Balance of planning land

本规划用地类型，主要分为建筑用地、道路广场用地、园林绿地、水体等四类

Land use in this plan are classified into four types: Construction Land Use, Roads and Square Land Use, Park and Greening Land Use and Water Body.

序号	用地类别	面积（万平方米）	比例（%）
1	总用地面积	48.1	——
2	建筑占地面积	3.27	6.8%
3	总建筑面积	7.14	——
4	道路广场水体面积	11.58	24.1%
5	绿化面积	32.22	67%
6	停车场面积	1.03	2.1%

Number	Type of Land Use	Area（hm^2）	Rate（%）
1	Total Area of Land Use	48.1	——
2	Occupied Area of Construction	3.27	6.8%
3	Total Construction Area	7.14	——
4	Water Area of Road and Square	11.58	24.1%
5	Greening Area	32.22	67%
6	Parking Area	1.03	2.1%

第二节 主要指标

多国佛教园——华藏世界各类用地数值一览表：

- ▲ 建筑密度：6.8%　Construction Density：6.8%
- ▲ 容积率：0.15　Volume Rate：0.15
- ▲ 绿化率：67%　Greening Rate：67%

小品布置示意图

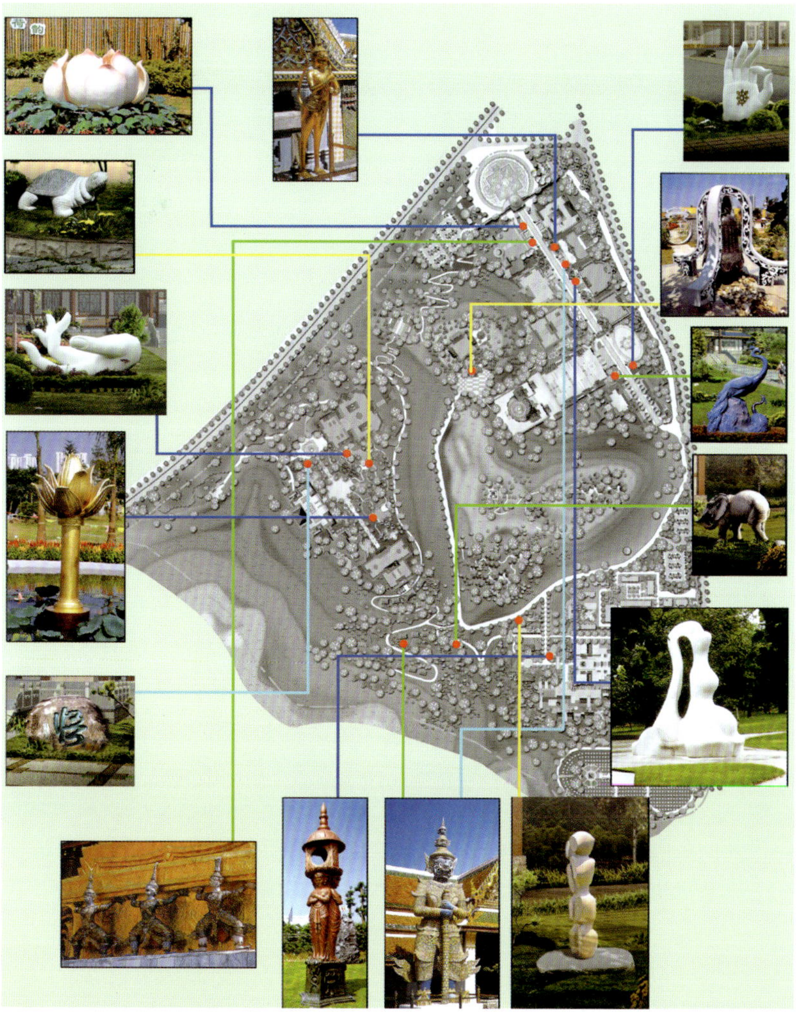

小品布置示意图

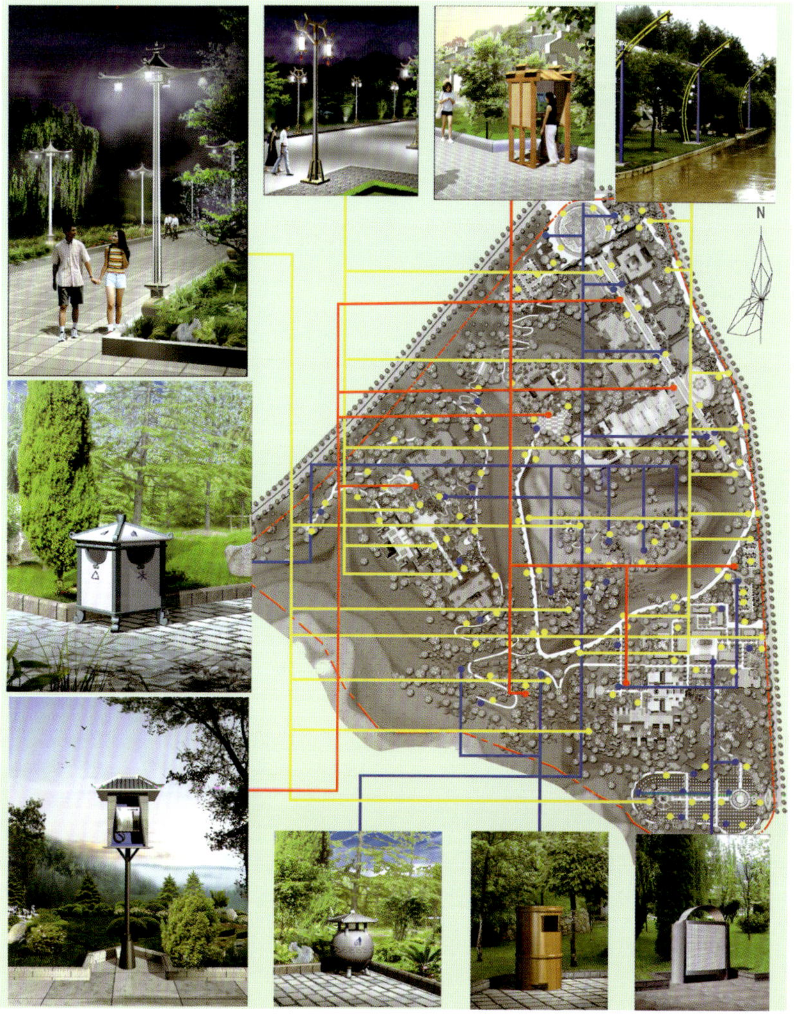

总体鸟瞰 Buddhism area

效果图件 Effective Pictures

Buddhism Park of Multi-Country

佛教园区 Buddhism Area

出入口

Entrance and Exit

主大门：采用一个融整个亚洲佛教建筑经典为一体的建筑形制，运用典型的印度建筑风格，并采用大量的“华严壁”。

Gate: It adopts a style which was mingled with whole architectural essence of Asia, the typical India architectural style and great many of "Huayan walls".

广场：客源集散地及文体活动场地。“华严壁”置于广场大门两侧，华严壁面采用佛教图案用浮雕手法展示。

Square: It is the place for visitors and public activities. "Huayan Stones", with the patterns of Buddha in sculptures, are set on the tow sides of the square gate.

①大门效果图 ②大门局部效果图 ③大门局部效果图 ④管理房效果图

①

②

③

④

①

东南亚各国佛教园区 Buddhism Parks of Asian Countries

①多国佛教街景 ②多国佛教街景

▲ 斯里兰卡佛教园：园内庙宇殿堂严格按照斯里兰卡的康提城中的佛牙寺建筑形制构建，佛牙寺以供奉佛祖释迦牟尼的佛牙而闻名。在斯里兰卡园当中，除了有三座最有特色的佛塔以外，护栏、大门也是由佛像构成。另外还希望通过佛协和中国佛协引取知名高僧的舍利放到塔里来供养，这样能增加整个斯里兰卡园的宗教氛围和可观赏性。

Sri Lanka Buddhism Park: Temples and palaces are constructed strictly according to the shape of Temple of Tooth Relic, Kandy, Sri Lanka, where tooth of Buddha is dedicated. In the park, the most three special pagodas are consisted of Buddha, so are the guardrails and the gate. In addition, schelli of eminent monks are expected to be introduced by Buddhism Committee and Buddhism Committee of China to the pagodas and dedicated, which can increase religious atmosphere and viewing value of the whole.

④

▲ 尼泊尔佛教园：园内庙宇殿堂严格按照尼泊尔最古老的佛教寺庙、著名的佛教圣地斯瓦扬布寺建筑形制构建。集中展示尼泊尔佛教文化的精髓。

Nepal Buddhism Park: Temples in this park are constructed seriously according to Swayambhunath, the most ancient temple of Nepal, to show concentratedly the essence of Buddhism.

⑤

①斯里兰卡佛教园鸟瞰 ②斯里兰卡佛教园局部 ③斯里兰卡佛教园局部 ④尼泊尔园佛教园鸟瞰 ⑤尼泊尔佛教园局部

▲日本佛教园：园内庙宇殿堂严格按照大和民族的大德寺建筑形制构建，在日本园里设置了禅茶区，有来自日本、中国的茶道，各个地方茶道的表演，很多游客可以在这里休息、品茶，成为游客的休息中心。

Japanese Buddhism Park: Temples and palaces are constructed strictly according to the structure of Daitokuji Temple of Japan. In this park, the tea area where Japanese and Chinese Tao of Tea and the performance of Tea Tao of many other places are available is the rest center for visitors to have a break and enjoy some tea.

①日本佛教园鸟瞰 ②日本佛教园大雄宝殿 ③日本佛教园局部

①

▲ 泰国佛教园：园内庙宇殿堂严格按照泰国最著名的也是目前泰国唯一一个没有和尚居住的寺庙，同时也是泰国三大国宝之一的玉佛寺形制而建造。园内庙宇按照玉佛寺内绘有178幅以印度古典文学《罗摩衍那》史诗为题材的精美彩色连环画，并附有泰文译诗的壁画长廊，是凸显泰国的独特建筑艺术，并与其佛教文化完美地结合在一起，集中展示泰国佛教文化的精髓。

Thailand Buddhism Park: Temples and palaces are constructed strictly according to the structure of Yufo Temple, the most famous and the only one temple with no monk living in in Thailand, and also one of the three national treasures of Thailand. There are 178 paintings elaborating the Indian Ancient historic poem of "Ramayana" hung in the houses in accordance with Yufo Temple, along with a gallery with version of this poem in Thai language on the wall paintings, which demonstrates the unique architectural arts of Thailand, combine perfect with Buddhism culture and show the essence of Thailand Buddhism culture.

①泰国佛教园鸟瞰 ②泰国佛教园大门 ③泰国佛教园主殿 ④泰国佛教园局部 ⑤泰国佛教园小品

▲ 韩国佛教园：园内庙宇殿堂严格按照韩国最大的寺院佛国寺建筑形制构建，内设多宝塔和释迦塔，这两座塔高 10.4m，分别耸立于大雄殿和紫霞门之间庭院的东西两侧。在韩国园区中我们可以安排一些有东北亚特色的佛乐的表演，有东北亚特色的佛教文化的展览。

Korea Buddhism Park: Temples in this park are constructed seriously according to the structure of Foguo Temple, the hugest temple of Korea. The Duobao Pagoda and Shijia Pagoda, both 10.4 meters in height, are standing separately the east and west sides between Daxiong Palace and Zixia Door. In the park, we can arrange some Buddhism musical performance and exhibitions of Buddhism culture of Northeast Asia.

①韩国佛教园鸟瞰 ②韩国佛教园大门 ③韩国佛教园局部 ④韩国佛教园局部 ⑤韩国佛教园局部

▲ 缅甸佛教园：缅甸被称为“万塔之国”，最出名的是它的佛塔。园内庙宇殿堂依照缅甸的仰光瑞达光大金塔建筑形制结合缅甸当地佛教建筑形式构建而成，庙宇殿堂内设置关于小乘佛教的慈悲、忍耐、修身、自我解脱以及积功德报来世为主要内容，集中展示缅甸佛教文化的精髓。在缅甸园区中可进入“禅浴中心”，它可以让游客在这里休息，放松一下筋骨，进行禅浴的一些活动。一方面增加游客行游的时间，另一方面能够亲身去感受佛教文化不仅仅是出世，在入世当中也有很多佛教文化的表现。

Burma Buddhism Park: Burma is famous for Buddhism pagodas, and called "a country with thousands of pagodas". Temples in the park are constructed imitating the structure of Shwedagon Paya of Burma and local Buddhism constructions. The houses of the temples are decorated with the theme of mercy of Hinayana Buddhism, tolerance, self-cultivation, and so on, which demonstrates the cultural essence of Burma Buddhism. In addition, this park is equiped with "bath center" where visitors can relax themselves and take bathes. It is a good way to make visitors' travel last in the park and visitors experience Buddhism culture by theirselves.

②

▲ 柬埔寨佛教园：园内庙宇殿堂按照柬埔寨著名的“银阁”佛庙结合柬埔寨当地的寺庙建筑特点的建筑形制构建。在柬埔寨佛教园中设置展示东南亚一带素斋、小吃等食品的商业区。

Cambodian Buddhism Park: Temple hall in the park combines the construction features of the famous "Silver Pagoda" in Cambodia and local temples. There is the commercial district set up in the park showing the foods of vegetarian and snacks in Southeast Asia.

③

①缅甸佛教园鸟瞰 ②柬埔寨佛教园鸟瞰 ③柬埔寨佛教园主楼

▲中国佛教园："中国佛教园" 地处佛教园区末端，与 "印度佛教园" 形成呼应，寓有佛教文化自印度发端，至中国发展到极盛的象征意义。"中国佛教园" 是多国佛教园——华藏世界的中心，集中展示中华佛教的灿烂文化，中国佛教园主体为一 99m 高的毗卢佛殿，佛殿内设毗卢殿、药师殿、弥陀殿、观音殿、地藏殿、文殊殿、普贤殿等。

Chinese Buddhism Park: It is at the end of this area with the meaning that Buddhism culture was from India and had its prime time in China. "Chinese Buddhism Park" is the center of the "Sino Tibetan World" fully displaying the splendid culture of Chinese Buddhism. The main building in Chinese Buddhism Park is Pilu Temple in 99 meters height, in which there are Pilu Palace, Pharmacist Palace, Amitabha Palace, Avalokitesvara Palace, Manjusri Bodhisattva Palace and Samantabhadra Bodhisattva Palace.

①中国佛教园鸟瞰 ②中国园大门 ③观佛广场

▲单体大佛像：自成一景，置于中国佛教园后方，高度为99m左右。佛有三生：有报身佛、有化身佛、有法身佛。毗卢遮那佛是佛教的法身佛，法身佛用入世的观点讲它代表的佛学的精神，佛的这个色身，外在的体现是非常多的，这也是佛教讲的"众生是佛"。毗卢遮那佛代表了佛教的统一宇宙观，代表了佛教的"空即是色，色即是空"，我们塑的就是毗卢遮那佛。同时我们要让它成为佛中有庙，庙中有佛。它不仅仅是一尊佛，在它的脚上、膝上都有建筑物，建筑物里依然还会有佛像。一方面让人感觉到一种耳目一新的感觉，另一方面感觉到更宽阔的空间感受。

Single Buddha Statue: It is behind the Chinese Buddhism Park with about 99 meters in height. There are three Kayas in Buddhism: Dharmakaya, Sambhogakaya, Nirmanaraya. Vairocana, which preach its Buddhism spirit with the concept of "go into the secularity", is the Dharmakaya in Buddhism where Buddha can be everybody in the world. Vairocana represents the concept of unity in Buddhism. And what we want to construct here is the Vairocana on whose feet and knees are constructions in which are Buddhas, we aim to reach the effects of Buddha contains temples, and vice versa, and giving tourists a broader space.

②

▲四大天王：中国园单体大佛前设置四大天王，意为守护毗卢主佛。四大天王为东方持国天王多罗咤，手抱琵琶，护持国土；南方增长天王毗琉璃，手持利剑，令众生善根增长；西方广目天王毗留博叉，臂缠一龙，以净眼观察，护持众生；北方多闻天王毗沙门，左手握银鼠为法器，右手执伞，代表福德，保护人民财富。他们所拿的法器，分别代表着"风、调、雨、顺"，都含有吉祥的意义，是人们对年景美好的愿望。

Four Deva-Kings: On both sides of the temple the four deva-kings are housed. The one in the East, the protector of the country, is holding a Pi-pa (a Chinese musical instrument) in his hands; the one in the South, representing proliferation, is holding a sword in his hand; the one in the West, known as the wide-eyed, is holding a dragon in his left hand and a pearl in his right; the one in the North is famous for his profound knowledge, and he is holding a pagoda in his right hand. The things they hold represents "wind", "harmony", "rain" and "stability" which are auspicious signs according to Chinese Tradition.

①单体大佛像 ②四大天王

▲ 登高塔：自成一景，以现代手法结合古代传统佛塔建造，上设观景台，游人、香客可登此鸟瞰“华藏世界”全景，远眺清镇市和红枫湖风景名胜区全景，既满足人们登高望远的心理需求，又营造出“华藏世界”立体空间效果。

Observation Pagoda: It is a single scene constructing in ancient traditional shape with modern ways with a observation plat on the top where visitor can get a bird's-eye view of the whole "Sino Tibetan World", and enjoy scenery of Qingzhen and Hongfenghu Lake scenic spot in the distance, which can not only meet people's mental needs of standing higher and looking farther, but also create three-dimensional spatial effect.

▲ 俗家食府：为游人、香客提供酒荤饮食享受的俗家食府。
▲ 素斋馆：为游人、香客提供素食斋饭的餐饮处。
Restaurant: It supplies visitor and pilgrims with meat dishes and wines.
Vegetarian Restaurant: It is the place which supplies visitors and pilgrims with vegetarian diet.

①钟楼效果图 ②素食街效果图

①

道场区 Rites Area

▲弘法道场：高僧弘法讲道的场所（可兼作会所用，高僧弘法讲道时，为弘法道场；有会议承接业务时，可作会议接待用）。

Rites of Preaching: It is the place where eminent monks preach. Also it can be used to undertake conferences.

①道场鸟瞰 ②道场局部 ③方圆楼效果图 ④方圆楼正面效果图

▲上客堂：上客堂主要是俗家弟子饮食起居的地方。除此以外，上客堂还是寺庙为重要顾客提供的休息、餐饮的场所，在上客堂游客可以更加直接的体验佛教博大精深的文化。

②

③

Distinguish Guest House: It is the place where the distinguish guests have rest and food, visitors can more directly experience the extensive and profound Buddhism culture.

①上客堂鸟瞰 ②上客堂大门 ③上客堂局部

①

佛学文化区

▲佛学院：讲经布道的场所，内设讲经堂、自习室及会所（会务接待）。

Buddhism Culture Area

Buddhism Institute: It is the place for lectures and business reception.

①佛学院鸟瞰 ②佛学院大门 ③佛学院局部

①

②

③

④

⑤

⑥

⑦

①～⑥佛学院局部 ⑦佛学院藏经阁

武

▲武僧馆：多国佛教园——华藏世界护寺“武僧”场所，护寺“武僧”分两类，一类为真正习武的僧人，一类为日常维护多国佛教园——华藏世界秩序的保安人员（该类人员也以“武僧”面貌出现）。

Monk House: It is the place where live guards of the temple of "Sino Tibetan World". Guards of monks are classied into tow types, one is the real monks mastering Wushu, the other is the security men in the costume of monk.

①武僧馆鸟瞰 ②武僧馆大殿 ③武僧馆练功房

①

禅修区

▲禅修斋：主要为游客提供一个长期度假或者休养的场所。

Buddhism Cultivating Area

Buddhism Cultivating House: It provides old people without abilities with living place.

▲居士寮：提供给一些高端的居士、德高的高僧来修行的地方。

Laymen Area: It provides some high-end people with location to preach.

①禅修斋鸟瞰 ②禅修斋局部 ③禅修斋局部 ④居士寮

多国塔林区 Multinational Pagodas Area

▲佛塔群：仿效东南亚各国塔林形制，建造具有各国特色的佛塔群。
Pagoda Group: To construct pagodas groups with different features of counties imitating the ones of every country of Southeast Asia.

▲功德碑林：立功德碑若干，铭刻多国佛教园——华藏世界投资捐款人的功德。
Forest of Steles of Virtue: To set some steles of virtue carving the names of investors and donors for "Sino Tibetan World".

③

▲登高塔：自成一景，以现代手法结合古代传统佛塔建造，上设观景台，游人、香客可登此鸟瞰多国佛教园——华藏世界全景，远眺清镇市和红枫湖风景名胜区全景，既满足人们登高望远的心理需求，又营造出多国佛教园——华藏世界立体空间效果。

Observation Pagoda: It is a single scene constructing in ancient traditional shape with modern ways with a observation plat on the top where visitor can get a bird's-eye view of the whole "Sino Tibetan World", and enjoy scenery of Qingzhen and Hongfeng Lake scenic spot in the distance, which can not only meet people's mental needs of standing higher, looking farther, but also create three-dimensional spatial effect.

①塔林局部效果图 ②塔林局部效果图 ③登高塔效果图

①

慈善公益事业区 The Social Charity Activities Part

▲孤儿院：为孤弱幼儿提供良好教育住宿生活的场所。

Orphanage: It offers orphans good education and lodging.

①慈善中心鸟瞰 ②慈善中心大门效果 ③孤儿院主楼 ④孤儿院局部 ⑤孤儿院局部

▲安养院：为年迈无自理能力的老人提供安享生活的场所。

Sanitarium: It provides old people without abilities with living place.

①～④安养院局部

溶洞观赏区

①水景鸟瞰 ②～⑤水景局部

公用设施及景观小品：由园区公共厕所、指示牌、休闲亭、休憩椅、灯饰、电话亭、垃圾箱、艺术景观小品等构成。

Infrastructure and Scenery Accessories: They are made of public washroom, guiding board, pavilion for resting and leisure, benches for rest, lamp, telephone booth, and little bin, art sketched.

Tourism R
旅

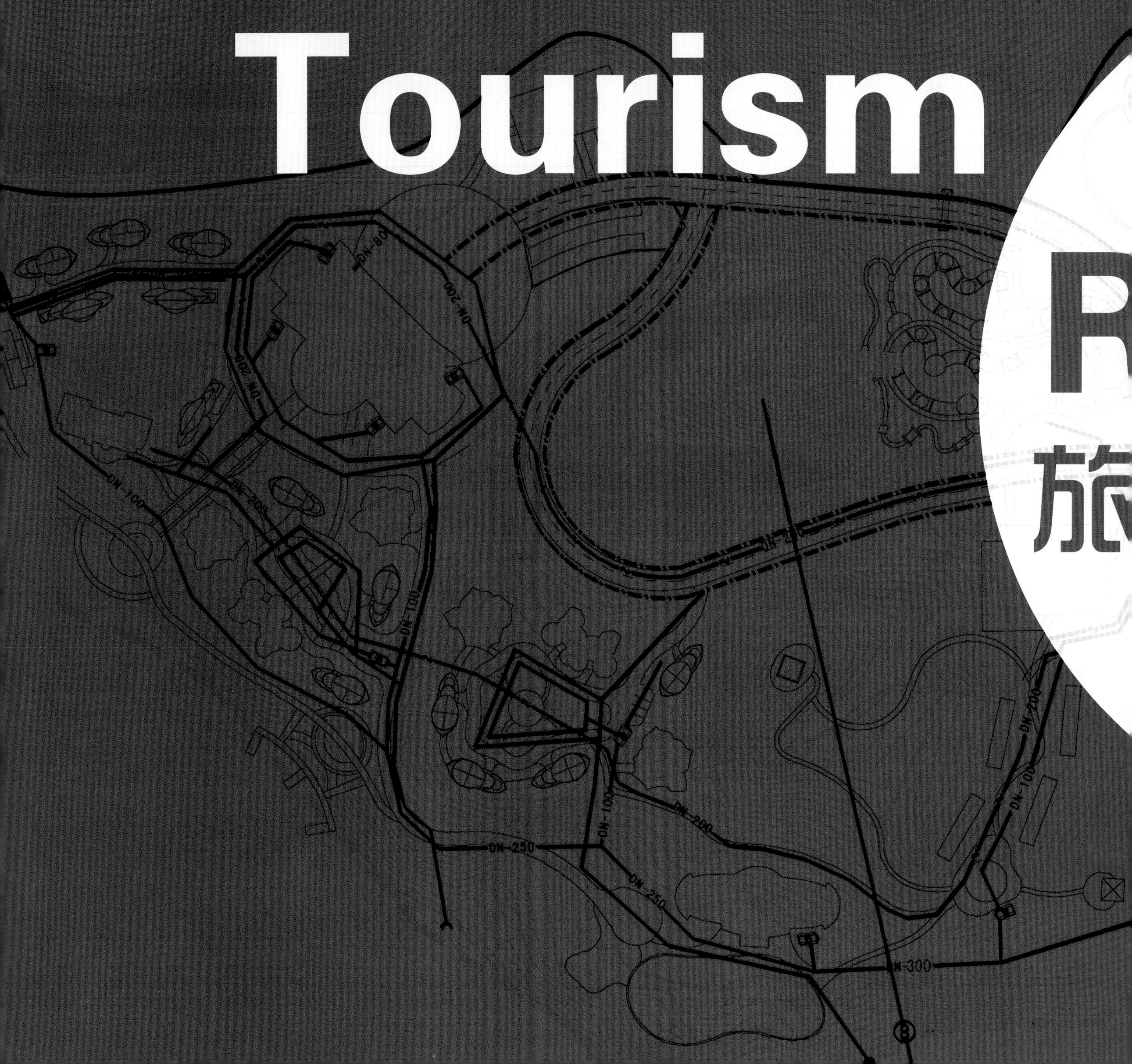
DN-80
DN-200
DN-200
DN-100
DN-200
DN-100
DN-250
DN-200
DN-100
DN-250
DN-200
DN-100
DN-300

esort

游度假区 ↘

1 Outline of Tourism Resort 认识旅游度假区

美国未来学家甘赫曼曾寓言“休闲时代”是继“信息时代”之后人类社会发展的第四次浪潮。据 WTO 预测，休闲产业将取代传统的主导产业和现代信息产业等跃居世界第一大产业的位置。

Herman Gan, the American futurist, had forecasted that "Leisure Time" is the forth wave of the development of human society after "Information Time". According to WTO projection, leisure industry will take the place of the traditional leading industries and modern information industries to be the first in the world.

在中国，这种预测正在成为现实，随着国民经济收入地提高以及国家对法定假日的调整，越来越多的人倾向于通过定期的休闲度假来放松身心，普通民众的旅游度假意识逐渐形成，定期度假需求越来越强烈。据全球私人咨询机构 AC 尼尔森公司 2007 年的调查显示，有超过半数（53%）的中国消费者计划把手中余钱花在度假上面，并有近 3/4 的受调查人员在过去的 12 个月里进行过一次或更多的休闲旅游。可见，中国人的旅游方式已由传统的观光型向休闲度假型转变，定期的休闲度假成为现代人追求的生活时尚。在此大背景下，旅游度假区大规模兴起。

This projection is becoming the reality in China. With the increase of people's income and the adjustment of national statutory holiday, more and more people are inclined to relax themselves through regular holidays. As a result, the tourism consciousness of general public has formed gradually, and the demand of regular holiday has become more and more strong. According to the investigation of AC Nielsen Company, the global private advisory institution, there are more than half (53%) of Chinese consumers plan to spend their left money on the vacation, and almost 3/4 investigated people had one or more times of leisure travel in the past 12 months. It is obvious that traditional tourism type of viewing has changed to be tour with entertainment and taking vacation has become the trend of modern people.

1.1 旅游度假区的发展历程 Development Course of Holiday Resort

度假始于公元初，当时罗马军团为了满足执政官的需要，在温泉地带建立了公共浴室和相应的饭店配套设施。14 世纪，比利时在温泉地建立欧洲第一处度假地——帕斯度假地。19 世纪，随着西方社交活动和赌博娱乐活动的活跃，出现了山地度假、海滨度假、湖泊度假和温泉度假等多种形式。20 世纪初，欧美一些国家又出现滑雪度假，度假的活动内容也日益丰富，目的已由单一的休闲转向休闲与健身、娱乐、会议相结合，旅游度假区也越来越趋于多样化和综合化。但直到这时，度假旅游仍然主要是上流社会和有产者的活动，平民大众很少涉足。进入 20 世纪 60 年代，伴随着旅游业的大发展，度假已成为发达国家大众性旅游活动，各类旅游度假区，特别是依托于“3S”资源的旅游度假区如雨后春笋般兴起。目前世界上较为著名的度假区有西班牙的加那利蓝岛、土耳其的南安塔利亚、多米尼加的波多普拉、墨西哥的坎昆、韩国庆州的波门湖和印尼的巴厘岛等。现在看来，度假旅游已经成为发达的休闲方式和生活方式，并且成为国际旅游市场的生力军。

Vacation can be dated from the beginning of the year when the Roman legions set up public bathrooms and the corresponding hotel facilities around the hot springs in order to meet the needs of the ruling palace. In 14th century, Belgium had established the first resort of Europe in the spas—Perth Resort. In 19th century, with the prevailing of the Western social activities and gambling entertainment activities, the various forms of resorts have come out including mountain resorts, beach resorts, lake reports and spa resorts, and so on. With the great tourism development, vacation has become the public tourism activities in the developed countries since 1960s. The variety of tourist resorts has sprung up like bamboo shoots after a spring rain, especially the ones depending on "3S" resources. At present, the better-known resorts in the world are: the Canary Blue Island of Spain, the Antalyaspor of Turkey, the Puerto Pula of Dominican Republic, the Cankun of Mexico, the Wave gate lake of Gyeongju of South Korea, and the Nusa Dua Bali of Indonesia, etc. In short, holiday resort has become developed leisure and lifestyle, and the main force of the international tourism market.

我国的旅游度假区最初的雏形是为满足帝王将相、皇亲国戚以及上流社会休闲度假需求而修建的皇家园林及私家园林。其中最著名的当属承德避暑山庄以及华清池，由此可以看出当时的度假旅游是一种极高端的休闲享受。新中国成立后，旅游度假区成为企事业单位职工修养疗养之地，这类旅游度假区的特点是“配额制”，一般不对外营业，其中具有代表性的是河北北戴河。度假旅游区在我国的真正兴起始于 1992 年，时值中国改革开放和经济建设的新一轮高潮，经过数年的发展，中国旅游业开始酝酿由单一的观光旅游模式向休闲度假旅游模式的转变，于是开始了旅游度假区的建设，当年国务院正式批准了 10 个国家级旅游度假区，如今我国有国家级旅游度假区 16 个，省级旅游度假区 100 余个。

The development of holiday resorts in our country began with the construction of royal and private gardens to meet the needs of leisure of the emperors, royal relatives and the high society. Among them, the most famous was undoubtedly the Chengde Mountain Resort and the Huaqing Hot Spring, from which we can see that spending vacation in holiday resorts in that time was a kind of extremely high-end tourism. After the founding of China, holiday resorts have become the places for institutional and business employees to recuperate. This type of tourism resorts is characterized by "quota system" and not for business generally, with a typical one of Beidaihe of Heibei. The real rise of holiday resort in our country began in 1992 when a new round of high tide of reform and opening up and economic construction were happening. With many years of development, China's tourism industry has begun to prepare to transform the single viewing to leisure tourism, thus construction of tourist resort started when the State Council officially approved 10 state-level tourist resorts. By now we have 16 state-level resorts and 100 provincial tourism resorts.

1.2 旅游度假区的特点 Features of Holiday Resort

1.2.1 休闲性

旅游度假区是度假旅游发展的必然产物，而度假旅游的实质便是为游客提供各类休闲服务，使游客在休闲的过程中得到身心上的放松与享受。随着休闲时代的到来，度假旅游不仅仅是一种时尚，还成为人们生活中不可或缺的一部分，旅游度假区作为集中了各种休闲项目的旅游目的地理所当然的具备了休闲性。

1.2.2 综合性

旅游度假区的综合性主要体现在两个方面，首先是旅游产品的综合性，旅游度假区是一个旅游功能相对完整独立的整体，某一个单纯的旅游产品都不能够支撑这一完整的体系，因为旅游者具有复合型的旅游需求，希望通过某一旅游目的地一次性获得康体健身、观光休闲、陶冶情操等多种满足。其次是设施的综合性，由于旅游度假区是为了多种旅游目的而设计的，因此需要不同的硬件设施来满足旅游产品的供给；另外旅游度假区是集吃住行游购娱于一体的旅游目的地整体，因此需要与此相关的基础配套设施及服务。

1.2.3 文化性

众所周知，文化是旅游的灵魂，这一不二法则同样适用于旅游度假区。文化是区别度假区的法宝，是构筑度假区特色的武器，是吸引旅游者，打造品牌的关键。地域文化与休闲度假文化地结合是打造旅游度假胜地的有效途径。印尼巴厘岛、美国夏威夷其自然条件与同纬度上的其他地区相差无几，但是他们却依赖浓郁的本土文化成为全球知名的度假天堂。

1.2.4 生态性

生态性是旅游者的需求，同时也是旅游度假区希望实现的目标。现代人每天面对钢筋混凝土的都市环境以及紧张压迫的都市生活，因此若旅游度假区有良好的生态环境就能使旅游者置身其中能够感受到久违的松弛。一个污水横流、垃圾遍野的旅游度假区是不能带给旅游者这种感受的。另外，通过旅游度假区良好的生态环境吸引旅游者，以此带动旅游目的地的社会效益及经济效益，从而实现旅游度假区的可持续发展，这是旅游度假区、当地政府及居民期望实现的良性循环局面。

1.2.1 Leisure

Holiday resort is the product of the development of holiday tourism which is essentially to offer tourists all leisure service, so that tourists can relax and enjoy themselves in leisure atmosphere. As leisure time is coming, holiday tourism is not only a kind of fashion, but also an indispensable part of people's life. Holiday resort, as the scene spot concentrated all kinds of leisure items, is of leisure for grant.

1.2.2 Synthesis

The synthesis of holiday resort reflects from two aspects: first, it is the synthesis of tourist products. Holiday resort is an independent unity with rationally completed tourism functions, any single tourist product can't support this united system as the tourists have comprehensive tourists requirement that they hope to satisfy themselves with; the other one is the synthesis of facilities. As holiday resort is designed for various tourism purposes, many different hardware facilities are needed to supply for tourist products. In addition, holiday resort is a unity including catering, living, traveling, walking, and shopping, thus related fundamental facilities and service are needed.

1.2.3 Culture

As we all know that culture is the soul of tourism, which is the same for holiday resort. Culture is the essence of the difference of holiday resorts, the represent elements of characteristics of resorts, and the key of attracting tourists and forming brand. The combination of regional culture and vacation culture is an effective means of creating holiday resort.

1.2.4 Ecology

Ecology is tourists' need, and also the object that holiday resort hope to achieve. Nowadays, what people face every day is the urban environment of steel and the pressed and tense urban life, thus their needs of going back to nature and relaxing themselves require tourism resorts should be with excellent ecological environment, so that tourists can get relaxed when they come in. In addition, to attract tourists with the excellent ecological environment is a good way to promote the local social and economic benefits, so that sustainable development will be achieved in the area of tourism resorts, which is the situation of good circle that local government and residents desire.

2 Exploration of Developing Tourism Resort in Our Country

把脉我国旅游度假区规划

经过十余年的发展，我国的旅游度假区取得了可喜的成绩，并为地方社会经济作出了卓越贡献，成为旅游业发展的一大热点。随着休闲度假旅游的进一步推进，旅游度假区的规划开发将迎来新一轮高潮。因此，天海规划在看到旅游度假区成绩的同时居安思危，将旅游度假区存在的问题进行归纳总结，寻求解决方案，以期在旅游度假区的规划中避免相同问题的再次发生。

With more than ten years of development, holiday resort has been enhanced a lot and contribute to local economy, thus it has become a hot spot of the development of tourism industry. With further promoting of holiday tourism, holiday resort will get a new round of high tide. Therefore, analyzing the situation of holiday resort, Tianhai Planning has summarized the accomplishments, as well as the existing problems, in order to seek solutions and avoid the same problems in the planning of holiday resort.

2.1 旅游度假区建设的盲目性 Blindness in the Construction of Holiday Resort

国家级旅游度假区、省级旅游度假区、区县级旅游度假区……度假旅游区如雨后春笋般兴起，然而大多数旅游度假区在建设初期都没有经过严密的论证，只看到整体大市场对旅游度假区的需求，却没有认真分析该地的资源、环境等条件是否适合开发旅游度假区，在这样的背景下开发的旅游度假区容易陷入两个误区：一是认为度假旅游还停留在供社会名流享受的阶段，一味地追求高端，全然不觉度假旅游已经走向了大众旅游发展的路线；二是认为旅游度假区可以脱离资源的束缚，仅仅依托康体度假休闲设施与人工造景就能取得成效，殊不知现代度假旅游具有综合性，观光、休闲一个都不能少。

Holiday resort has sprung up like bamboo shoots after a spring rain, such as national tourist holiday resort, provincial tourist holiday resort, prefectural tourist holiday resort, and so on. However, most of the planning of tourism resorts haven't been proved rigorously with only knowing the market of tourism holiday resorts, but never analyzed seriously whether local conditions such as resources and environment etc. are suitable to develop holiday resort. There are two errors: one is that the social notable was still oriented as their target market without knowing holiday tourism has become common; the other is that tourism resorts were developed with too much man-made landscape.

2.2 欠缺文化内涵，缺乏自身特色 Lack of Cultural Content and Characteristics

目前我国旅游度假区给人的整体印象是似曾相识，大体雷同，缺乏特色。造成这一现象的主要原因除了上面提到的旅游度假区盲目建设、仓促上马之外，其深层次的原因是文化内涵的缺失。基础配套设施是旅游度假区的体魄，而文化才是旅游度假区的灵魂，但是现在对旅游度假区开发的认识却存在偏差，认为旅游度假区的精髓在于硬件设施的建设，在度假区的开发中只强调其功能性，缺乏文化内涵的注入，致使大江南北的旅游度假区无论从功能还是外形上都趋于一致，无法体现差异性。

没有文化内涵的旅游区是空洞乏味，无法引起旅游者共鸣，无法吸引旅游者眼球，无法得到旅游者青睐，尤法长久立足的。

At present, the overall impression that China's tourism resorts leave to people is similarity without features respectively. Besides constructing blindly and hastily that mention above, the causes of this phenomenon is the lost of cultural connotation. Infrastructure is the body of tourism resort, while culture is the soul. Nevertheless, the current recognition of development of tourism resort is biased that the construction of facilities were regarded as its essence. And only the functionality was emphasized in the development, but cultural connotation was ignored, which made the tourism resorts wherever they are tend to be that same either the functions or the appearances.

Any scenic spot without cultural connotation is dull, it is unable to arouse the tourists' interest and attract tourist that it can't stand for a long time.

2.3 休闲度假氛围不明显 Unobvious Leisure Atmosphere

旅游度假区是为旅游者提供休闲度假旅游产品的综合体，其实质是通过旅游度假区这一载体实现休闲文化氛围的营造及传播。但是如今的旅游规划开发更多注重的是一种技术性、功能性的要求，而没有站在旅游需求的角度上考虑度假区的规划开发，致使有些旅游度假区有向高档住宅区发展的趋势，更有甚者还出现了浓郁的小城镇气息。这些与旅游者渴望回归自然，寻求身心松弛的愿望都是背道而驰的。这里需要提出的是度假区休闲氛围的营造不见得要高端，而在于一种温馨、浪漫、清新、朴实的氛围的营造。

Tourism resort is the synthesis that provides tourists with tourism products for leisure, and its essence is to realize the creation and transmission of the atmosphere of leisure culture with the carrier of tourism resort. However, the current tourism planning stresses the requirement of tech and function, rather than the tourists' real demand, which make some tourism resorts have the tendency of upmarket residential areas, and even some towns. These are opposite to the tourists' desire of returning to nature and seeking physical and mental relaxation. What we want to point out is the leisure atmosphere of tourism resorts is not necessarily high-end, but should be warm, romantic, fresh, and simple.

2.4 对周边特色资源挖掘利用不足 Deficiency of Excavating and Using of the Surrounding Special Resources

旅游度假区是一个统一的整体，吃住行游购娱貌似样样齐全，但是由于受到规模的限制不可能样样都精，尤其是“游”这一项，“风景名胜区管理条例”明确规定风景名胜区内不能兴建度假区，致使度假区的自然及人文旅游资源没有亮点，使得旅游度假区的品质大打折扣，因此需要同周边的资源联动发展，形成观光、度假分流的联动局面。世界著名的旅游度假区南非太阳城便是一个典范：太阳城为旅游者提供休闲度假的所有设施，但是却将旅游者输送到外围的著名景点如好望角参观游览（在旅游景点周围却没有任何的休闲度假设施），这样既实现旅游度假区与旅游景点的共赢，又确保了旅游资源受到外界最小程度的侵扰及破坏，保持旅游资源的原汁原味。但是目前我国的旅游度假区在规划开发中往往只将眼光放在旅游度假区自身的建设上，忽略甚至从来不考虑与周边资源的联动开发，这种将本土特色资源拒之门外的做法不仅造成了资源的浪费还使得旅游度假区放弃了树立自身特色的机会。事实上合理利用周边各种自然、人文资源，就能有效地增加旅游度假区的本土特色，形成自身特色。

Tourism resort is a united whole which includes audition, lodging, traveling, walking, shopping and entertainment, but confined by scales, it is impossible that every aspect is good enough, especially "traveling". Holiday resort is not permitted in scenic spot according to the "Scenic Area Management Regulations" that holiday resorts were lack of natural and cultural bright spots and the quality of them has been greatly reduced. Thus it is necessary to unite holiday resorts with the surrounding resources to form a connective situation of viewing and taking entertainment. Sun City of South Africa, the world famous tourism holiday resort, is a good example: the Sun City offers the tourists all facilities of entertainment and relaxation, while transports them to the famous attractions of periphery like the Cape of Good Hope for sightseeing, and there is no facilities around the scenic spots. In this way, both tourism resort and tourism attraction get benefits, and the interference and damage to tourism resources reduce to be minimal, so that the origin of tourism resources will be maintained. However, the current tourism resorts of China have the orientation of constructing the resorts without uniting the surrounding resources to develop together. Actually, using reasonably the various surrounding natural and cultural resources can enrich local features and form characteristics for tourism resorts.

3 Strategy of Developing Tourism Resort

为旅游度假区规划开方抓药

3.1 全面分析，明确方向 Determine Orientation

天海规划认为旅游度假区的开发要拒绝盲目性，就要对规划地进行全方位的分析，了解规划地是否具备旅游度假区的开发条件，明确规划地的开发方向。第一是对旅游背景的分析。这样才能确定规划区的发展方向。第二是资源条件的分析。应该跳出旅游度假区重人工建筑，轻资源禀赋的怪圈，旅游度假区需要的是优美的环境、清新的空气、独特的人文。第三是区位条件的分析。一般的旅游目的地多集中于较为繁华城市市区，这样地安排得益于市区成熟的基础配套设施体系，但是旅游度假区却不同，它一般远离城市处于山林、海湖之滨，同时还需要具有一定的封闭性以确保旅游度假区的安全，这些条件都具备后还需要便捷的可进入性。

Tianhai Planning has always believing that developing holiday resorts should reject blindness, but analyze the planned place completely to know whether it was endowed the conditions of developing holiday resort, and determine the orientation of development. Firstly, the analysis of tourism background: it is to determine the orientation of the planned area. Secondly, the analysis of resources conditions: what the holiday resort need is beautiful environment, fresh air, and special culture, thus the phenomenon of stressing man-made constructions, but ignoring natural resources should be deleted. Third, the analysis of geographic conditions: generally, the tourism destinations concentrate in the cities because of the mutual urban infrastructure system, but holiday resort is different, it is often located in maintains or sea and lake shores, and with certain closeness to ensure safety, besides, the convenient accessibility is needed.

天海规划在进行《黔西化屋旅游度假区规划》编制的过程中首先对规划地——黔西东风湖进行了全方位的分析，我们认为其具备成为旅游度假区的先决条件：

In the process of compiling "Planning of holiday resort of Dongfeng Lake (Qianxi)", Tianhai had analyzed completely the planned place—Dongfeng Lake, Qianxi, and we conclude that Dongfeng Lake of Qianxi has the prerequisites of developing holiday resort.

首先从旅游背景分析：

(1) 贵州省两大湖泊旅游区（红枫湖、百花湖）作为水源地被保护，旅游发展受限，为东风湖湖泊旅游资源的开发带来机遇。

(2) 国务院对国家法定节假日制度作了调整，同时出台了《职工带薪休假条例》,新的休假制度实行,预示着我国公民休闲度假旅游时代已全面到来。

(3) 观光旅游同质化严重，旅游者对个性化、多元化、精品化休闲度假产品需求更为迫切。

(4) 广大旅游者在消费理念上越来越与国际接轨，对休闲度假旅游的认知越来越成熟,出游动机也越来越理性。人们出游的目的已不再是盲目跟风,不再以“到此一游”为乐,开始由单纯的游山玩水、追求猎奇性向追求舒适型、享受型的方向转变。

Tourism Background:

(1) As the two major lake scenic spot, Hongfeng Lake and Baihua Lake are protected as water sources, which constrain the tourist development and bring opportunities to Dongfeng Lake to develop.

(2) The State Council has adjusted the National Holiday system and issued a "workers paid leave regulations", which declared that the era of leisure tourism holiday has come in our country.

(3) Sightseeing tend to be saturation, what the tourists pursue is the personalized, diversified, boutique-oriented leisure tour.

(4) Most tourists have had more and more international consumption concept, mature understanding to leisure holiday tour, and rational traveling motive. People have not toured just for following the tendency blindly, but turned to pursue comfort and enjoyment.

其次从资源条件分析：

化屋度假区所依托的东风湖资源条件优越,除了“山似三峡而水胜三峡，水似漓江而山胜漓江”的东风湖湖域以及两岸林地风光外，独树一帜的“歪梳苗”文化也是区域内不可多得的。自然风光湖滨、山地、森林以及少数民族村寨，化屋村拥有构筑休闲度假区的所有“原材料”。

Resource Conditions:

The resource conditions of Dongfeng Lake are superior. The mountains, the waters, all scenery around Dongfeng Lake are splendid; besides, the unique culture of the "Waishu Miao Ethic Group" is rare within the region. Huawu is endowed with all the "original materials" to construct tourism holiday resort with natural lakeshore, mountains, and forest.

最后从区位条件分析：

(1) 东风湖处于乡村、山林和海湖之滨，距贵州省省会贵阳 100 余公里，贵毕高等级公路穿越县域全境，距黔西县城 30km 处设有钟山贵毕高速公路匝道，可与化屋村相连；东风湖距黔西县城 50km，已开通柏油公路。另外，随着成贵铁路的建成通车，以及清镇——黔西快速通道的建成将更加加强东风湖的可进入性。

(2) 相对封闭的区域环境适合度假旅游区的开发。度假旅游区的一个先决条件是安全性，由于东风水库的修建，化屋村片区多数居民已经迁居，片区滨湖靠山，只有一条公路与外界相连，在此修建度假区便于管理，安全系数较高。

Geographic Conditions:

(1) Dongfeng Lake is located in the villages and woods, and about 100 kilometers from Guiyang, the capital of Guizhou. With the completement of Chenggui railway and the fast-track from Qingzhen to Qianxi, Dongfenghu Resort will be more easily to visit.

(2) Relatively closed regional environment is suit to develop tourist resorts, as one of prerequisites of developing tourism resorts is security. The residents around Huawu region have been moved due to the construction of Dongfeng reservoir, this region face to lake and back with mountains, and with only one high road connected with the outside world, which provides safe environment to construct tourism resort in this area.

3.2 因地制宜，展示文化 Displaying Culture

文化因地域的不同而具有明显的差异性，因此展现当地文化是实现旅游度假区的重要途径。这种文化要通过旅游度假区的建筑、景观以及旅游活动的设置向旅游者展现出来，使旅游者产生直观的印象。

Culture is different obviously because of geographic differences, thus manifesting local culture is an important way to accomplish tourism resort. This culture should be through architectures in tourism resort, landscape and tourism activities to show tourists who will get visual images then.

例如天海规划在进行《黔西化屋旅游度假区规划》编制的过程中便是通过具象的建筑将当地文化展现得淋漓尽致。化屋度假区的建筑大体上呈现出两种风格：一种是以杜鹃花为创意元素的现代风格；另一种是以东风湖旅游区所在地化屋村的建筑为原型的原生态风格。之所以这样考虑是出于两方面的原因：第一，东风湖所在地黔西县拥有“中国杜鹃花都”的美誉，其“百里杜鹃”是地球上同纬度地区规模最大，数量最全，姿态最美的，可以说杜鹃花是黔西的核心资源，我们也将之引申为度假区的创作原型；第二，东风湖所在地化屋村具有浓郁而独特的民族特色，将这一特色运用于度假区的建筑风格中，是对地方文化内涵的诠释。另外从景观上来说，大到入口大门，小到垃圾桶、路灯、休闲凳都延续了杜鹃花以及民族风格，事无巨细地体现了蕴含其中的地方文化特色。最后根据当地特色设置旅游产品，体现地方文化与休闲度假文化。

In the "Planning of Dongfeng Lake (Qianxi)", Tianhai has manifested the local culture completely through concrete constructions. Architecture in Dongfeng Lake Holiday Resort generally has two styles: Modern style with the azalea as creative element, and primeval style with the prototype of architectures of Huawu village in the area of Dongfeng Lake Holiday Resort. There are two reasons that we gave this orientation: On one hand, Qianxi County where Dongfeng Lake located enjoys the fame of "Azalea Capital of China", where "Baili Azalea Belt" is the widest and most complete and beautiful compared with the other places at the same latitude in the world, so we can say that Azalea is the core resource of Qianxi that we take it as the creative prototype; on the other hand, Huawu village where Dongfeng Lake located is of thick and unique national features, thus to inject such features into the architectural style of this resort is the best elaboration of local cultural connotation. In addition, on the aspect of landscape, either big or small, such as the gates of entrance, or trash cans, street lamps and leisure benches are along with azalea and national style to embody local cultural characteristics. The last step is to display local culture and the culture of leisure holiday by determining tourism products according to local flavor.

化屋旅游度假区旅游产品一览表

旅游产品		旅游项目
观光旅游产品	东风湖风光游	观光游轮、东风湖沿岸风光等
	东风湖—织金洞水路游	观光游轮、东风湖—织金洞水域风光、织金洞风光、游轮餐会等
民俗旅游产品	化屋苗族风情游	化屋歪梳苗板凳舞、芦笙拳舞、无伴奏多声部合唱、化屋苗族特色饮食、手工艺品等
度假旅游产品	东风湖滨湖度假游	六星级（三星级）度假酒店、度假别墅、滨湖酒吧、咖啡吧、茶楼、游艇、小型豪华游轮、水上摩托艇、快艇、冲浪池、电玩城、儿童乐园、歌影剧院、棋牌娱乐、垂钓、篝火广场、歌舞广场、风情街等
康体健身旅游产品	化屋康体健身游	天然森林氧吧、摇曳、旋转器械、弹跳跨越器械、自行车、游泳、推拿、桑拿等
	化屋拓展训练游	攀岩、拓展训练器械等
疗养保健旅游产品	化屋疗养保健游	SPA、理疗、按摩、推拿、中药保健、药膳等
商务公务旅游产品	化屋商务游	度假酒店、会议中心、康体会所、歌影剧院等

Tourism Product		Tourism programs
Tourism products for sightseeing	Travel in Dongfeng Lake	Viewing ferries and scene along the banks of Dongfeng Lake
	Water line of Dongfeng Lake–Zhijin Cave	Viewing ferries, water scene from Dongfeng Lake to Zhijin Cave, scenes in Zhijin Cave and feasts on ferries
Folk tourism products	Travel in HuaWu for Miao flavor	Bench dance, Lusheng dance, polyphonic chorus without musical accompaniment, special food and crafts of Huawu, and so on
Toutism Products for holiday	Holiday travel in HuaWu	Six–star (three–star) resort hotels,resortvillas,Lakesid bars,cafes,tea houses,ships,small–type great ship, aquatic motorboats,yachts,surfing pool,electronic game center,Children's playground,opera house,cards entertainment,fishing,bonfire square,songs and dances square,ethnic street and so on
Tourism products for healthy and fit	Travel of fitness in HuaWu	Natural forest oxygen bar,swinging and swirling equipments,jumping equipments,bicycles,swimming, massage,sauna and so on
	Travel of expending training in HuaWu	Rock climbing,equipment of expending training and so on
Tourism products for recuperation	Recuperation Travel in HuaWu	SPA,physical therapy,massage,health–care with Chiness medicines,medicine diet and so on
Tourism products for business	Business travel in Dongfeng Lake	Resort hotels,meeting center,fitness club,opera house and so on

3.3 完善设施，营造氛围 Improving Facility

首先，硬件上要有完善的休闲娱乐设施。旅游度假区是一个整体，提供的是整个休闲度假产品体系，包括吃住行游购娱各个方面，除此之外还需要根据休闲度假旅游的特质设置休闲娱乐设施。黔西化屋旅游度假区配备有中高端度假酒店、度假别墅、影视电玩城、康体休闲会所、民族风情街、民族活动展示区、冲浪池、户外拓展基地、游艇、客船、水上运动设施、儿童游乐设施及动物观赏园等休闲娱乐设施，可以说形成了一个完整的休闲度假体系。这其中值得提出的是儿童游乐设施及动物观赏园的设置，事实上儿童游乐设施都是很简单的攀爬设施，动物观赏园也只是诸如兔子、鸡鸭等家禽，但是它却是在考虑到休闲度假旅游以家庭为单位出行的特点而专为儿童提供服务的休闲度假设施，这一区域场地相对封闭、安全，有专人管理，让儿童拥有自己的小天地，锻炼独立能力的同时，解除家长的后顾之忧，专职于自身的休闲放松。

其次，通过有形设施营造休闲氛围。休闲强调的是一种感受，那么感受从何而来呢？这种感受来自于有形设施对旅游者感官上的刺激。同样以"黔西化屋旅游度假区"项目为例，度假区从建筑外观上运用木材、茅草等当地建筑元素，进行艺术手法上的再加工，加上黔西水西彝族常用的尖角装饰，力图打造出粗犷古朴又具有当地特色的原生态建筑，使旅游者通过建筑外观就能体会到野趣十足的松弛氛围，同时度假区还通过对杜鹃花花蕾、花蕊、花瓣以及花叶进行抽象创意，从内容和形式上体现黔西特色的同时，营造度假区温馨浪漫的整体氛围，丰盈和完善度假区的休闲松弛感受。

First of all, the complete facilities for leisure and entertainment are indispensable. Holiday resort is a whole, and what it offers is the product system for leisure vacation, including diet, lodging, traveling, shopping and entertainment. Moreover, it is necessary to set up facilities for leisure and entertainment according to the idiosyncrasy of leisure vacation for tour. There has formed a integrated system of leisure vacation in Dongfeng Lake Tourism Holiday Resort with the facilities for leisure and entertainment including resort hotels, resort villas, film & electricity animation game center, health & leisure club, custom street, performance area of ethnic activities, surfing pool, outdoor expansion base, yachts, boats, water sport facilities, children playground equipments, and animal park, etc. Among them, we would like to talk about children playground equipments and animal parks. Actually, children playground equipments are simple climbing, and there are some poultry in the animal park, such as rabbits, chickens, ducks and so on, which, however, are special for the families with children. This area is relatively closed, safe, and taken charge of certain people, so that children can have their own small world, which is good for their ability of independence, and also offers their parents free times to relax.

Second, it is an effective way to create leisure atmosphere through the physical facilities. What leisure stresses is a kind of feeling, and what does this feeling come from? It is from the sensory stimulation that the physical facilities to tourists. Also take the project of "Dongfeng Lake Tourism Holiday Resort of Qianxi" as example, in order that the tourists can feel the leisure atmosphere full of wild charm, the primeval architectures with rough and simple flavor, and local characteristics will be created by using the local architectural elements such as wood, thatch, etc., applying artistic means, and adding the common cusp decorations of Yi ethnic of Qianxi. Meanwhile, the abstract creative on the basis of flower buds, stamen, and petal of azaleas will be used to embody characteristics of Qianxi on the content and form, create the overall atmosphere of warm and roman in the tourism resorts.

3.4 周边联动，实现共赢 Accomplishing Mutual Benefits

旅游度假区的综合性以及旅游者需求的多样性注定了旅游度假区不仅仅是为旅游者提供食宿的场所，它还需要兼具观光游览、增长知识、探古寻幽、参与体验等其他旅游功能。将这些所有的要素都包含在一个旅游度假区内显然是有难度的，这时候就需要将度假区周边的资源统一纳入旅游度假区中来统筹安排。

天海规划在进行《黔西化屋旅游度假区规划》编制之初了解到规划地以西30km处是素有"织金归来不看洞"美誉的国家级风景名胜区织金洞，并且规划地所依托的东风湖延伸至织金洞水域可供游船通行，由此我们意识到这一条水域交通线是化屋旅游度假区的亮点所在，因为贵州著名的旅游景区的交通设施大多数为陆上交通，而且由于受到地形的限制山路崎岖，交通不便，外地游客多会出现身体上的不适，将大量时间耗费在交通上不仅限制了客源市场的可进入性，也会使已进入的游客游兴大减。织金洞的资源品味不言而喻，但是其交通状况却存在以上提到的问题。因此我们在化屋旅游度假区的开发中利用这条水域交通线，与织金洞联动发展，以东风湖景区的雄奇秀美为依托，以各色游船为载体，以丰富的游船娱乐活动为补充，借用织金洞的名气、水域交通线的创意，打造东风湖——织金洞度假观光项目，丰富旅游产品，增加游客停留时间，从而实现资源的共享，利益的双赢。

Tourism holiday resort is destined not to provide accommodation for tourists only due to the comprehensiveness of tourism holiday resort and the diversity of tourists' needs, but other tourism functions are needed such as sightseeing, enriching knowledge, exploring history, and participating activities, etc. It is obviously that a tourism holiday resort containing all these functions is difficult. In this case, we need to integrate the surrounding resources into the resort.

At the beginning of compiling "Planning of Dongfeng Lake Tourism Holiday Resort of Qianxi", we had known that there is "Zhijin Cave", which is a state-level scenic spot and famous as "no other cave is worth to a glance after you've been to Zhijin Cave", on the west of Dongfeng Lake 30 kilometers away. Moreover, the boats for visitors on the waters extending from Dongfeng Lake to Zhijin Cave are available, which makes us realize that this traffic line on waters is a bright spot of the Tourism Holiday Resort of Dongfeng Lake. Most ways of transportation of the famous scenic spots of Guizhou are land vehicle. And the traffic is inconvenient due to the rugged terrain that foreign tourists often suffer physical sick and spend a lot of time on transportation, which decreases not only the accessibility of tourist market, but also the tourists' traveling interesting. The resources taste of Zhijin Cave is self-evident, but traffic conditions are constrained by the problems we mentioned above. Therefore, in the development of the Tourism Holiday Resort of Dongfeng Lake, we made good use of this traffic line of waters in order to unite Zhijin Cave to accomplish the project of Dongfeng Lake—Zhijin Cave sightseeing & resort with the dependence of magnificent and beautiful scenery of Dongfeng Lake, the carriers of various boats, the supplement of rich cruise entertainment, the fame of Zhijin Cave and the creative of this water traffic line, so as to enrich tourism products and increase visitors' stay, and finally, achieve resource sharing and mutual benefit.

Tourism Resort
旅游度假区

Huawu Resort 化屋度假区

HuawuResort

前期策划 Prophase Planning

1 引言 Introduction

黔西位于贵州省中部偏西北、乌江中游鸭池河北岸，是毕节地区东大门，素有“一枝花”的美誉，是贵州省规划的全省6个特色旅游区涉及县之一，县内旅游资源十分丰富，分布广、种类多。主要有百里杜鹃国家级森林公园、国家级文物保护单位观音洞旧石器时期古文化遗址、乌江源百里画廊东风湖等3处国家级旅游资源，有中国古彝圣水之称的贵州第一大湖支嘎阿鲁湖、贵州境内三大淡水湖群之一的柯家海子群；有省“苗族歌舞之乡”化屋基苗寨、吸引王阳明亲临写下《象祠记》并载入《古文观止》的九龙山象祠、清代兵部尚书李世杰墓院等省级文物保护单位……

黔西众多资源中尤其值得一提的是乌江源百里画廊东风湖景区，可谓“山似三峡而水胜三峡，水似漓江而山胜漓江”，景区兼具三峡的雄壮及漓江的秀丽，实属难得；同时在东风湖景区高峰峻岭，碧波秀水之中掩映着“苗族歌舞之乡——化屋基苗寨”更使景区平添几分民族文化气息；此外东风湖向西可通达贵州省著名风景区织金洞，构筑了一条别具特色的水路交通线，总之整个景区雄浑大气、深邃清幽、风情浓郁。

在认识到资源优势的同时，我们也必须客观的认清资源的不足之处：

1. 品牌资源季节性过强，致使传统的观光旅游市场优势不明显

“黔西百里杜鹃国家级风景名胜区”无疑是黔西打得响的品牌，然而每年2～3个月的花期限制了品牌效益的进一步扩散。

2. 人文资源品位高却缺乏市场

黔西观音洞是中国古人类发祥地之一，考古学界有“北有周口店，南有观音洞”之说。然而由于缺乏具体实物的支撑以及专业性过强等原因，致使对该类产品的需求量不足，无法将之开发为旅游产品。这样的问题并不具有偶然性，北京人遗址、重庆巫山人遗址都存在这样的问题，拥有世界级的资源和无数风光头衔却鲜少有旅游者问津。

3. 景点分散、交通不便利、特色不突出，无法形成规模市场

黔西县旅游资源丰富，但却呈点状分布，遍布黔西各地，景区交通多为乡级道路，较为不便，加上旅游资源不具备绝对优势，替代性较强，旅游者没有必要非去不可，因此无法形成规模市场。

基于以上分析，黔西资源有特色，但是也有缺陷，在开发中要做的便是最大限度的彰显优势掩盖劣势，那么优势如何彰显呢？我们拒绝闭门造车，拒绝就资源谈开发，我们要做的是以市场为前提，以资源为载体，以人的主观能动性为依靠，实现黔西旅游业的跨域发展。

QianXi area is located in the north-west of GuiZhong Province, at the north bank of YaChi River, middle reach of WuJiang River, which is natural north gate of Bijou Area. It is honored as "a bloom flower", one of the 6 areas of Gui Zhou Province features (planning) in tourism. It is rich in tourism resources, which allocated vast and variety in species. With Cuckoo main national forest parks, national heritage conservation units in the Cave Paleolithic period of ancient cultural sites, and Hundred-Miles-Gallery of Wujiang Dongfeng Lake. Three resources of Nation Class, namely Relic of Paleolith Age, Hundred-Miles-Gallery of Wujiang Origin, ZhiGaALu River the biggest river in GuiZhou Province, called Holy Water of Ancient Yi Minority, I Miao Tribe, Huawu Area, called Origin of Miao's Dancing and Singing in GuiZhong Province, XiangCi of Jialing Mountain, which is mentioned in XiangCi Ji by Wang YangMing, recorded in GuWenGuanZhi, the grave of LiShiJie, Military Minister of Qing Dynasty, etc.

Among the famous sites, one remarkable is the Hundred-Miles-Gallery of Wujiang Origin, which renowned to the world "mountain scenery alike those in Three Gorges, but water scenery superior to those in Three Gorges, water scenery alike those in Li Jiang, but mountain scenery superior to those in Three Gorges". The scenery possesses both the sturdiness of Three Gorges, and elegance of Li Jiang, which is really rare and one in a thousand. Meanwhile, the Ji Miao Tribe, Huawu Area, called Origin of Miao's Dancing and Singing in GuiZhong Province, covered by the rivers and mountains in DongFeng travel Resort, adding a tinge of minority culture. Besides that, the DongFeng Lake leads to JingZhi Cave in west Guizhou Province, forms exotic water transportation. After all, this resort is pageantry and grand, concealed and peaceful, full of exotic scenery.

In recognition of resources advantages, we must objectively recognize the inadequacy of resources:

1. Brand Resource is rather seasonal, which fails to make the classic tourism market distinctive

The Cuckoo Natural Forest Park of Nation, undoubtly is the key brand

Of QianXi Area, however, the not-so-long 2~3 months blooming season prevents the further expanding of the brand effect.

2. High grade of human & culture resources is in lack of market

Guanine Cavity of QianXi is one of the cradles of Chinese ancient man, which equals to ZhouKouDian in Beijing, as a saying of archaeology, "North ZhoukouDian, South GuanYin Cavity". However, due to lack of specific kind of support, and also too professional, the demand for the products is deficient, and fails to be developed into tourism product. This problem is not by chance, Peking Man Site, Chongqing Wushan Man site also exist such problem, with the resources and numerous world-class beauty titles but there were fewer tourists cares.

3. Scattered spots, inconvenient transportation, non-prominent feature, resulting failing to form market into large scale

The rich tourism resources are dotted randomly in Qian Xi. The transports between the tourist sites are inconvenient. Furthermore; the tourism resources have no absolute advantage, and they are replaceable, not a must to the visitors. In a word, it is hard to form to market-scale.

Based on the above analysis, tourism in QianXi has both advantages and disadvantages. In the process of development, advantages have to be utmost highlighted to make up the disadvantages. But what to do with it? We should refuse to act blindly, refuse to develop tourism at the cost of ruining natural resources. What we should do is realize the cross-region development of tourism in QianXi, with market as premise, resources as carrier, and man power as support.

2 市场背景分析 Market Background Analysis

中国旅游自改革开放以来，经历了 30 年的快速发展，国人旅游的愿望也越来越强烈，在旅游业发展如火如荼之际，旅游者却发出这样的疑问“名胜大川我们都看完了，现在该看什么？”是的，30 年间旅游者的观光需求得到了满足，然而旅游者永远是“不知足”的，旅游市场永远是变化发展的，随着旅游者的旅游品味提高、闲暇收入与闲暇时间的增长、旅游观念的成熟，旅游者不再满足于视觉上的享受，他们开始追求能使身心得到彻底放松的休闲度假旅游。

Since the reform and opening up policy, Chinese tourism has undergone rapid development for 30 years. The desire to travel for Chinese people becomes more and more intensive. While the tourism is on the heyday now, the tourists still have a doubt "where should I travel to after having visited all the famous mountains and rivers in China?" It is true that tourists are satisfied with their travel requirements, but they always want to see more and more and the market is changing with every second passing by. As tourism tastes increased, the growth of leisure time and increase of income, the concept of travel matured, Tourists are no longer satisfied with the enjoyment of the visual. They pursue a leisure travel which could release them from busy life mentally and physically.

(一) 休闲旅游的发展是大势所趋

1. 国务院对国家法定节假日制度作了调整，同时出台了《职工带薪休假条例》，新的休假制度实行，预示着我国公民休闲度假旅游时代已全面到来。

2. 观光旅游趋于饱和，旅游者对个性化、多元化、精品化休闲度假产品需求更为迫切。

3. 广大旅游者在消费理念上越来越与国际接轨，对休闲度假旅游的认知越来越成熟，出游动机也越来越理性。人们出游目的已不再是盲目跟风，不再以“到此一游”为乐，开始由单纯的游山玩水、追求猎奇性向追求舒适型、享受型的方向转变。

I Leisure Tourism Development is the Trend of the Times

1. State Council made adjustment on the National Holiday system, also announced a "Paid vacation regulations" to implement the new vacation system, indicating an era of public leisure travel of our citizens has come.

2. Tourism tends to saturation. The requirements on personal, multiple and refined travel products are more and more pressing.

3. As the consume concept of travelers is integrated with international ones, they are more and more familiar with the know-how of leisure travel. The motive of travel becomes more and more rational. The aim to travel is no longer to follow blindly, nor to travel just for fun as I-have-been-here. The simple sightseeing and seeking novelty change into a comfort and enjoyable journey.

（二）黔西发展休闲度假旅游是旅游业发展高起点高标准的体现

II Qianxi Developing Leisure Tourism is a Manifestation of the Development of a High Standard Starting Point of Tourist Industry

黔西旅游要实现真正意义上的高起点，高标准，就要跳出自身资源谈开发，要认识到资源是无法决定旅游产业发展的，只有市场才能；要认识到资源并非休闲旅游的决定要件；要认识到资源是有限的，智慧是无穷的，休闲产业从某种意义上来讲是人类智慧的结晶。

If we want to develop Qian Xi tourism from a high standard and based on a high level, we must realize that natural resources cannot determine the development of tourism, only market can. Natural resources are not a must for leisure travel. Natural resource is limited but the wisdom is unlimited. Leisure travel industry in some extent is the outcome of human wisdom.

首先，从旅游市场发展来看，休闲旅游占据了越来越重要的地位，大力发展休闲旅游是旅游业发展的大趋势，黔西要在旅游业上走在前列，必须具备发展休闲度假旅游的远见卓识。

First, from the perspective of tourism development, leisure travel takes a more and more important position on the market. Sparing no efforts on developing leisure travel is a macro trend. We must have a long vision and judicious judgment on the development of leisure travel.

其次，从黔西旅游资源看，虽然遍地开花但是都存在这样或那样的问题，导致旅游业全面发展存在难度，因此我们要跳出现有资源，根据市场发展旅游，休闲度假旅游对资源不具有绝对的依赖性，大力发展休闲度假旅游是黔西的必然之选。

最后，从智慧层面上说，人具有很强的主观能动性，人类的智慧向来是不容怀疑的，在县委县政府的带动和重视下，凝结各方智慧的黔西休闲度假旅游业必将扬帆远航。

Secondly, from the perspective of Qian Xi Tourism resources, the travel sites scattering over the province causes problems and difficulties here and there. Thereby, we should put aside the natural resources, sparing no efforts on leisure tourism development according to the market, based on the principle that tourism doesn't have to totally rely on the natural resources and leisure travel development is the best choice for Qian Xi.

Last but not least, from the view of wisdom, human being is strong in subjective initiative. The wisdom of human being is undoubted all along. Under the driving and leading of County Government and Communist Committee, with all the wisdom from all parts, the sailing of developing leisure travel in QianXi must be going farther and farther.

3 优势分析 Advantages analysis

（一）政策优势

无论是党中央国务院、地方各级政府对旅游业作出的战略性发展方针，还是国务院对国家法定节假日制度作出的调整，以及出台的《职工带薪休假条例》都为旅游业尤其是休闲旅游业地进一步发展铺平了道路。

1. 金融危机下旅游业受到前所未有的关注，使得旅游业在社会经济中的地位越发突出。

2. 旅游业成为贵州省国民经济新的增长点，并被确立为后续支柱产业，为贵州旅游业的发展铺平道路。

3. 黔西县全力打造旅游业的宏伟战略。

Ⅰ Policy Advantage

CPC Central Committee State Council, local governments at all levels have published guiding lines for strategy of tourism development and State departments regulated national holidays, published Regulation on Vacation with Pay. All of which lays a broad and flat way for the development of leisure travel.

1. With unprecedented attention to the tourism industry under financial crisis, Tourism socio-economic status is more prominent.

2. Tourism has become a new growth point of national economy of GuiZhou Province, and it is confirmed as the pillar industry in the future, which paves the way for tourism development in Guizhou.

3. Strategy on developing tourism in Qian Xi County with all efforts.

（二）区位优势 Ⅱ Geographic Advantages

从交通区位上看黔西是毕节的东大门，贵毕高等级公路穿越县域全境，县城东距贵阳 117km，西离毕节 115km，交通便利。

From the view of traffic, Qian Xi is the East gate of BiJie. The GuiBi High Way goes through Qian Xi, whose county seat is 117km west of Gui Yang, 115km east of BiJie, boasting a convenient traffic.

从旅游区位来说黔西是贵州省规划的 6 个特色旅游区涉及县之一，处于黔西——织金特色旅游区，境内的东风湖可从水路通往织金洞，是省内最具特色的水域风光交通线。

Geographically speaking, Qian Xi is one of the 6 areas of Gui Zhou Province features in tourism, located in the special travel area of QianXi-ZhiJin. From Dong Feng Lake located in Qian Xi, Zhi Jin Cave is accessible by water transport, which claims the unique scenery of water.

（三）成贵铁路过境 Ⅲ Chengdu-Guizhou Rail Transit

成贵铁路在黔西境内设站，届时从黔西到贵阳只需要 20min 车程，而从成都到黔西的车程也将大大缩短。

The station of Cheng Gui railway will be set in Qian Xi, then it is only about 20 minutes' train from Qian Xi to Gui Yang. The time from Cheng Du to Qian Xi will also be greatly shorted.

另外，成贵快速铁路是与贵广快速铁路连接的，届时黔西通往珠三角的交通也将被打通，为黔西开拓省外市场带来了巨大优势。

Moreover, ChengGui Express is joint with GuiGuang Express. Then, Qian Xi will lead to ZhuJiang Triangle Area by railway, which benefits QianXi greatly when developing the out-province market.

4 定位 Market Position

以化屋东风湖——织金洞水路黄金线为特色，以化屋片区山地森林、苗族村寨为依托，集节庆会议、康体疗养、民族风情、观光于一体，同时以此为核心辐射黔西其他旅游景点景区的休闲度假集散地。

Tourism in Qian Xi should be characterized in Golden Water Line, HuaWu Dongfeng Lake–ZhiJin Cave, supported by forest of HuaWu Mountain area and Miao Tribe, integrated with holiday ceremony meetings, resort sanatorium, ethnic customs, and sightseeing. It should be leisure travel center radiating to travel sites out of GuiZhong Province.

5 目标 Target

将之打造为贵州真正的旅游度假区。

Making Gui Zhou Province into a real travel area.

利用旅游度假区的产品、服务、设施优势，形成黔西旅游集散地，形成吃、住、娱乐休闲在度假区，观光在外围辐射区的整体效果。

We should taking advantages of products, services, facilities of holiday resorts to forming food, housing, entertainment and leisure in the resort, tourism in peripheral areas of the overall effect of radiation.

（一）近期目标

化屋旅游度假区要做到的便是将旅游者留下至少住宿一夜，游玩两天，形成周边市场周末度假地。

（二）中远期目标

完善化屋度假区特色项目，以此作为核心吸引力，成为辐射周边省市的休闲度假区，黔西旅游集散地。

Ⅰ Short–term Goal

We should try to accommodate travelers at least for 1 night and 2 days in Huawu Travel resort, which could also form a weekend resorts around it.

Ⅱ Long–term Goal

We should perfect special holiday items in HuaWu Holiday Resorts, to make it the core attraction, radiating leisure resorts around by, and becoming travel center in Qianxi.

6 初步构想 Initial concept

（一）总体构想

1. 资源本位的时代已经过去

视觉上的享受看一次就够了，心灵上的放松 + 视觉上的享受却是无穷的，休闲度假旅游的发展是未来旅游业发展的必然趋势。

2. 黔西发展休闲度假旅游大有可为

化屋是黔西休闲度假旅游的首选之地，计划在化屋发展休闲度假旅游，将化屋打造为辐射黔西的休闲度假集散地，主要是基于以下几个原因作出如此考量：

Ⅰ Overall Concept

1. Resource–based era is over

For visual enjoyment, once is enough. While visual enjoyment with relaxation of hearts is boundless. The development of leisure travel is a inevitable trend of tourism.

2. Developing leisure travel in Qian Xi is promising

Huawu is the first choice as the leisure travel site in Qian Xi. Developing leisure travel as planned would make Haw a travel center for leisure travels. The reasons are as following:

第一，旅游度假区一般应该远离城市，处于乡村、山林或海湖之滨，化屋远离贵阳，距黔西县城30km，且地处苗族村寨、山林以及东风湖滨，符合度假区的区位特征。

First, travel resorts should be far away from city, located in country, forest, mountain, beach and bank of lakes. Huawu is far away from Gui Yang, 30KM away from Qian Xi County Seat, located among Miao Tribe, mountain forests and Dong Feng Lake. In a word, it geographically meets the requirements of a leisure travel resort.

第二，化屋具有黔西最具特色的旅游水域交通：东风湖——织金洞。贵州绝大多数旅游景点景区之间都是以陆路交通连接，以这样一条风光旖旎，雄奇秀美的水域风光将黔西与织金洞串联起来，势必会为黔西旅游带来新的特色，是织金洞风景名胜区确定的旅游服务站（旅游服务村）。

Second, HuaWu possesses a unique water transport for travelling: Dongfeng Lake– JinZhi Cave. Most of the travel resorts are connected by roads and highways. While DongFeng Lake and JinZhi Cave are connected by such a charming and magnificent water route, which without a doubt will bring new feature to tourism in Qian Xi. Huawu is set as the service station (travel service village) by Jin Zhi Cave Resort.

第三，东风湖与化屋苗寨良好的开发基础。

Thirdly, DongFeng Lake and Miao Tribe in HuaWu have a very good base for development.

第四，相对封闭的区域环境适合度假旅游区的开发。度假旅游区的一个先决条件是安全性，由于东风水库的修建，化屋片区多数居民已经迁居，片区滨湖靠山，只有一条公路与外界相连，在此修建度假区便于管理，安全系数较高。

Fourthly, a relatively closed and isolated environment is fit for travel development. The premise of travel resort is security. Due to the construction of DongFeng reservoir, most of the residents in HuaWu have moved out. Lakes there are against mountain, the only link of which with outside is a highway. Resorts set up here will be easily to manage, of course, more secure.

第五，化屋片区滨湖、临山的地理环境、开阔的地势是发展滨湖度假旅游、山地森林度假旅游的理想场所。自然风光、外围旅游区、湖滨、山地、森林以及少数民族村寨，化屋拥有构筑休闲度假区的所有“原材料”，现在的关键就在于能工巧匠的加工组装，这项工作将由我们完成。

Fifth, Lakes there, a Mountain–facing environment, board and flat, which is an ideal place for setting up HuBin resorts and Leisure travel in Mountain forest. Natural scenery, travel resorts around, lakes, mountains and minority tribes, build the leisure travel resort as all needed raw material. Now the key to success is craftsmen who can process and assemble them together, which will be accomplished by us.

（二）分区布局构想 Ⅱ Partition Layout Idea

1. 乌江源百里画廊——织金洞水域风光带
2. 滨湖疗养度假区
3. 山地康体度假区
4. 化屋民俗度假区

1. Hundreds–miles–gallery of Origin Wujiang—JinZhi Cave Water Scenery
2. Lakeside wellness Resort
3. Mountain Sports Resort
4. Folk Custom Resort in Huawu Village

（三）建筑风格构想 Ⅲ Construction Style

基于穴居形式的原生态建筑风格，建筑形式有地面建筑，半地下建筑，建筑顶部以圆顶与椎体造型为主，建筑材料大量使用传统的木材、石材以及茅草，意在从外观上营造一种充满野性的原生态氛围，向旅游者传达此处便是释放野性、卸下伪装完全自我的休闲空间。

The construction style should be based on the troglodyte style, the original construction style, the form of which could be on ground, half–burying, the top of the building is in the shape of dome and cone. The materials should be original elements of local residents, i.e.: wood, stone and couch grass, all of which are meant to build a wild atmosphere. That will convey to visitors here is the place one can release his or her ferity and offload the disguise.

建筑外观古朴野性，但是内部装潢却要体现现代的方便及舒适。

The out of the building is of primitive simplicity and wild, but inner of it should display comfort and convenience of modern life.

7 总体布局 General Layout

（一）东风湖——织金洞水域风光带

黔西旅游的亮点所在，这一亮点不是体现在乌江源百里画廊的雄奇秀美上，也不是体现在织金洞“织金归来不看洞”的美誉中，而是体现在水域旅游交通上。

Ⅰ Dong Feng Lake-Zhi Jin Cave, Water Scenery

The light spot of travel in QianXi lays not on Hundreds-Miles-gallery of WuJiang origin, nor on the JinZhi Cave which is called in a saying "no cave visiting after visiting JinZhi Cave" ,but on the water transport.

贵州著名的旅游景区的交通设施大多数为陆上交通，而且由于受到地形的限制山路崎岖，交通不便，外地游客多会出现身体上的不适，长时间的将时间耗费在交通上不仅限制了客源市场的可进入性，也会使进入的游客游兴大减。织金洞的资源品味不言而喻，但是其交通状况却存在以上提到的问题。开发一条水域交通线，以东风湖景区的雄奇秀美为依托，以各色游船为载体，以丰富的游船娱乐活动为补充，借用织金洞的名气、水域交通线的创意，打造东风湖——织金洞度假观光项目，丰富旅游产品，增加游客停留时间。

The transport among famous travel resorts rely on roads and path, which is limited by landform thus inconvenient. The visitors will feel uncomfortable after long-time travel, which is a time waste and limits visitor's access to the resorts; also reduce the interest of visitors. The resources of JinZhi Cave speak for itself, but the transport encountered the same problem above-mentioned. We should explore a water route, relying on the imposing and beautiful DongHu water scenery, carrying out with variety of yachts, complimented with abundant recreational activities. With fame of JinZhi Cave and creation of water transport, DongFeng Lake-JinZhi Resorts Project could be accomplished, which enriches the travel products, and makes visitors linger for longer.

旅游者早上从化屋码头登船——沿途欣赏东风湖风光——下午抵达并游览织金洞景区——傍晚返回游船——用餐——参加游船狂欢派对——夜游东风湖风光（东风湖沿岸增设夜景工程）——午夜返回化屋度假区继续度假之旅。

Visitor go on board at HuaWu Dock in the morning-enjoy the water scenery of DongHu -Arrive at JinZhi resorts in the afternoon-Get back to the yacht in the evening-dinner -participate in carnival party on the yacht-night journey on Dong feng Lake (Night Scenery Project should be added along Dong Feng Lake) -Return Huawu Resorts and continue his travel.

（二）滨湖疗养度假区

滨湖度假是旅游度假区中较为常见的一种，初步计划以迷人的水上风景，温馨舒适的住宿环境、设施，刺激安全的水上运动，丰富多样的娱乐活动以及特色疗养为其主要特色，将之打造为整个化屋度假区的中心旅游区，开展高端休闲度假旅游的同时，借地区第三届旅游发展大会在化屋举办的东风，大力推广及发展会展旅游市场，形成贵州著名的休闲旅游、会议旅游目的地。

Ⅱ Lakeside Wellness Resort

Lake-side resort is relatively common among holiday resorts. The initial plan is to build it into the centre of the entire Huawu resort featuring fascinating water scenery, cozy accommodation, thrilling but safe aquatic activities, plenty and colorful recreational events and special convalescence. While developing high-class leisure tourism, we should seize the opportunity of the regional tourism development convention held in Huawu and vigorously promote and develop convention tourist market to make our county the leisure tourist and convention tourist destination of Gui Zhou Province.

1. 核心特色 Core Features

蹦极：目前贵州市场上没有一处蹦极设施，在此区域引入蹦极设施，形成贵州唯一，增强旅游区的娱乐项目的刺激性。

Bungee jumping: Currently, there is no Bungee facility in Gui Zhou market. Our county will be the first and the only one who introduce Bungee jumping into this region, making the recreational events more thrilling.

观景度假酒店：采用后退阶梯式格局，尽量让所有房间都能看到湖光山色。

Lake-viewed holiday hotel: layout of receding stair-way adopted, making it possible to see the beautiful view of the lake and mountains from every room.

儿童游乐园：专为儿童提供服务，场地相对封闭、安全，专人管理，让儿童拥有自己的小天地，锻炼独立能力的同时，解除家长的后顾之忧，专职于自身的休闲放松。

Children's playground: Children only. Relatively closed, safe, with hand care. Children can have their own little world there and learn to be independent. Parents don't have to worry about their children and can totally concentrate on their own pleasure and relaxation.

特色 SPA 养生馆：苗药水疗、苗族推拿、苗族药膳。

Special SPA sanatorium: SPA with Miao group medicine, Miao Group Massage, Miao Group tonic cuisine.

2. 常规项目 Regular Events

会议中心：设施先进，能够承接大中型会议
剧院：电影放映、室内演出
滨湖露天演出场：广场演出、篝火晚会
滨湖休闲吧：包括咖啡店、酒吧
露天游泳场：游泳池、躺椅、太阳伞
水上项目：垂钓、游艇、独木舟、皮划艇、摩托艇、水上茶坊等

Convention Center: with advanced facilities, adequate for large and medium conventions
Theatre: Movies and Plays
Open–air Lake–side Show Plaza: Plaza shows, bonfire Night Party
Lake–side lounge bar: Cafeterias, Bars
Open–air swimming pool: swimming pool, chaise longue, parasol
Aquatic events: Fishing, Yachting, Canoeing, Kayaking, motor boating, Water–side Teahouses

（三）山地康体度假区 Ⅲ Mountain Sports Resort

山地森林旅游度假区也是当今旅游度假市场的重要组成部分。山地康体度假区主要依靠连绵起伏的山地景观以及良好的森林植被为旅游者提供优美的视觉景观、清新的空气，辅以丰富的康体健身活动，使旅游者在汗流浃背的运动过程中，远离城市的喧嚣，享受身与心的彻底放松。

Mountain forest holiday resort is also an important part of the current tourist market. Relying on rolling mountains and forests, Mountain fitness resort provide tourists with beautiful visual scenery and fresh air, as well as plenty fitness activities. In the sweaty exercises, tourists will find themselves away from the hustle and bustle of the city and totally relaxed, both physically and mentally.

旅游项目：依山而建的别墅、帐篷营地；拓展训练营；烧烤场；山地自行车、网球、羽毛球、壁球、乒乓球、攀岩、滑翔翼等。

Travel program: Villas and camps that built along the hill; out–ward bound camps; Barbeque site; Mountain bike riding, tennis, badminton, racket ball, ping–pong, rock–climbing, hang–glider.

（四）化屋民俗度假区Ⅳ Folk Custom Resort in Huawu Village

乡村旅游、民俗旅游是休闲度假旅游的有机组成部分。化屋村被命名为贵州省歌舞之乡，化屋歪梳苗歌舞技艺精湛，在苗族中独树一帜，但是由于修建水电站，化屋原有民居村落被湖水淹没，现有建筑大部分建造时间在20世纪八九十年代，缺乏科学系统的建设规划和统一管理，建设散、乱。现有村民住宅，结构形式以草木、石木、砖石为主。建筑形式、材料和色彩混乱，缺少特色，整体形象较差。因此化屋民俗度假区在外形上对化屋现有建筑进行改造，并新建民居客栈；在内容上丰富、充实苗族文化，增强游客的参与性，形成乡村旅游与度假旅游并举的休闲度假旅游发展局面。

旅游项目：化屋民族歌舞表演，民俗工艺表演，民族商品购物，民俗工艺体验，民居客栈体验等。

Travel featuring Villages and Folk Custom is an integral part of leisure travel. Huawu Village is entitled as the home of singing and dancing art of Gui zhou Province. People of the ethnic group, Waishumiao, in the village are outstanding singers and dancers among the Miao group. However, the former village is buried under water after the construction of the Hydroelectric Power Station. Most buildings in the village were built around 1980s to 1990s, disorderly scattered, without systematic planning and centralized management. The residences of the village people are mostly constructed with random materials like weeds, wood, stone and bricks in unflattering shapes and colors with no characteristics, making a bad impression as a whole. That's why Huawu folk custom resort improves its appearance by making renovation on the buildings and setting up new folk taverns, and enhances its competence by enriching the culture of Miao Group and encouraging tourists' participation so as to build itself into a leisure tourist destination that is suitable for both village travel and holiday travel.

Travel Program: Huawu ethnic dancing and singing show, folk custom craft show, shopping for ethnic novelties, experiment on folk custom craft, and experiment on folk taverns.

HuawuResort

项目规划 Item Planning

1 基础概况 Outline

第一节 自然地理

一、地理位置

旅游区所属的化屋村位于新仁乡南部，鸭池河大峡谷、东风湖北岸，属二水（鸭甸河、六圭河）交汇、三县（黔西县、织金县、清镇市）连界的河谷地带，东与清镇市隔河相望，南与织金县隔河相邻，西与羊场乡以哈冲沟为界，北与群益村接壤，距乡政府驻地 9km，距黔西县城 50km，距毕节市 80km。

二、地质地貌

该区地貌特征属喀斯特地貌，地处低纬度、高海拔的高原山地，地势西高东低，由西北向东南倾斜，海拔 870 ~ 1360m，山高坡陡，河谷深切，六冲河在此于左岸汇入乌江。域内出露地层以二迭系、三迭系为主，全为石灰岩分布地区，岩溶较发育，溶洞、暗河、封闭式盆地、海子较多。

三、气候水文

该区位于中亚热带湿润季风气候区，素有春迟夏短、秋早冬长、水热同季、干湿分明等气候特征。常年气候宜人，冬无严寒、夏无酷暑，雨量充沛。年平均日照 1348.9h，无霜期 264d，年平均气温 14℃，雨日 188d，降雨量 1005.2mm。

Section Ⅰ Land

Ⅰ Land

Huawu Village, where the tourist site is located, lies on the south of Xinren County, the north bank of Yachi River and Dongfeng Lake, and the valley of the cross of two rivers (Yadian River and Liugui River) and the boundary of three counties (Qianxi, Zhijin and Qingzhen), about 9 kilometers from its county government, 50 kilometers from Qianxi City, and 80 kilometers from Bijie Prefecture. The land extends north to Qunyi Village, and to the other three directions of east, south and west, there are rivers respectively dividing it from Qingzhen Prefecture, Zhijin County and Yangchang Town.

Ⅱ Terrain

This are is uplands and mountains with karst geomorphologic features. The average election is 870~1 360 meters and the overall land trend is high in west and low in east. The exposed lands are limestone region and mainly in Triassic or Dyas Systems with karst caves, subsurface streams, closed basins, and wet lands.

Ⅲ Climate

This region is in subtropical monsoon climate with comfortable climates and plenty of rain. The average sunlight times per year are 1 348.9 hours, 246 frost-free days and 188 rainy days, rainfall 1 005.2 mm.

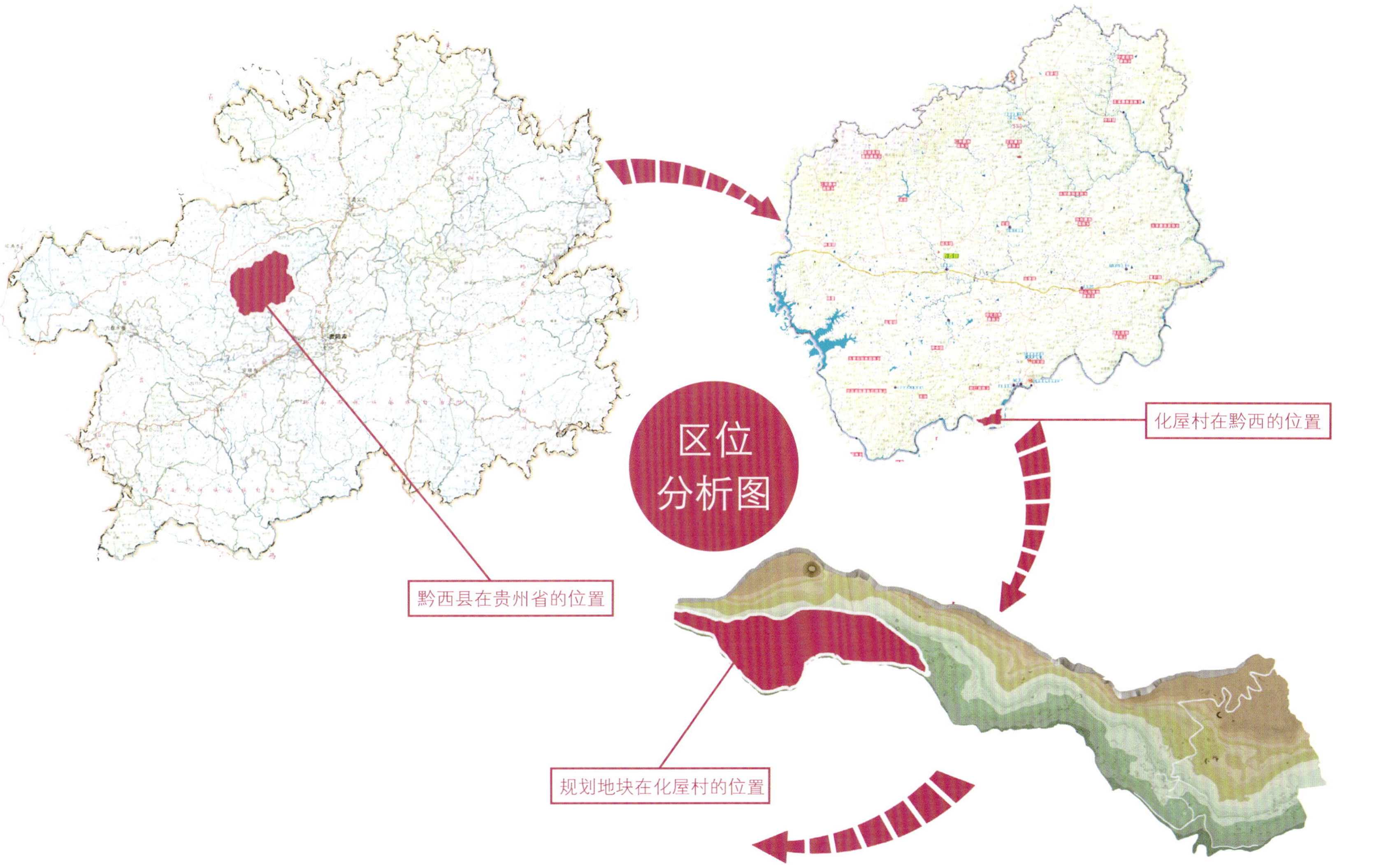

主要技术经济指标一览表

总规划用地面积		247161.82㎡
建筑总占地面积		14420.50㎡
其中	民俗风情街	3204.80㎡
	风情酒店	1720.84㎡
	六星度假酒店	3257.88㎡
	度假别墅	2892.48㎡
	影视电玩城	1278.00㎡
	康体会所	983.49㎡
	啤酒屋	261.27㎡
	管理用房	821.74㎡
总建筑面积		45684.64㎡
其中	民俗风情街	6782.31㎡
	风情酒店	8762.41㎡
	六星度假酒店	15987.27㎡
	度假别墅	7307.60㎡
	影视电玩城	2556.00㎡
	康体会所	2950.47㎡
	啤酒屋	261.27㎡
	管理用房	1077.31㎡
建筑密度		5.83%
容积率		0.18
绿地率		42.5%
停车位		108 个
其中	酒店停车位	58 个
	社会停车位	50 个

第二节 人文环境

一、历史沿革

化屋村虽为片壤之地，然开发较早。南朝梁、陈时期，昆明夷闽支卢鹿部，沿江东进，占据此地；唐代，先后划入琰州之琰川县和龚州（今黔西）；宋为“罗氏鬼国”；元代，先属鸭水县，后归鸭水长官司；明代，先后隶属贵州宣抚司和水西宣慰司；清初，吴三桂平水西时被困之于三岔河，战火将这里毁成废墟，沦为荒野，后有4户苗民来此，开荒狩猎，重建家园，遂又逐步发展，划归黔西安德里三甲。民国6年（1917年）置特编保（乡级），属南一区。民国21年与新仁乡合并，称仁化乡。民国27年分开置化屋特编联保，下辖3个保。民国30年同新仁、慕义合并称仁慕乡，直至解放。1953年5月置化屋乡，隶属大关区。1957年6月，贵州省人民委员会批准为苗族乡。1958年9月改称人民公社。1959年改称管理区。1961年3月复称公社。1984年6月，恢复化屋苗族乡建置。1991年11月，建并撤后划为新仁苗族乡，称化屋村。

Section II Culture & Environment

I History

Although Huawu was a remote village, it was developed fair early. Lulu tribe of Kunming ethnic groups marched along the river towards east and occupied this land in Liang and Chen Dynasty; it was controlled by Yanchuan County of Yan State and then by Gong State (Qianxi Country now) in Tang Dynasty, named "Ghost Nation of Luo" in Song Dynasty, and at the beginning of Qing Dynasty, when Wu sangui was troubled fighting for Shuixi in Sanchahe, this region was destroyed by the war. Then four Miao families came and lived here, and this area were rebuilt and redeveloped and was affiliated wit Andeli, Qianxi. The sixth year of the Republic (1917), it was listed in the Nanyi Town as a county, and combined with Xinren Country named Renhua Country three years later until twenty-seventh year (1938) when Huawu country resumed and governed three districts, then united Xinren and Muyi country to form another country named Renmu till the liberation. In May 1953, Huawu country was set under the attached to Daguan Town; in June 1957, it was authorized by People's Committee of Guizhou Province to be Miao County, and in November 1991, it was labeled Xinren Miao Town and named Huawu Village.

高程分析图

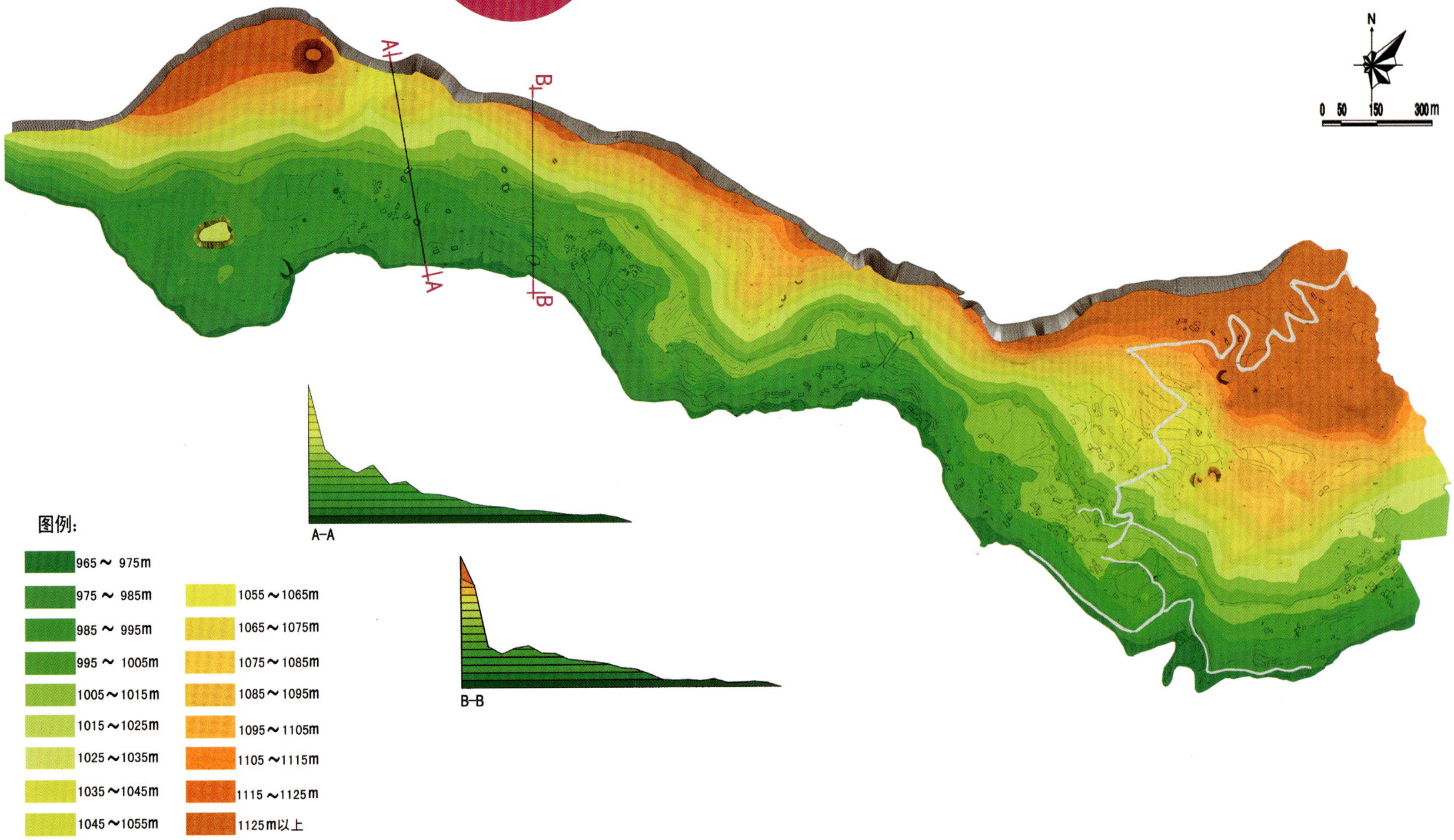

二、社会经济

全村辖 7 个村民小组，总户数 197 户，总人口 1035 人，少数民族 986 人。全村地域面积8.2km^2，耕地面积417亩，山多地少，土质条件差，水土流失严重，经济发展较为滞后，生产力水平较为低下。全村主要产业为种植业和养殖业。2007 年人均纯收入 1560 元，人均生产粮食 424kg。

Ⅱ Social Economy

Villagers in Huawu are divided into seven groups, totally 197 families, 1 035 villagers, among them the population of ethnic is 986. The land occupies 8.2km^2, only 417 mu cultivated field. The poor fields with many mountains and hills but few fields and serious soil and water lost had hampered badly the development of county economy. In 2007, the average income is RMB 1 560, and average grain 424 kilogram.

三、基础设施

建成了 14km 的通村油路，路宽 6m，结束了化屋不通公路的历史。但该路弯多坡陡且路况舒适性较差，一旦遇上雨季及极端天气，可进入性较差。村庄由农村电网供电，基本能满足村民生活用电需要，有少量道路照明设施；村内生活饮用水引自哈冲山泉水，水质和水量基本能达到要求，但输水管道年久失修；无污水处理设施，生活污水经简单收集后自然排放或农灌；有少量垃圾箱、公共厕所等环卫设施，垃圾收集处理设施不完善。

Ⅲ Infrastructures

The completion of the road, 14 kilometers length and 6 meters width, has ended the history of traffic deficiency of Huawu. However, it is dangerous for traffic to go through the road in the rain seasons and bad weather, due to many turnings and steep slops. Rural electric net supplies power to the villagers and few facilities of road illumination; spring water from Hachong Mountain, with good quality and quantity, was introduced to supply them for daily use; the facilities of hygiene and environment like trash cases and public toilets are few, and the disposal facilities of sewage are unsound.

坡度分析图

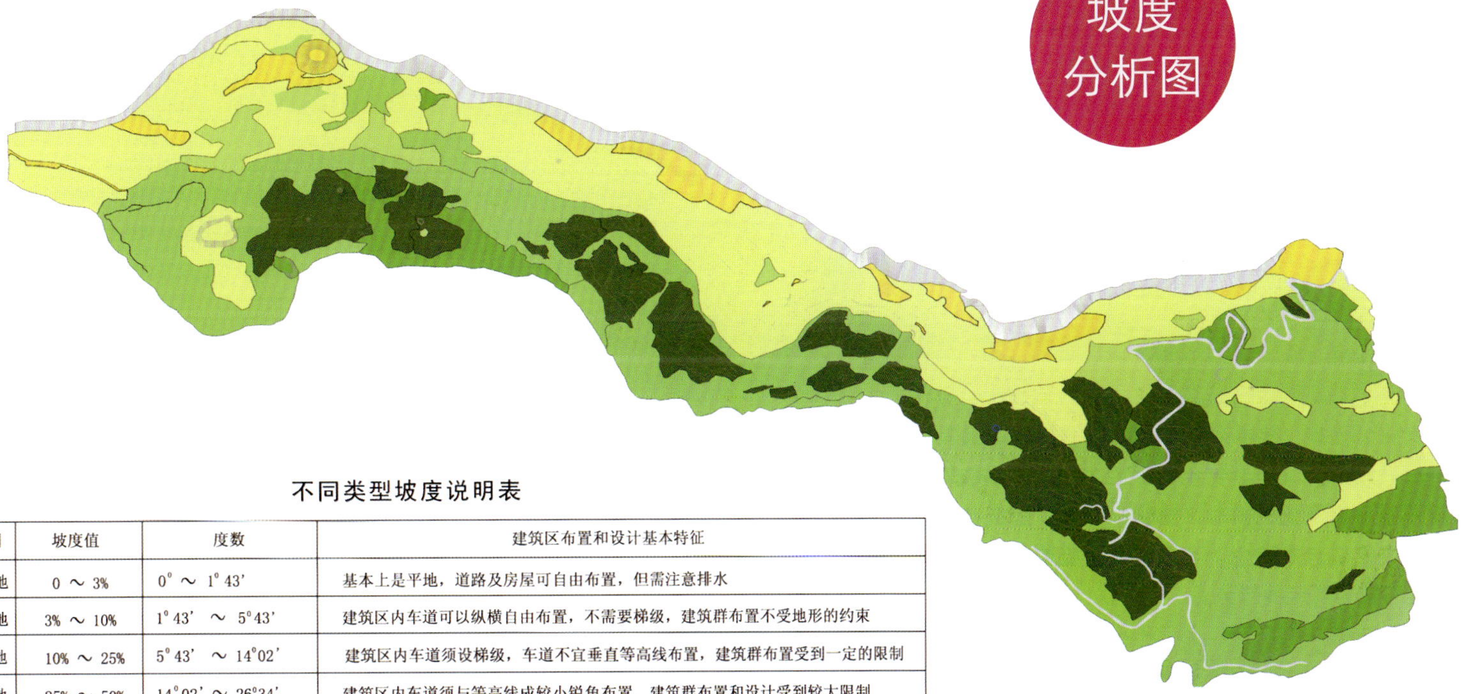

不同类型坡度说明表

图例	类别	坡度值	度数	建筑区布置和设计基本特征
	平坡地	0 ～ 3%	0° ～ 1°43′	基本上是平地，道路及房屋可自由布置，但需注意排水
	缓坡地	3% ～ 10%	1°43′ ～ 5°43′	建筑区内车道可以纵横自由布置，不需要梯级，建筑群布置不受地形的约束
	中坡地	10% ～ 25%	5°43′ ～ 14°02′	建筑区内车道须设梯级，车道不宜垂直等高线布置，建筑群布置受到一定的限制
	陡坡地	25% ～ 50%	14°02′ ～ 26°34′	建筑区内车道须与等高线成较小锐角布置，建筑群布置和设计受到较大限制
	急坡地	50% ～ 100%	26°34′ ～ 45°	车道须曲折盘旋而上，梯道须与等高线成斜角布置，建筑设计需特殊处理
	悬崖地	大于100%	大于45°	车道及梯道布置极困难，修建房屋工程费用大，一般不适合作建筑用地
	水 体			

2 背景分析 Background Analysis

第一节 旅游资源分析

一、资源概述

1. 自然资源

旅游区自然资源主要是以山水秀丽的东风湖为主线沿湖分布，主要为地文景观类资源，其中具有代表性的主要有以下景点：天竹奇峰、大鹏展翅、水西姑娘、水西坐佛、神扇灭蝇、龛壁借光、人面狮身、天鹅戏水、红运天书、灵猫戏鼠、天蟾静卧、绝壁栈道、五彩飞瀑、哈冲燕城、峰壁云海等。

2. 人文资源

化屋人文资源主要体现为当地的歪梳苗民族风情。极具特色的蜡染刺绣服饰，优美动听的多声部民歌，奇特的芦笙舞、板凳拳舞、打鼓舞等舞蹈无不体现了苗族历史的源远流长和深厚的文化内涵。

其中，化屋苗族芦笙拳舞、多声部民歌曾在北京、内蒙古、广西南宁等地表演并多次获奖。1993 年，被贵州省文化厅命名为“苗族歌舞之乡”；2006 年被毕节地区摄影家协会授牌确定为“毕节地区摄影家协会创作基地”；2007 年 5 月，“化屋苗族文化空间”被贵州省人民政府列为省级民俗非物质文化遗产。

二、资源评价

1. 自然资源壮阔秀美

东风湖旅游区自然资源主要依托东风湖水体以及沿岸山地景观，东风湖与周边山体共同构成了旅游区雄伟壮阔的峡谷景观，水秀山雄，有“山似三峡而水胜三峡，水似漓江而山胜漓江”的美誉。可以说东风湖旅游区自然资源是兼具了雄壮与秀丽为一体的。

2. 人文资源独具特色

化屋歪梳苗是贵州数十支苗族分支中的一支，该村居住着苗、彝、汉三个民族，其中苗胞占 95%，是黔西县最具代表性的苗族村寨之一。其极具特色的手工艺品及苗族歌舞在继承了传统苗族歌舞精髓的同时又结合、融入了歪梳苗以及化屋当地风格，形成自身特色，可以说以歪梳苗为特色的化屋人文资源具有较为鲜明的特色。

3. 资源组合相得益彰

首先，自然资源的组合相得益彰。秀丽的湖水与雄伟的峡谷山峰相映成趣，山与水完美结合，形成山水交相辉映的高峡平湖景观。

其次，自然资源与人文资源组合相得益彰。这样的资源组合丰富了资源类型，丰盈了旅游内容。

Section Ⅰ Analysis of Tourism Resources

Ⅰ Overall of Resources

1. Natural Resources

Wonderful peaks, lush green forests, colorful waterfalls and seas of clouds contribute to the natural resources of landscape along Dongfeng Lake in this fine spots.

2. Cultural Resources

The cultural resources of Huawu are mainly embodied by the folk flavor of local Waishu Miao ethnic group. The special wax painting and embroidery costumes, beautiful polyphonic folk song, fantastic Lusheng Dance, Bandengquan Dance and Drum Dance, all of them are reflecting the long history and profound cultural connotation of Miao people.

Among them, the Lusheng Dance and polyphonic folk song had performed and won many prizes in Beijing, Inner Mongolia, and Nanning of Guangxi. In 1993, Huawu was named by Guizhou Cultural Department "The Home of Miao Sings and Dance", and was appointed by the "Photographers Association of Bijie Prefecture" as the "Creation Base of the Photographers Association of Bijie Prefecture" in 2006; and in May 2007, "Huawu Miao Cultural Space" was rated among the provincial folk intangible heritage by People's Government of Guizhou.

Ⅱ Evaluation of Resources

1. Magnificent Natural Resources

Natural Resources of this fine spot, magnificent and beautiful, are mainly the water of Dongfeng Lake and the mountains landscape on two sides of the bank that it enjoys the fame of "With the similar mountains, but more magnificent water than the one of the Three Gorge; with the similar water, but more beautiful mountains than the one of the Lijiang River".

2. Unique Cultural Resources

Waishu Miao is one of ten branches of Guizhou Miao ethnic groups, living along with the other two nations of Yi and Han in Huawu which is the most representative village of Qianxi County. With unique handicrafts and special songs and dances of Miao people which had inherited the essence of traditional songs and dances of Miao and combined and integrated the local style and the one of Waishu Miao people, Huawu village is covered by cultural resources with distinct features.

3. Perfect Combination of Resources

First, it is the perfect combination of natural resources, which forms a wonderful landscape of high mountains and clear waters with beautiful lake reflecting magnificent valleys.

Second, it is the perfect combination of natural resources and cultural resources, which even adds the types of resources and enriches tourism contents.

现状分析图

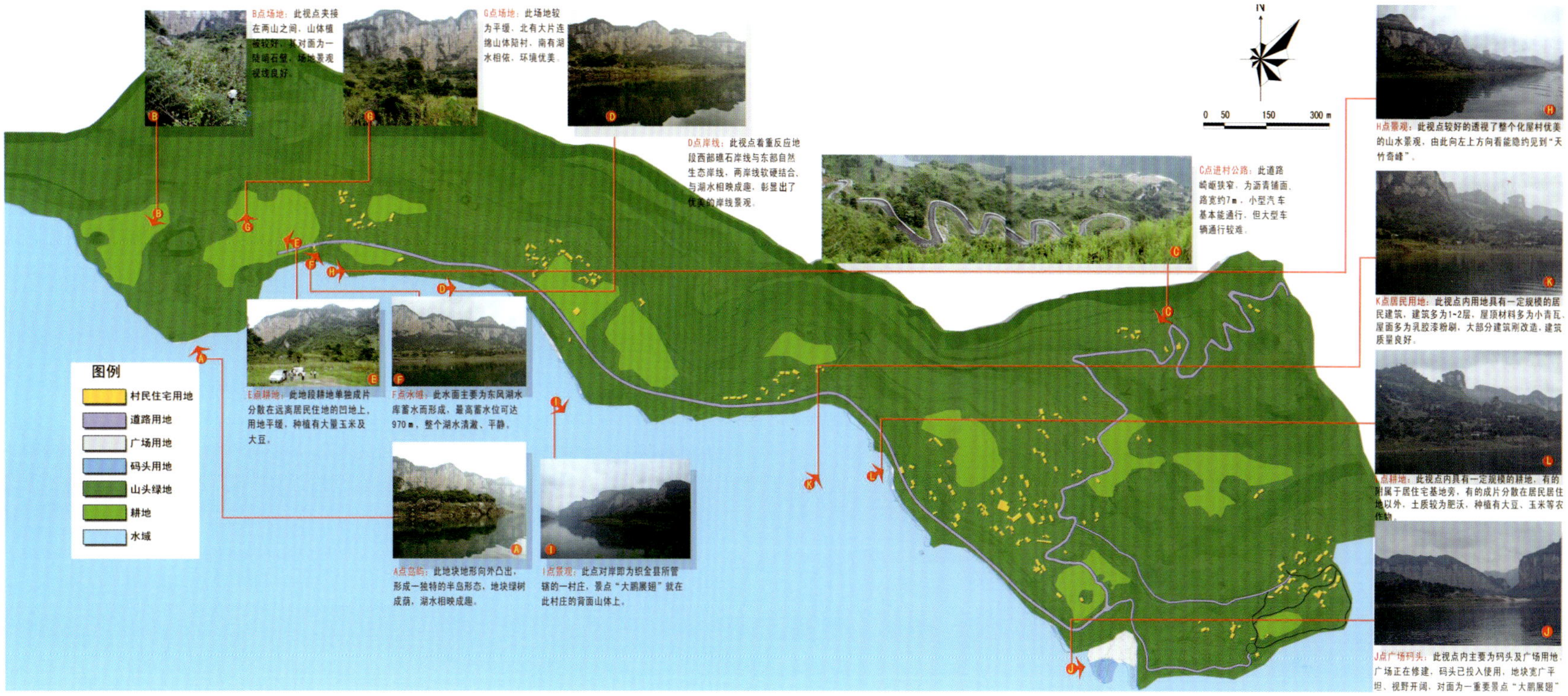
B点场地：此视点夹接在两山之间，山体植被较好，其对面为一陡峭石壁，场地景观视线良好。
G点场地：此场地较为平缓，北有大片连绵山体陪衬，南有湖水相依，环境优美。
D点岸线：此视点着重反应地段西部礁石岸线与东部自然生态岸线，两岸线软硬结合，与湖水相映成趣，彰显出了优美的岸线景观。
C点进村公路：此道路崎岖狭窄，为沥青铺面，路宽约7m，小型汽车基本能通行，但大型车辆通行较难。
N
0 50 150 300 m
H点景观：此视点较好的透视了整个化屋村优美的山水景观，由此向左上方向看能隐约见到“天竹奇峰”。
K点居民用地：此视点内用地具有一定规模的居民建筑，建筑多为1~2层，屋顶材料多为小青瓦，屋面多为乳胶漆粉刷，大部分建筑刚改造，建筑质量良好。
L点耕地：此视点内具有一定规模的耕地，有的附属于居住宅基地旁，有的成片分散在居民居住地以外，土质较为肥沃，种植有大豆、玉米等农作物。
J点广场码头：此视点内主要为码头及广场用地，广场正在修建，码头已投入使用，地块宽广平坦、视野开阔，对面为一重要景点“大鹏展翅”
图例
村民住宅用地
道路用地
广场用地
码头用地
山头绿地
耕地
水域
E点耕地：此地段耕地单独成片分散在远离居民住地的凹地上，用地平缓，种植有大量玉米及大豆。
F点水域：此水面主要为东风湖水库蓄水而形成，最高蓄水位可达970m，整个湖水清澈、平静。
A点岛屿：此地块地形向外凸出，形成一独特的半岛形态，地块绿树成荫，湖水相映成趣。
I点景观：此点对岸即为织金县所管辖的一村庄，景点“大鹏展翅”就在此村庄的背面山体上。

第二节 旅游市场分析

一、中国旅游市场发展趋势

1. 中国旅游市场前景巨大

进入21世纪以来，我国综合国力的进一步提高，经济社会的全面进步，使得旅游业及其相关产业加快了与国际接轨的步伐，不断向纵深发展，在拉动内需、带动相关产业发展、提高人民生活质量方面发挥了突出的作用。

2009年上半年，在全球旅游消费低迷的背景下，中国的出境旅游消费依然保持着接近12%的增长，突显着中国国际旅游地位日益上升的事实。据世界旅游组织预测，到2020年我国将成为世界第一大旅游目的地国家和第四大客源输出国。由此可见，中国旅游者的数量还将会快速增加，中国旅游客源市场前景巨大。

2. 休闲度假成为生活时尚

随着国人经济收入的提高以及国家对法定假日的调整，闲暇时间增加，越来越多的人倾向于通过定期的休闲度假来放松身心，普通民众的度假旅游意识逐渐形成，定期度假需求越来越强烈。“中国旅游族日益壮大：勤劳的中国人开始享受旅游休闲生活”。可见，中国人的旅游方式已由传统的观光型向休闲度假型转变，定期的休闲度假成为现代人追求的生活时尚。

3. 高端旅游逐步兴起

近年来，旅游业在经历了数量扩张阶段后，向以质量为本的新战略转变，再加上中国高收入群体的不断增加，整个旅游市场也在悄然变化，旅游产品逐渐摆脱“价格低、质量次、匆匆跑、曾到过”的最初阶段，开始追求愉悦体验的更高层次，高端旅游市场逐步兴起并形成一定规模，展示出蓬勃生机和活力。

二、省内旅游市场分析

在省内旅游市场方面，通过对近年来省内居民的闲暇时间、可随意支配收入、旅游项目选择、旅游满意度、旅游消费水平的调查分析可以看出，省内旅游市场呈现以下特点：

(1) 有休闲度假愿望的城市居民快速增加，近程和中程潜在客源市场呈扩大趋势。

(2) 省内实际旅游者的数量也在不断攀升。

(3) 贵州的休闲旅游产品处于更新换代阶段。以贵阳为代表的各大中心城市普遍具有浓厚的休闲地域文化，而目前的休闲产品难以满足市民日益变化的需求，有出行愿望的人越来越崇尚自驾游、野外拓展等参与性更强的高层次和时尚型的休闲活动。

(4) 贵州客源有超前消费的冲动，旅游者的消费水平逐渐提高。以贵阳市为例，据统计部门的相关数据显示，月收入低于或接近全市平均水平的人群中，每月有休闲消费支出的人数接近50%。这一比例大大高于西安、沈阳等同等收入水平的地区。

Section Ⅱ Analysis of Tourism Market

Ⅰ Tendency of Tourism Market of China

1. Bright Prospects of Chinese Tourism Market

With the promotion of comprehensive national strength and completed economic and social progress of China since 21st century, tourism and the relative industries have increased developing profoundly to keep pace with the international situation, which has played a prominent role in expanding domestic demands, accelerating development of relative industries and improving people's life.

With the background of low tourism consumption of the world in the fist half year of 2009, China has increased in abroad tourism consumption by 12%, which proved that the position of international tourism of China has been promoting gradually. By 2002, according to the World Tourism Association, China will be the first among the tourist destinations and the forth in the quantity of tourists output. Thus, the number of Chinese tourists will increase quickly and the market resource of tourism in China is prosperous.

2. Fashion of Leisure Travel

As the income of Chinese people increased and the national holiday adjusted, people have had more spare time and they have trended to have leisure travel to relax themselves and the common people have formed tourism consciousness gradually and ask for relaxing stronger and stronger. Therefore, Chinese has changed their tourism form from traditional sightseeing to leisure travel.

3. Rise of High-end Tourism

In the recent years, after the stage of the increase of quantity, tourism has changed the strategy to promote quality. In addition, with more and more high-incomes groups in China, the whole tourism market has changed that the tourism products gradually left the primary stage of "low price and quality, little value of experience", but pursued a higher stage of pleasant experience. Since then the high-end tourism rose gradually and formed a certain scale to show its prospect and vigor.

Ⅱ Analysis of Tourism Market in the Province

We can get the following features of tourism market in the province by analyzing people's spare time and income, choice of traveling program, tourism satisfaction and consumption.

(1) The citizens with leisure travel have increased rapidly, the potential short-range and medium-range customer sources tend to expand.

(2) Tourists in the province are continuous increasing.

(3) Leisure travel products in Guizhou are renewing that every great center city, represented by Guiyang, is with profound regional culture. However, the present leisure products can not meet the changing needs of citizens as many travelers prefer to drive and seek thrill in the wild.

(4) Tourists of Guizhou have the impulse of overconsumption and the tourists' consumption has risen gradually. Take Guiyang as example, among the groups with monthly income beneath or close to the average level, about 50% of them spend money on leisure and relaxing every month. This proportion is greatly higher than the regions with the same income of Xi'an and Shenyang.

3 SWOT 分析 SWOT Analysis

第一节 优势分析

一、资源优势

1. 化屋湖光山色的原始风光、独树一帜的民族文化深藏于大山深处，由于开发较晚，环境及资源品质未受到破坏，保存了原生态的自然风貌，给人予心旷神怡感受的同时，为下一步高水平、高起点的规划开发提供了良好的基础。

2. 化屋具有黔西乃至贵州最具特色的旅游水域交通：乌江源百里画廊——织金洞。贵州绝大多数旅游景点景区之间都是以陆路交通连接，以这样一条风光旖旎，雄奇秀美的水域风光将黔西与国际级的旅游景点织金洞串联起来，势必会为黔西旅游带来新的特色。

二、区位优势

1. 化屋位于清镇、织金、黔西交汇处，贵毕高速公路在距化屋 30 千米处设有匝道，由通村油路连接；化屋西有东风湖水路直通黔西，东有公路连接清镇：另外，县城距化屋村 50km，已开通柏油公路。

2. 从旅游区位来说黔西是贵州省规划的全省 6 个特色旅游区涉及县之一，处于黔西——织金特色旅游区，化屋村境内的东风湖可从水路通往织金洞，是省内最具特色的水域风光交通线。

三、政策优势

1. 金融危机下旅游业受到前所未有的关注，为旅游业的发展带来机遇

(1) 党中央国务院下拨 4 万亿人民币紧急救市，提振内需。

(2) 国家将旅游作为拉动消费，扩大内需的战略产业和重要突破口。

(3) 城乡居民收入继续增长，居民旅游消费需求潜力依然巨大。

(4) 政府对旅游业发展采取了积极关注的态度。

(5) 国家旅游局提出了“大力发展国内旅游”的战略思想。

2. 贵州旅游业继续阔步向前

(1) 旅游业成为贵州省国民经济新的增长点，并被确立为后续支柱产业。

(2) 中国农业银行贵州省分行为我省旅游业提供了 60 亿元授信额度，扶持旅游项目的开工建设。

(3) 贵州省政府下拨 200 万资金拉动内需，资助短途旅游者、自驾旅游者。

Section Ⅰ Analysis of Advantage

Ⅰ Resources Advantage

1. Huawu, with original scenery of lakes and mountains and the unique national culture hiding in the mountains, can leave people pleasant impression and, moreover, provide good fundament for next development with high-level planning.

2. Huawu has the most special water traffic line for tour in Qianxi County and even in Guizhou province: from Hundred miles of Scenery along Dongfeng River to Zhijin Cave. Since most scenic sites in Guizhou are connected with each other by road, this charming view of water line linking Qianxi and Zhijin Cave, an international scenic spot, must inject new vigor to tourism of Qianxi.

Ⅱ Position Advantage

1. Huawu is located at the confluence of Qingzhen, Zhijin and Qianxi, 30 kilometers away is the entrance to Guibi Highway connecting Huawu with a blacktop; in the west of Huawu, water line of Dongfenghu runs towards Qianxi, and in the east blacktop expanding to Qingzhen, in addition, there is blacktop road from Qianxi City to Huawu.

2. Qianxi, as the special tourism region of Qianxi-Zhijin, is one of the six special tourism regions in the tourism scheme of Guizhou. Also, Dongfeng Lake of Huawu Village is the most special water traffic line in the province.

Ⅲ Policy Advantage

1. Opportunities from Financial Crisis

(1) The Central Committee of the Party and the State Council have given RMB4000 billion to save market and increase domestic demands.

(2) Tourism has been determined as the strategic industry and important outbreak of expanding domestic demands and driving consumption by our country.

(3) The income of urban and rural residents is growing constantly so that residents' potential consumption on tourism is still enormous.

(4) The government positively pays attention to the development of tourism industry.

(5) The National Tourism Bureau has pointed out the strategic thought of "greatly developing domestic tourism".

2. Tourism Moving Ahead in Guizhou Province

(1) Tourism has become the new economic increasing point of Guizhou province and was appointed as the back-up industry.

(2) The China Agriculture Bank (Guizhou) has provided 6 billion yuan of line of credit to support the construction of tourist programs.

(3) Guizhou Government has supplied a capital of 2 million yuan to expand domestic demands and fund the excursionists and drivers.

3. 黔西县全力打造旅游业的宏伟战略

2006年，黔西县第十次党代会在认真总结“旅游基地”建设的基础上，从县域经济发展战略全局的高度，确立了“旅游活县”的发展思路。经过近三年的积极工作和精心打造，全县旅游业得到持续发展，证明了“旅游活县”的战略思路是正确的，今后必将坚定不移地贯彻执行这一战略思想，为化屋东风湖度假区的开发提供政策上的支持和保障。

3. Strategy of Developing Tourism in Qianxi County

On the basis of concluding the construction of "tourism base" on the tenth Party Congress in 2006, the development thought of "Better Tourism, Better city" was determined considering the overall regional economic development. With almost three years of hard work and exquisite construction, tourism has been continuously developing, which has proved this strategy is proper, so this strategic thought seriously must be fulfilled to provide support and insurance of policy.

第二节 劣势分析

一、基础设施较为落后

化屋旅游区旅游业起步较晚，目前只有一家度假酒店，接待能力有限，且景区连接道路等级较低，这在一定程度上制约了旅游者的进入，但是将起步晚转化为起点高的优势势必能够形成超越。

二、旅游开发较为初级

化屋旅游业的发展仅仅局限于自然资源的观光以及民族风情的简单观赏，随着旅游者对旅游产品品质要求的不断提升，这样的产品无法继续满足旅游者的需求，势必会进入衰退阶段，急待新的旅游产品丰盈及支撑化屋旅游的发展。

Section Ⅱ Analysis of Weakness

Ⅰ BackWard Infrastructure

With only one hotel and law class roads in this region at present, the capacity of reception and tourists visiting are confined. The tourism region of Dongfeng Lake has been developed relatively late. However, if this disadvantage of late development could be transformed to the advantage of high-level beginning, the surpassing development can be achieved.

Ⅱ Primary Tourism Development

Tourism of Huawu has been only sightseeing of natural resources and simply feeling of folk flavor, which, as the continuous growth of tourists' demands for tourism products, can't meet the needs of the tourists and must destine to get recession, so new tourism products are needed to enrich and support the tourism development of Huawu.

第三节 机遇分析

一、良好的开发机遇

1. 贵州省两大湖泊旅游区作为水源地被保护，旅游发展受限，为东风湖湖泊旅游资源的开发带来机遇。

2. 休闲旅游时代的来临为化屋东风湖旅游度假区的开发带来了绝佳的机遇。

(1) 国务院对国家法定节假日制度作了调整，同时出台了《职工带薪休假条例》，新的休假制度实行，预示着我国公民休闲度假旅游时代已全面到来。

(2) 观光旅游趋于饱和，旅游者对个性化、多元化、精品化休闲度假产品需求更为迫切。

(3) 广大旅游者在消费理念上越来越与国际接轨，对休闲度假旅游的认知越来越成熟，出游动机也越来越理性。人们出游目的已不再是盲目跟风，不再以“到此一游”为乐，开始由单纯的游山玩水、追求猎奇性向追求舒适型、享受型的方向转变。

二、成贵铁路过境

2010年成贵铁路将建成通车，届时将在黔西境内设站，

Section Ⅲ Analysis of Opportunities

Ⅰ Good Opportunity for Developing

1. Being tourism region, the two great lakes of Guizhou Province are preserved as the water sources, which leave the opportunities to Dongfeng Lake to developing its tourism resources.

2. The time of leisure travel has created an excellent opportunity for developing Dongfeng Lake Resort of Huawu.

(1) The State Council has adjusted the national holiday rules and, meanwhile, worked out 《Paid Leave of Employees》. The fulfillment of the new holiday rules indicate the time of holiday travel of residents in our country is coming.

(2) Sightseeing travel trends to be enough that tourists more urgently demand holiday travels of personality, diversity and excellence.

(3) Vast visitors have more and more fashionable consumptive concept, mature cognition on holiday tourism and rational motivation of travel. People have not blindly followed trend of travel and "have a visit here", but they have changed their tourism from monotonous travel and exploring to pursuing comfort and enjoyment.

到那时从黔西到贵阳只需要20min车程，而从成都到黔西的车程也将大大缩短。另外，成贵快速铁路是与贵广快速铁路连接的，届时黔西通往珠三角的交通也将被打通，为黔西开拓省外市场带来了巨大优势。这为拥有高品质资源的东风湖旅游区旅游业的发展提供了先导条件，带来了极大的机遇。

三、清镇——黔西快速通道的建设

清镇至黔西高速公路启动项目前期工作，进行工程可行性研究。

清镇——黔西高速公路起于清镇，修建特大桥跨越鸭池河，进入黔西境内。届时，从贵阳经清镇进入化屋只需1h左右的车程。

第四节 挑战分析

一、周边县市对东风湖资源的同质开发

由于地域及行政区划的关系，规划区所依托的东风湖湖域遍及黔西、织金、清镇三地，由于东风湖水域风光秀丽，两岸景观状况，织金县、清镇市都将旅游发展眼光投向东风湖旅游区，对于旅游区的共同开发势必会构成同质竞争。

但是只要黔西先于他人，形成品牌效益，不仅能够靠自身吸引客源，还能分享织金县、清镇市客源市场。

二、前期初级开发可能带来的消极影响

由于前期对化屋的旅游产品只做了初级开发，可能会使旅游者对化屋旅游留下初始印象，对化屋旅游停留在简单的观光及“农家乐”阶段，短期内难以扭转这一印象，对化屋旅游发展不利。

Ⅱ Chenggui Railway

Chenggui railway will be completed and a station will be set in Qianxi in 2010 when people can only spend 20 minutes from Qianxi to Guiyang and much less time than the present from Chengdu to Qianxi. In addition, traffic from Qianxi to the Pearl River Delta will be convenient as Chenggui railway links to Guiguang railway, which brings great advantages for Qianxi to expand tourism market of other province.

Ⅲ Construction of the Express Route

Before starting the project of the highway from Qingzhen to Qianxi, feasibility study of this project had been fulfilled.

The express route from Qingzhen to Qianxi will be constructed beginning from Qingzhen and with a great bridge spanning over Yachi River towards Qianxi region. It will only take one hour from Qingzhen to Guiyang after the express route completed.

Section Ⅳ Analysis of Weakness

Ⅰ The Development of the Sane Resources from Surrounding Counties

The region of Dongfeng Lake covers three counties of Qianxi, Zhijin and Qingzhen because of administration division. With beautiful scenery of Dongfeng Lake and its banks, the other two counties will make investment on tourism development of Dongfeng Lake, which will bring competition with the same resources.

However, if Qianxi strive to develop first and form their own brand, they can attract visitors of themselves, as well as the ones from these two counties.

Ⅱ Possible Negative Influence from the Primary Development

The primary development of Huawu may leave the visitors an impression of simply sightseeing and Joyous Farmer's House that can't be changed in a short time, which will be bad for Huawu to develop tourism.

4 规划总则 Overall Principle of Planning

第一节 规划依据

《中华人民共和国环境保护法》
《中华人民共和国土地管理法》
《旅游规划通则》GB/T 18971—2003
《旅游发展规划管理暂行办法》
《旅游安全管理暂行办法》
《旅游区（点）质量等级的划分与评定》GB/T 17775—2003
《中国旅游业发展“十五”计划和 2015 ~ 2020 年远景目标纲要》
《风景名胜区规划规范》GB 50298—1999
《风景名胜区管理条例》
《村庄和集镇规划建设管理条例》
《贵州省旅游管理条例》
《贵州省旅游发展总体规划（2003 ~ 2020）》
《贵州旅游业发展“十一五”规划纲要》
《贵州省委贵州省人民政府——关于加快旅游业发展的意见》
《毕节地区旅游发展总体规划（2003 ~ 2020）》
《黔西县国民经济和社会发展第十一个五年规划纲要》
《黔西县旅游发展规划（2005 ~ 2020）》

Section Ⅰ Regulars

Environmental Protection Law of the People's Republic of China
The Land Administration Law of the People's Republic of China (2004 Revised Edition)
Tourism Planning Rules(GB/T 18 972—2003)
Provisional Regulations for Tourism Development Planning and Management
Provisional Regulations for Tourism Security and Management
Classification and Evaluation on Quality of Scenic Areas (Spots)
Fifteen Years Plan of Tourism Development of China and Outline for Long-term Goal in 2015 and 2020
Rules of Scenic Area Planning(GB 50 298—1999)
Management Regulations of Scenic Area
Management Regulations of Construction of Countries and Towns
Tourism management regulations in Guizhou Province
General Plan of Tourism Development of Guizhou Province (2003~2020)
Outline for Eleventh Five Year Tourism Plan in Guizhou Province
Advice on Quickening the Development of Tourism Industry issued by Guizhou Provincial Committee of the CPC and the People's Government of Guizhou Province
General Plan of Tourism Development of Bijie Prefecture (2003~2020)
National Economy and Social Progress Outline for Eleventh Five year Program in Qianxi County
Plan of Tourism Development of Qianxi (2005~2020)

第二节 规划指导思想

一、生态优先的原则

化屋旅游区是集水上游览观光、水岸休闲度假、滨湖游乐活动为一体的综合性高端旅游区，良好的生态环境对于化屋旅游区发展的重要性更是不言而喻。只有保持东风湖的清澈、沿岸植被的繁茂、空气的清新，才能打开化屋旅游区的知名度，吸引更多的游客。因此，在规划当中应由始至终贯彻“生态优先”的规划指导思想，以保护、彰显当地的生态环境特色为第一要务。

二、高标准高规格原则

化屋旅游区是以有高端旅游需求的人群为目标市场的，所以无论是从住宿、餐饮、娱乐、交通工具等旅游硬件设施还是导游接待服务、旅游行程安排、游玩深度等旅游软件条件，与普通旅游目的的设施标准有所区别，而都应以高标准高规格为原则进行规划。

三、全面协调的原则

今天的旅游开发已由原有的景区（点）的开发转向为以目的地建设为主的开发，这就意味着旅游开发不再是单一的独立的开发，而是多元的综合的开发，化屋旅游区的发展，应站在足够的宏观高度去认识，全面协调区域内的相关资源，形成合力，引导化屋周边乡镇的经济社会和谐发展，谋求区域经济的不日腾飞。

Section Ⅱ Guideline

Ⅰ Principle of Ecologic Priority

Dongfeng Lake is a comprehensive high-end tourism region with the programs of sightseeing on the lake, leisure on the banks and playing on the lakeshores. It is self-evident that good ecological environment is very important to the development of Dongfeng Lake Resort. Therefore, only keeping Dongfeng Lake clear, the mountains verdant and air fresh can we expand the fame of Dongfeng Lake Resort and attract more visitors. Thus, we have to implement the guideline planning "priority ecology", so as to preserve and highlight the local features of ecological environment.

Ⅱ High Standard and Specification

The target market of Dongfeng Lake Resort is the groups with high-end tourism demands, thus, either the hardware facilities of lodging, board, entertainment and traffics, or the other conditions of reception service, schedule of tour, travel programs and so on should be different from the ordinary tour sites, but should be schemed in high-standard and scale.

Ⅲ Overall Coordination

Now, tourism development has transformed from development of scenic spot to construction of tourist site, which means that tourism development has not been a single project, but a comprehensive one. The development of Dongfeng Lake Resort should be considered with a macroscopic vision to coordinate the relevant resources in the region in order to conduct economy and society to develop harmoniously.

5 规划定位及目标 Planning Orientation and Object

第一节 战略定位

以化屋东风湖——织金洞水路黄金线为特色，以化屋片区山地森林为依托、苗族文化为支撑，集观光、会议、康体健身、民族风情于一体的中高端休闲度假中心。

第二节 形象定位

花舞蝶飞，梦境天堂

黔西是杜鹃花都，可以说，杜鹃花是黔西的代表。而化屋度假区是未来黔西旅游的代表，因此将二者结合起来便是化屋度假区的形象来源，以杜鹃花漫天飞舞的景象成就了“花舞”之说：“花香自有蝶飞舞”，有花有蝶不仅应景，还以蝴蝶象征游客，寓意化屋旅游区将会以自身高远的定位，高品质的产品及服务吸引旅游者纷至沓来；与此同时“花舞蝶飞”还营造了置身于化屋旅游区的浪漫氛围。

“闲折两枝持在手，细看不是人间有”，唐代大诗人白居易曾这样赋诗赞美杜鹃花。“不是人间有”的杜鹃花以各种形式绽放在化屋旅游区中不就代表了化屋是梦境中的天堂吗。“梦境天堂”这一主题形象旨在说明化屋旅游区美得如梦如幻，如临仙境；此外“梦境天堂”想要传达的另外一层意思是旅游区的游客在这里都会得到优质称心的服务。

第三节 目标定位

一、总体目标

将之打造为贵州真正的旅游度假区。

利用旅游度假区的产品、服务、设施优势，形成集吃住行游购娱于一体的中高端旅游度假中心。

二、近期目标

利用化屋旅游区的打造以及东风湖——织金洞水域风光线，形成中高端周末旅游度假地，吸引旅游者至少住宿两夜，游玩两天。

三、中远期目标

完善化屋旅游区特色项目，以此作为核心吸引力，营造整体的休闲度假氛围，使旅游者的停留不局限于周末，形成真正意义上贵州中高端度假中心，从而形成黔西乃至贵州度假旅游的代表。

Section Ⅰ Strategic Orientation

The prime water line of Dongfeng Lake—Zhijin Cave is the specialty, reflected by the mountains and forests of the region of Huawu and supported by Miao culture to create center for leisure and taking vacation.

Section Ⅱ Image Orientation

Flying flowers and butterflies, Dreamlike Paradise

Qianxi is the azalea capital, so we can say that azalea is the representation of Qianxi. And Dongfeng Lake Resort of Huawu is the future tourism representation of Qianxi. Therefore, the combination of the two representations is the derivation of the image of Dongfeng Lake Resort. The image of azalea whirling in the sky elaborates "flying flowers"; and "where there is flower fragrance, there is butterflies flying around", which, on one side, is the common sense, on the other side, indicates that Dongfeng Lake Resort will attract more and more visitors with its high-level orientation and high-quality products and service as the butterflies symbolizes visitors. Moreover, "flying flowers and butterflies" creates a romantic atmosphere in Dongfeng Lake Resort.

Bai Juyi, the great poet in Tang Dynasty, had praised azalea in his poem "Fondled in hands are two twigs, which I do not think we have on earth". The azalea, "can't be found on earth" indicates that Huawu is the dreamlike paradise. The thematic image of "Dreamlike Paradise" aims to elaborate the magnificence and beauty of Dongfeng Lake Resort; in addition, it also means that visitors can get pleased service here.

Section Ⅲ Objection Orientation

Ⅰ Overall Objection

We aim to create it be a real holiday resort of Guizhou province and make good use of the advantages of products, service and facilities to form a medium and high-end tourist center combining with board, lodging, traveling, shopping and entertainment.

Ⅱ Short-term Goal

The short-term goal is to form medium and high-end holiday travel in weekends so as to attract visitors and keep them for two days at least by Dongfeng Lake Resort and the scenery line of Dongfeng Lake Zhijin Cave.

Ⅲ Medium & Long-term Goal

The medium and long-term goal is to improve the special project of Dongfeng Lake Resort and create the overall leisure atmosphere to attract tourists' coming not only in the weekends and form a real medium and high-end holiday center, so as to make Dongfeng Lake Resort to be the tourism representation of Qianxi County, and even Guizhou Province.

6 总体布局 Overall Layout

东风湖旅游区主要分为两个部分，十三个功能区。

一、高端区

第一部分为高端区，包括六星级酒店服务区、度假住宿区、滨河游览区。

六星级酒店服务区及度假住宿区

向高端旅游者提供住宿服务，是化屋旅游区的核心所在。其中六星级酒店周边点状布置别墅群，供游客住宿。

影视电玩区

该片区是丰盈化屋休闲、度假胜地的配套设施。电子游乐区针对儿童电子游艺和高端客户的成人电子游艺，是商务人士闲暇放松，儿童娱乐的重要场所。

商务休闲区

该片区主要为商务人士打造。它集商务会务、商务洽谈、康体疗养等众多功能为一体，是成功人士享受假期，处理公务，保养休闲的重要场所。

滨河游览区

滨河游览区主要体会的是亲水性，它分为陆地亲水和游船亲水两部分。游客可以在闲暇时光漫步于滨河步道欣赏湖光山色，亦可在码头登船欣赏沿江美景。其次，滨河游览沿江布置了啤酒屋及咖啡吧，供游客休闲娱乐。

The tourism region of Dongfeng Lake is divided into tow parts, thirteen functional areas.

Ⅰ High–end Area

The first part is high–end area, including six–star hotel service area, lodging area and walking area.

The Six–star Hotel Service Area and Lodging Area

Providing lodging for the high–end visitors is the core of Dongfeng Resort. Some villas are spread among the six–star hotels to provide the visitors with lodging.

Electronic Game Area

This area, with the matched facilities to enrich and promote leisure and entertainment of Huawu, is an important place for children and high–end customers playing electronic games and relaxing.

Business & Entertainment Area

This area is mainly for businessmen. With numerous functions like business meeting, negotiation, and sanitarium, it's the important place for successful people enjoying vacation, dealing with business and getting sanitarium.

Sightseeing Area

Visitors can feel and enjoy the wonderful scenery water in this area both on the banks and boating on the lake. Also, along the river there will be beer houses and coffee bars on the banks for visitors relaxing and entertainment.

二、大众区

第二部分为大众区，包括大众停车场、民族风情街区、三星酒店服务区、儿童游乐及动物观赏区、冲浪体验区、民族活动展示区、户外拓展区、电子游乐区、商务休闲区。

大众停车场

位于度假区入口处，用于进入景区车辆的停放。

民族风情街区

该片区是展示黔西民族文化的主要窗口。它结合了东风湖所在地化屋的"歪梳苗"和黔西彝族于一体。通过民族歌舞、民族技艺、民族手工艺品的集中展现，体会浓郁的黔西民族特色。

风情酒店服务区

该片区主要为中端客户和大众消费游客提供住宿服务。

Ⅱ Public Area

The second part is public including park lots, ethic streets, three–star hotels and children's playgrounds, zoos, surfing experience area, demonstration activities area, wild expanding area, electronic games area and business leisure area.

Public Park Lot

At the entrance of the tourism report for parking vehicles entering the resort.

Ethnic Streets Area

This is the main window demonstrating national culture. With the two ethnic groups of "Waishu Miao" and Yi, visitors can feel thick specialty of Qianxi ethics through their songs and dances, crafts and technology.

Ethnic Hotels

This area mainly provides medium visitors and public consumers with lodging.

总平面图

儿童游乐及动物观赏区

专为儿童提供服务，场地相对封闭、安全，专人管理，让儿童拥有自己的小天地，锻炼独立能力的同时，解除家长的后顾之忧，专职于自身的休闲放松。

Children's Ground and Zoo

This is for children, so the place is relatively enclosed, safe, and has special people taking charge of it that children can have their own world for playing and relaxing, which is good for improving their independent ability and removing the parents' worry.

冲浪体验区

该区域旨在增强旅游项目的参与性、刺激性，更能营造休闲度假的整体氛围，让游客真正地领略休闲、度假游的魅力。

Surf Experience Area

It is aimed to strengthen participation and thrilling of the program and create an entire atmosphere of leisure and making vacation, so that visitors can real admire leisure and travel.

民族活动展示区

民族活动展示区是化屋旅游区打造民族特色的又一亮点。它可以让游客在此观看婀娜的彝苗舞蹈，倾听甜美的彝苗歌声。特别是那享誉中外的化屋多声部苗歌。

Demonstration of Ethnic Activities

This is the other highlight of tourism region of Dongfeng Lake, where visitors can watch the wonderful Yi and Miao dance, enjoy the beautiful Yi and Miao songs especially the world-known polyphonic Miao songs of Huawu.

户外拓展区

该区域向游客提供攀岩等拓展训练项目，满足个人或企业开展户外拓展训练培训。

Wild Expanding Area

In this area visitors could try some training programs like climbing rocks, so it is a place to provide individual or enterprises with wild expanding training.

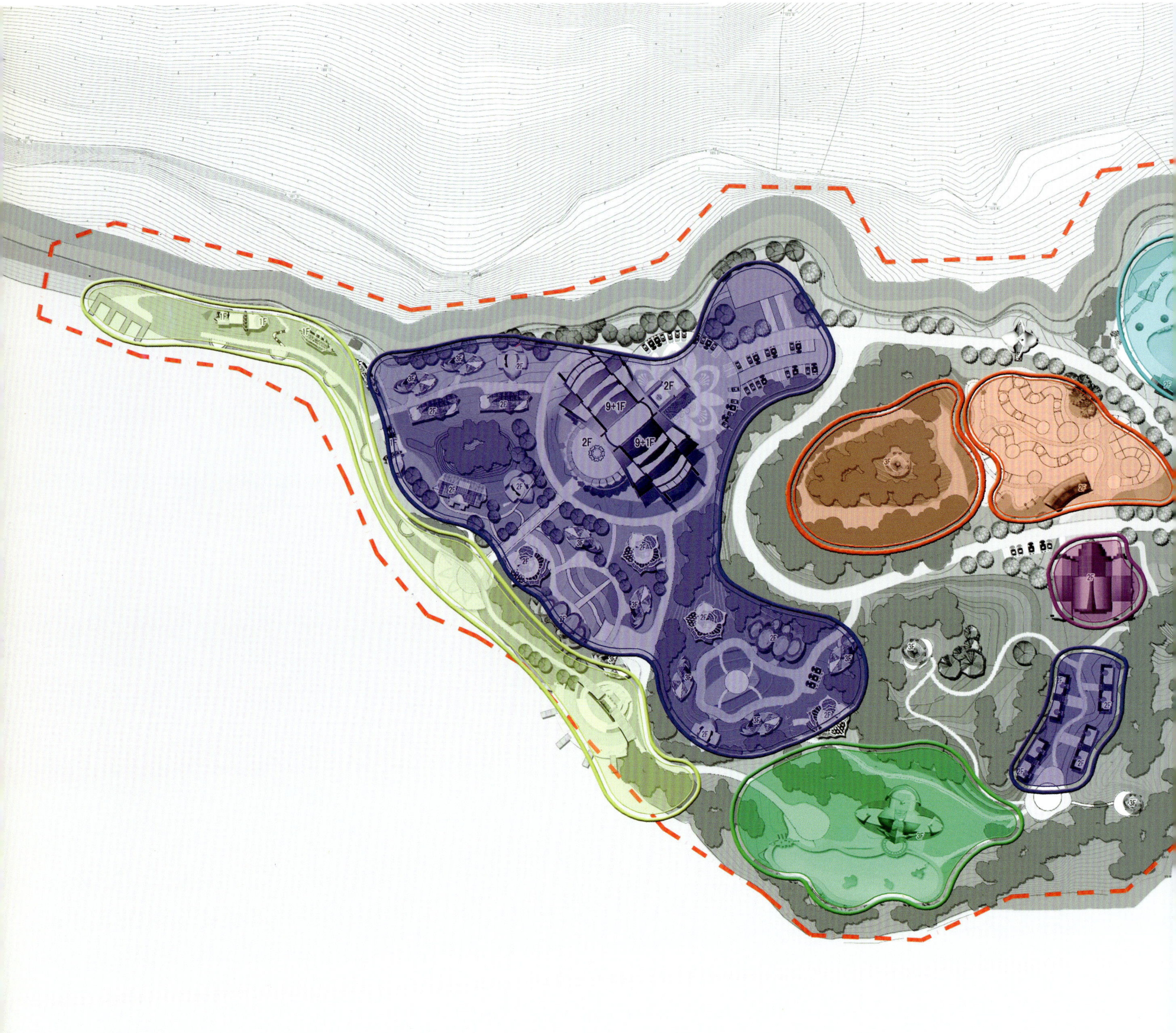
1F
2F
9+1F
3F

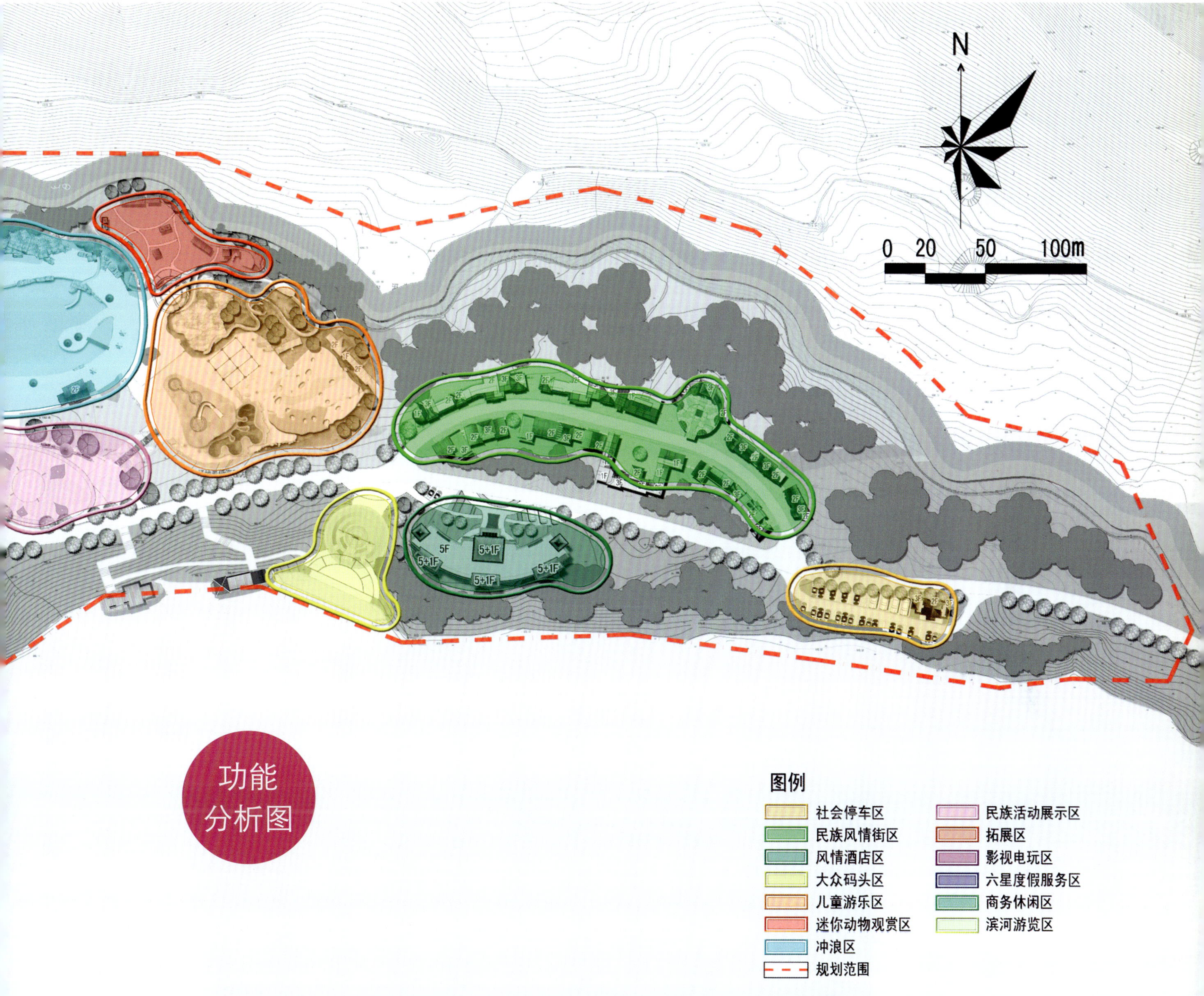
N
0 20 50 100m
2F
5F
5+1F
3F
1F
图例
社会停车区
民族风情街区
风情酒店区
大众码头区
儿童游乐区
迷你动物观赏区
冲浪区
规划范围
民族活动展示区
拓展区
影视电玩区
六星度假服务区
商务休闲区
滨河游览区

功能分析图

7 主体工程规划 Planning of Major Project

第一节 设计理念

化屋度假区的设计大体上呈现出两种风格：以东风湖旅游区所在地化屋村的建筑为原型的原生态风格以及以杜鹃花为创意元素的现代风格。

度假区应该营造休闲放松的氛围，又要运用当地元素，塑造当地形象。因此在设计中，我们根据功能布局，主要运用了两种不同的风格，运用设计风格对功能区进行巧妙的划分。

大众区的设计主要体现化屋当地的特色，运用木材、茅草等当地建筑元素，进行艺术手法上的再加工，加上黔西水西彝族常用的尖角装饰，力图打造出粗犷古朴又具有当地特色的原生态建筑，使旅游者通过建筑外观就能体会到野趣十足的松弛氛围。高端区的设计元素来源于黔西著名的杜鹃花。

黔西是杜鹃花之都，这无疑是黔西的核心资源之一，化屋度假区不仅仅局限于化屋，它还需要站在黔西整体范围内来考虑，因此将代表黔西的杜鹃花作为创意元素顺理成章，通过对杜鹃花花蕾、花蕊、花瓣以及花叶进行抽象创意，从内容和形式上体现黔西特色的同时，营造度假区温馨浪漫的整体氛围，丰盈和完善度假区的休闲松弛感受。

第二节 旅游产品策划

旅游产品		旅游项目
观光旅游产品	东风湖风光游	观光游轮、东风湖沿岸风光等
	东风湖—织金洞水路游	观光游轮、东风湖—织金洞水域风光、织金洞风光、游轮餐会等
民俗旅游产品	化屋苗族风情游	化屋歪梳苗板凳舞、芦笙拳舞、无伴奏多声部合唱、化屋苗族特色饮食、手工艺品等
度假旅游产品	东风湖滨湖度假游	六星级（三星级）度假酒店、度假别墅、滨湖酒吧、咖啡吧、茶楼、游艇、小型豪华游轮、水上摩托艇、快艇、冲浪池、电玩城、儿童乐园、歌影剧院、棋牌娱乐、垂钓、篝火广场、歌舞广场、风情街等
康体健身旅游产品	化屋康体健身游	天然森林氧吧、摇曳、旋转器械、弹跳跨越器械、自行车、游泳、推拿、桑拿等
	化屋拓展训练游	攀岩、拓展训练器械等
疗养保健旅游产品	化屋疗养保健游	SPA、理疗、按摩、推拿、中药保健、药膳等
商务公务旅游产品	化屋商务游	度假酒店、会议中心、康体会所、歌影剧院等

Section Ⅰ Design Concept

Generally, there are two styles of designs of Dongfeng Lake Holiday Resort of Huawu: one is the original ecological style based on the architecture of Huawu Village, the other one is the modern style with azalea as the creative element.

Leisure and relaxing atmosphere should be created in the resort, also local elements should be used to shape local image. Therefore, we mainly apply two different styles and use design styles to divide the functional areas masterfully according to functional layout in the design.

The public area is designed mainly embody the specialty of Huawu. We use local architectural elements like wood and thatch etc. and horn decoration of Yi ethnic group of Shuixi to create original ecological constructions, rough and primitive but with local features with our artistic design, so that visitors can feel the wild relaxing air just by the appearance of architectures. The design element of high-end area derives from the famous azalea of Qianxi.

Being the capital flower, azalea is undoubtedly the core resources of Qianxi. And Dongfeng Lake Resort of Huawu does not only represent Huawu Village, but Qianxi County. So it is reasonable to take azalea as one of the creative elements to design the buildings with the image of the buds, stamens, and petals of the azalea, so as to embody Qianxi flavor from contents and shapes abstractly and also create a romantic atmosphere and enrich and improve the relaxation and leisure of the whole resort.

Section Ⅱ Scheme of Tourism Products

Tourism Product		Tourism programs
Tourism products for sightseeing	Travel in Dongfeng Lake	Viewing ferries and scene along the banks of Dongfeng Lake
	Water line of Dongfeng Lake–Zhijin Cave	Viewing ferries, water scene from Dongfeng Lake to Zhijin Cave, scenes in Zhijin Cave and feasts on ferries
Folk tourism products	Travel in HuaWu for Miao flavor	Bench dance, Lusheng dance, polyphonic chorus without musical accompaniment, special food and crafts of Huawu, and so on
Toutism Products for holiday	Holiday travel in HuaWu	Six-star (three-star) resort hotels,resort villas,Lakeside bars,cafes,tea houses,ships,small-type great ship,aquatic motorboats,yachts,surfing pool,electronic game center,Children's playground,opera house,cards entertainment,fishing,bonfire square,songs and dances square,ethnic street and so on
Tourism products for healthy and fit	Travel of fitness in HuaWu	Natural forest oxygen bar,swinging and swirling equipments,jumping equipments,bicycles,swimming,massage,sauna and so on
	Travel of expending training in HuaWu	Rock climbing,equipment of expending training and so on
Tourism products for recuperation	Recuperation Travel in HuaWu	SPA,physical therapy,massage,health-care with Chiness medicines,medicine diet and so on
Tourism products for business	Business travel in Dongfeng Lake	Resort hotels,meeting center,fitness club,opera house and so on

第三节 旅游线路策划

一、化屋周末游

周五晚上抵达化屋旅游区，入住六星级杜鹃花度假酒店（或度假别墅、四星酒店），晚间可欣赏度假区夜景，并感受滨湖酒吧、咖啡吧、茶楼的梦幻氛围。

周六早上乘坐游轮（或租用游艇）沿途观赏东风湖——织金洞水域风光，中午抵达织金洞，在船上用完午餐后，参观天下第一洞——织金洞，参观完毕返回游轮，于游船上享用晚餐，然会参加游船狂欢 party。夜间返回度假区，宿六星级杜鹃花度假酒店（或度假别墅、三星酒店）。

周日全天感受度假区的休闲氛围，包括风情街、拓展训练基地、冲浪池、康体会所、电玩城、水上项目，儿童还可进入儿童乐园享受属于自己的娱乐空间，晚餐后乘车返回。

二、化屋深度游

第一天：中午抵达化屋旅游区，入住六星级杜鹃花度假酒店（或度假别墅、三星酒店），下午体会攀岩及拓展训练的惊险刺激，之后游览风情街，并于风情街用晚餐，晚上到篝火广场参加篝火晚会。

第二天：全天感受度假区的休闲氛围，包括康体会所、电玩城、水上项目，儿童还可进入儿童乐园享受属于自己的娱乐空间。晚间可欣赏度假区夜景，并感受滨湖酒吧、咖啡吧、茶楼的梦幻氛围。

第三天：体会度假区的景点项目：水路游览织金洞。早上乘坐游轮（或租用游艇）沿途观赏东风湖——织金洞水域风光，中午抵达织金洞，在船上用完午餐后，参观天下第一洞——织金洞，参观完毕返回游轮，于游船上享用晚餐，然后参加游船狂欢 party。夜间返回度假区，享受六星级杜鹃花度假酒店（或度假别墅、三星酒店）的完美设施及服务，次日早晨退房离开。

次日：退房离开或继续体验东风湖旅游区度假之旅。

Section Ⅲ Scheme of Tourism Line

Ⅰ Weekend Travel in Huawu

Visitors can reach Dongfeng Lake Resort and check in the Six-star Azalea Hotel (or villas and four-star hotels), and admire night scenery and feel the dreamlike atmosphere in the bars, cafes and tea houses in the night.

On Saturday morning, visitors can take ship from Dongfeng Lake to Zhijin Cave viewing along the lake, and arrive at Zhijin Cave in the noon. After lunch in the ship, they will enter the cave for viewing. In the evening, visitors return to the ship and have supper, and then join in the fantastic party on the ship. In the night they will go back to the six-star Azalea Hotel (or three-star hotels and villas) in Dongfeng Lake Resort.

The whole day of Sunday, visitor will experience leisure programs of the resort including Folk Streets, Expanding Training Base, Surfing Pool, Healthy Club, Electronic Game Center, and programs on the water, children can also come into Children's Playground to enjoy the space belonged to them. After supper, they will go back by bus.

Ⅱ Tour Huawu

The first day: Visitors arrive at Dongfeng Lake Resort and check in six-star hotels (or villas, three-star hotels) in the noon, and in the afternoon, experience thrilling of rock climbing and expanding training and enjoy walking along the ethnic streets and have supper there in the evening, in the night, they are arranged to participate bonfire party on the bonfire square.

The second day: Visitors feel thoroughly the leisure atmosphere of the resort including programs in sanitarium club, electronic center and boating on the river, children can even enter the Children's playground to enjoy their own space. In the night visitors can appreciate the night scenery of the resort and feel the dreamlike atmosphere of bars, coffee houses and tea houses along the river.

The third day: Visitors will enjoy the scenic sites of the resort. They can take ship from Dongfeng Lake to Zhijin Cave viewing along the lake, and arrive at Zhijin Cave in the noon. After lunch in the ship, they will enter the cave for viewing. In the evening, visitors return to the ship and have supper, and then join in the fantastic party on the ship. In the night they will go back to the six-star Azalea Hotel (or three-star hotels and villas) in Dongfeng Lake Resort, and leave the second morning.

The second day: visitor can check out or continue their holiday in Dongfeng Lake Resort.

8 基础设施规划 Planning of Infrastructure

第一节 公共服务设施规划

根据景区游客需要及用地性质、规模以及所应配备的设施，按照合理、就近、方便的原则，规划布置相应公共服务设施。本规划区公共服务设施包括住宿、餐饮、银行、商店、公厕、网球场、公共停车场、露天游泳池、公共电话亭、垃圾箱等。其中，酒店 2 处，餐饮服务点 2 处，银行 2 处，商店 9 处，码头 3 处，公厕 3 处，网球场 1 处，露天游泳池 1 处，室外生态停车场 2 处，垃圾箱 33 处。(具体详见《公共设施布置图》)

公共设施

Section Ⅰ Planning of Public Service Facilities

The matched service facilities should be set according to visitors' demands and feature of the planning land, and also the principles of ration, closeness and convenience. The public service facilities in this planning area including accommodation, restaurants, banks, public toilets, tennis courts, public park lots, swimming pools, public phone boxes, and trash case and so on. Among them, there are two hotels, two diet service centers, two banks, nine shops, three harbors, three public toilets, one tennis court, one outdoor swimming pool, two outdoor ecological parking lots, and thirty-three trash cases. (Specification is in《Layout of Public Facilities》)

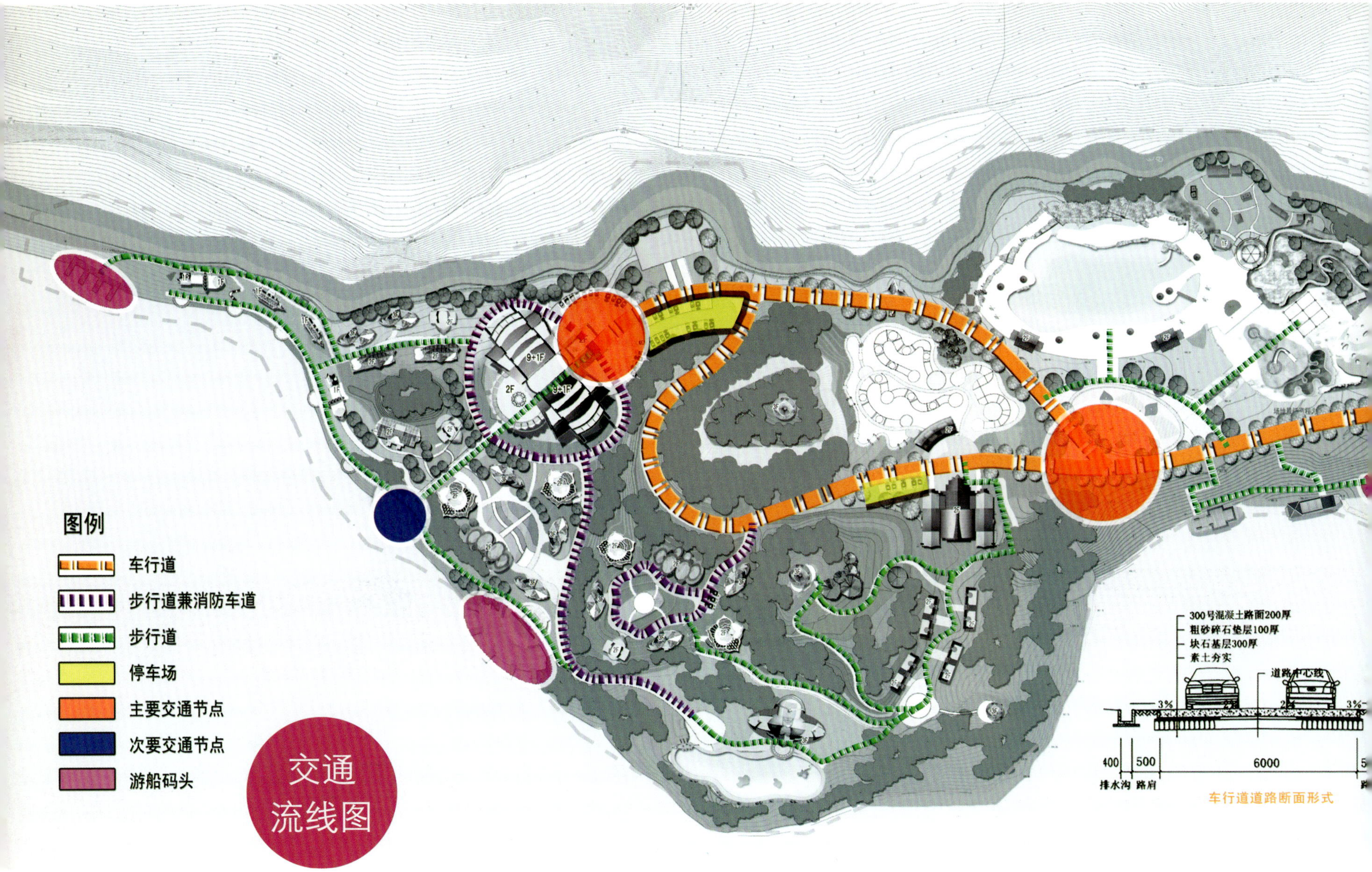

第二节 道路交通规划

一、道路分级

化屋旅游区道路主要由车行道、步行道兼车行道、步行道三级构成。景区的主干道是双向单车道的车行道，宽度为 7m；景区的次干道是步行道兼车行道，宽度为 4m；其余道路是步行道，宽度为 1.2 ~ 2.5m。主干道与次干道兼做消防车道使用。

二、交通组织

景区以沿东风湖的东西向车行道为主干道，该道路东至歌舞广场与化屋村的交叉路口，西连杜鹃花酒店，贯穿整个景区；风情街及酒店度假别墅群内部道路以步行道兼消防车道路与上述主干道相连，是景区的次干道；其余功能区、广场、码头、绿化带等区域均以步行道与主干道或次干道相连。东风湖沿岸东、中、西部分别布设了 3 个游船码头，供水上游船停靠使用，形成陆上、水上两个交通系统，交通较为便捷。景区还分布有 2 个主要交通节点和 3 个次要交通节点。（详见《道路系统分析图》）

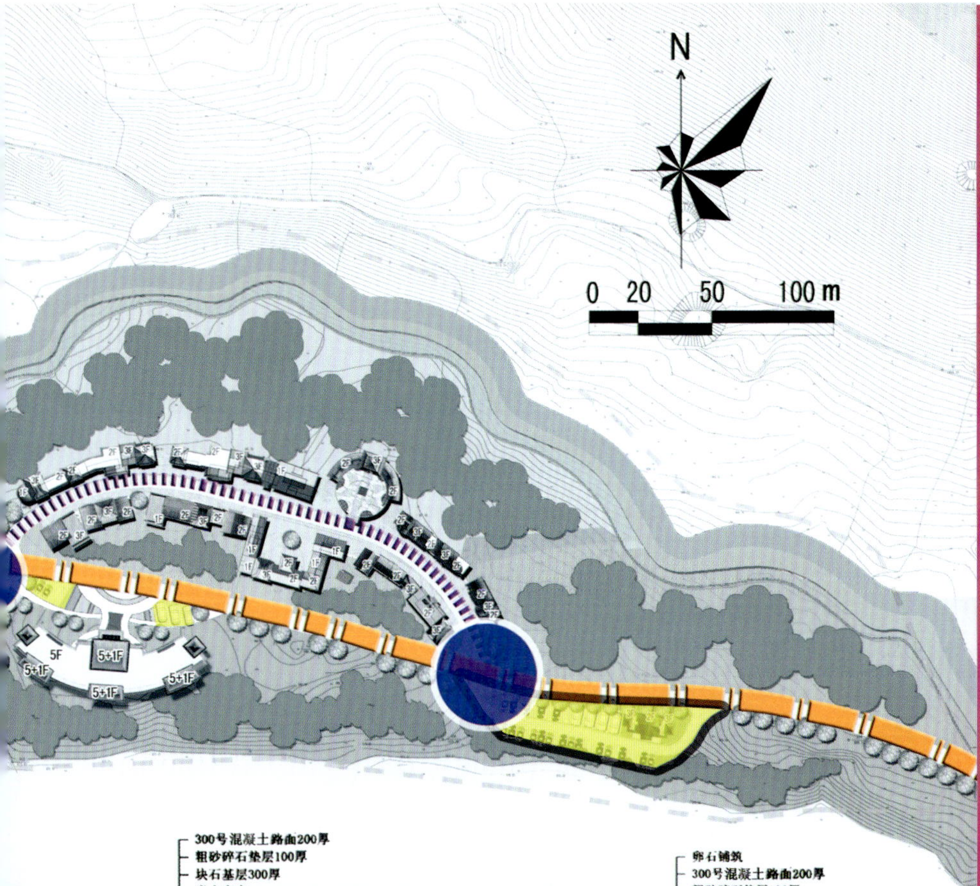

Section Ⅱ Scheme of Traffic

Ⅰ Classes of Road

Road web of Dongfeng Lake Resort consists of driveway, walkway and road for both driving and walking. The main road is a single driveway, 7 meters in width for vehicles of both directions; the secondary main road is for both driving and walking, 4 meters in width, the others are for walking, 1.2~2.5 meters in width. Both the main road and secondary main road can be used as fire lanes.

Ⅱ Traffic

The main road is from east to west along Dongfeng Lake, extending through the whole region from the intersection of Dancing and Singing Square and Huawu Village to the Azalea Hotel. Roads of folk streets and in the villas groups and hotels are the secondary main road linking the main road, for both walking and driving. The other functional areas, squares, harbors and green lines are connected with the main and the secondary main roads by walkway. There are three harbors respectively in the east, middle and west part of the bank of Dongfeng Lake for the ships docking. With the traffic systems both on water and land, Dongfeng Lake Resort is with a relatively convenient traffic. Also, there are two main traffic nodes and three secondary traffic nodes. (Specification is in《Analysis Map of Road System》)

Ⅲ Parking

Parking areas in the resort are outside in two large-scale ecological parking lots and three small-scale parking lots, totally can hold 108 vehicles. The large-scale parking lots are located by the east entrance (36 vehicles) and near the Azalea Hotel (51 vehicles) respectively, the three small-scale parking lots are separately in the area of the Folk Hotel (7 vehicles), the Electronic Game Center (10 vehicles), and the villa groups (4 vehicles).

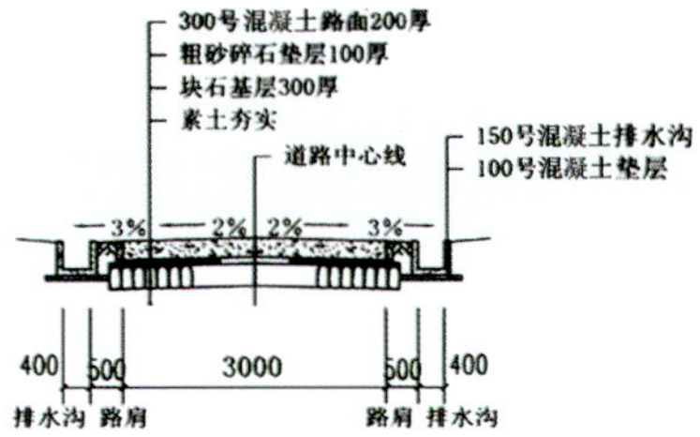

步行道兼消防车道道路断面形式

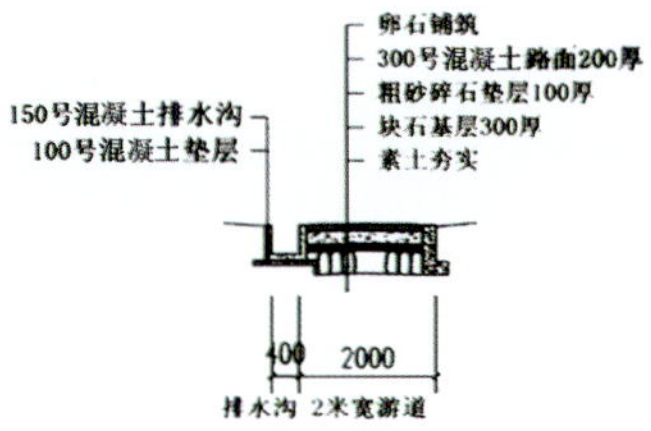

步行道道路断面形式

三、停车场地

景区停车方式主要以室外停放为主，设置了两处大型的室外生态停车场和三处小型停车场，共有车位 108 个。大型停车场分别位于景区东入口和杜鹃花酒店两处，景区东入口可停车辆 36 辆，杜鹃花酒店停车场可停车 51 辆；三处小型停车场分别位于风情酒店、影视电玩城及别墅群，其中风情酒店停车场可停车 7 辆，影视电玩城停车场可停车 10 辆，别墅群停车场可停车 4 辆。

第三节 给水排水及电力规划

一、给水工程

规划区给水排水工程包括生活给水、消防给水、污水和雨水处理三个部分。

1. 生活用水

(1) 给水源头：景区的饮用水和非饮用水采用不同的给水水源。饮用水给水由附近的龙坝寨给水管道引入。非饮用水给水（消防、绿化、清洁、洗车等用水）直接从东风湖抽取至山上蓄水池，经过处理后使用。

(2) 给水量：综合生活用水按建筑物用水 100m^3/ 万 m^2，绿化、道路、广场用水按 2L/（m^2 • d），未预见水量按上述之和的 20%计算，则该片区用水量为 619.452m^3/d。

(3) 给水管网系统采用给水母管、支状相结合的形式，给水主干管沿道路埋地铺设，以支管接入各功能区，整个给水管网的管径采用 DN50 ~ 400。

2. 消防给水系统

(1) 根据规划区规模以及《建筑设计防火规范》GB 50045—95，项目消防系统平面分区与功能分区一致，分设消防水池、消防泵房及消防给水管网系统。按同一时间内火灾次数为一次，一次灭火时间为 2h，灭火用水量按 20L/S 计算，消防用水总量为 144m^3。

(2) 在酒店以及其他公共场所另设置室内消火栓系统。同时，根据《建筑灭火器配置设计规范》在要求部位配套设置灭火器，并在不通车地段设置灭火器。消火栓沿主干道每隔 120m 设置，消火栓最小管径为 DN200，以保证消防用水安全可靠。

Scheme of Water System and Power

Ⅰ Water Supply

Water supply and drainage system in the planning place including three parts of domestic water, firefighting water and disposition of waste water and rain.

1. Domestic Water

(1) Water Source:Drinking water is introduced by pipes from the nearby Longba Village and non-drinking water, for firefighting, irrigation, cleaning and washing vehicles, is the water drew directly from Dongfeng Lake to reservoir on a mountain where it is dealt with for using.

(2) Capacity of Water Supply: Suppose 10 000 square meter of construction needs 100 cubic meter water, 1 square meter 1 day needs 2 cubic meter water for irrigation, road and square, and the unpredictable demands of water is 20% of the sum of above, then water demands for this area will be 619.452 cubic meter per day.

(3) System of service pipes web is consisted with major service pipes and branches. Major pipes are buried along the road with branches linking every functional area. All the pipes are DN50 to DN400.

2. Fire Water System

(1) Fire system of the project has its plane division accordant with functional division according to the scale of planning area and《Norm of Building Fire System》(GB50 045—95). Suppose it takes two hours to put out a fire and 20 cubic water per second, and only one fire happens at the same time, the total water consumption for fire is 144 cubic meters.

(2) To set indoor fire hydrant system in hotels and other public places, as well as the other places needed according to《Norm of Building Fire Hydrant Design》and the places without smooth traffic system. Fire hydrants are set every 120 meters and with the minimum DN200 to make sure water use safe and smooth.

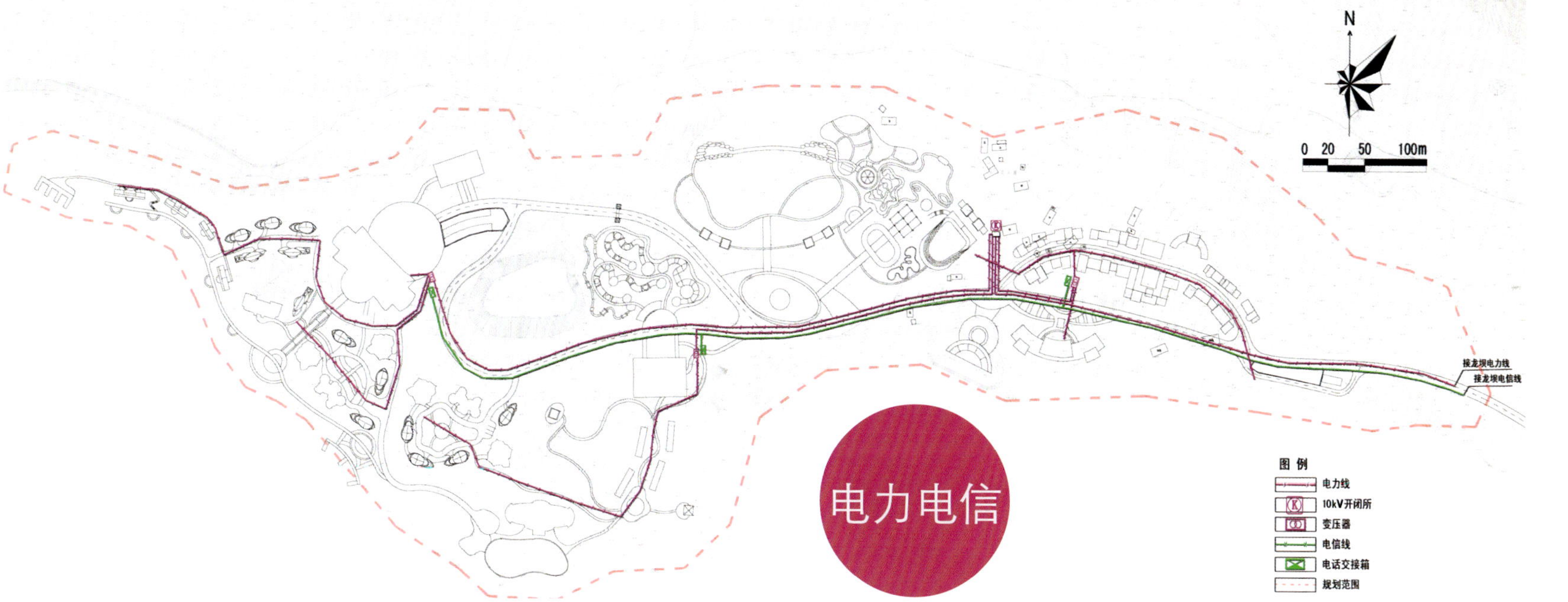

二、排水工程

本项目排水系统采用雨、污分流排水系统。

(1) 污水处理：污水量按总用水量的 80 % 计，为 553.452m^3 / d。餐厅污水需经隔油池预处理，粪便污水需经化粪池处理后排入各功能区的污水管，收集进入排污母管，最后经过污水处理站处理并达到排放标准后，可就近排入东风湖。

(2) 雨水处理：各功能区场地雨水经雨水管收集后分别就近汇入东风湖中。雨水排水采用暗沟与管道结合的方式进行排放。

(3) 排洪处理：为防止山洪、泥石流、滑坡等灾害，在规划用地北侧临坡面由西向东规划一截洪沟，在景区东部、中部、西部分别布设一个排水口，就近排入东风湖。

三、电力工程

1. 供配电：按照《总体规划》要求，该片区供电依托 110kVA 变电站，通过 10kV 输电线路经开闭所形成三个回路接入支线后输入规划区箱式变电器，再用电缆向各个功能区输电。

2. 负荷预测：采用单位面积负荷指针法，按建筑面积负荷指针进行预测规划区的用电量。整个景区总建筑面积为 46751.34m^2，按 80 W/(m^2 • d) 计算，景区规划用电量为 3740.1 kW/d。

3. 电网规划：景区总用电负荷为 5058.8kVA，供电电压等级为 10kV，规划区内需采用单回 10 kV 电力线路供电。由箱式变压器向各建筑物供电。

4. 景区内的变压器采用箱式变压器或结合建筑物设置户内式变电柜，形成良好的景观环境。

Ⅱ Drainage Project

Discharging rain and waste water separately will be adopted in this project.

(1) Disposal of waste water：Suppose the capacity of waste water is 80% of the total used water, it is 553.452 cubic meters per day. Waste water from restaurants must be disposed first in the oil-separating tank and processed in the water disposal station to reach the standard and released into Dongfeng Lake.

(2) Disposal of Rain：Rain in every functional area will be collected by rain pipes and released into Dongfeng Lake. Both underground ditches and pipes are adopted to drain water of rain.

(3) Disposal of Flood：In order to avoid the damage of mountain flood, mud-rock flow and land sliding, we design a ditch from west to east on the mountain side of the planning place and drainage outlets in the east, west and middle respectively for flood flow into Dongfeng Lake.

Ⅲ Power Project

1. Power supply: According to the "Overall Planning", this region equipped with a transformer of 110kVA, electric current runs through a 10kV transmission lines to form three loops and link to branch lines to flow into box type transformer, and then supply every functional area by electric cables.

2. Load Forecasting:Load pointer per unit area should be adopted to predict power consumption in the planning place. The scenery site has a total construction of 46751.34 square meters, power consumption in the scheme should be 3 740.1 kW/d if the average is 80 W/(m^2 • d) .

3. Power Network Planning:The total power load in this area is 5 058.8kVA, and power voltage class is 10kV. 10kV single electric circuit and box type transformer should be adopted to supply power in this planning area.

4. Box type transformer or inner cabinet type transformer setting in the buildings should be adopted to form a good landscape environment.

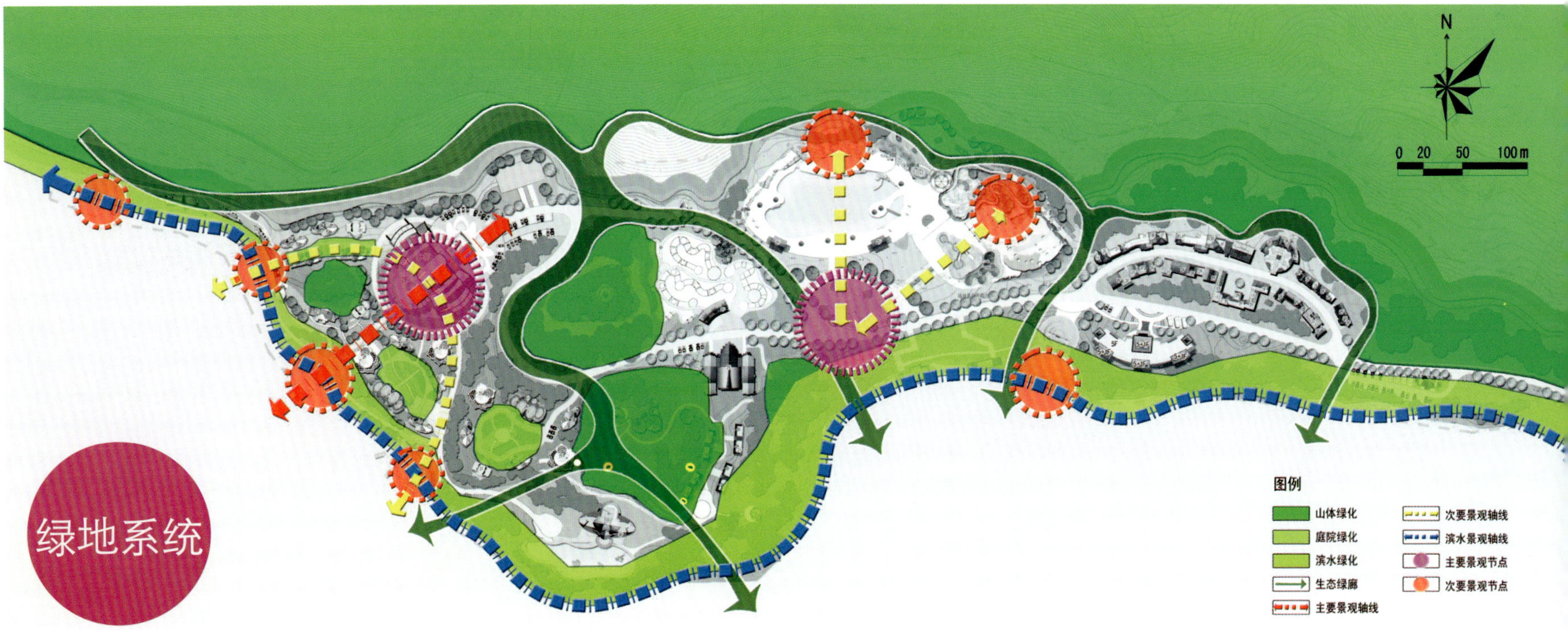

第四节 景观系统及植物配置规划　Section Ⅳ Green Land System and Plants

一、景观系统 I Green Land System

主要景观轴：以六星级酒店——滨河广场轴线作为整个景区的主景观轴线。

Main landscape Line: Six-star Hotel—Lakeshore Square.

次要景观轴：以六星级酒店——垂钓乐园、六星级酒店——高级码头、儿童游乐场——苗家篝火——电玩城、冲浪区——影视电玩城四条景观轴为次要景观轴线。

Secondary Landscape Line: Six-star Hotel—Fishing Paradise, Six-star Hotel—Great Harbor, Children's Playground—Bonfire of Miao—Electronic Game Center, Surfing Area—Electronic Game Center, totally four lines.

滨水景观轴：以快艇码头——垂钓乐园——滨河广场——高级码头——康体会所——大众码头所形成的空间序列为滨水景观轴。

Lakeshore Landscape Line: Swift Boat Wharf—Fishing Paradise—Lakeshore Square—Great Wharf—Healthy Club. beauty of environment, let the architecture hidden into deep as well as lush forests.

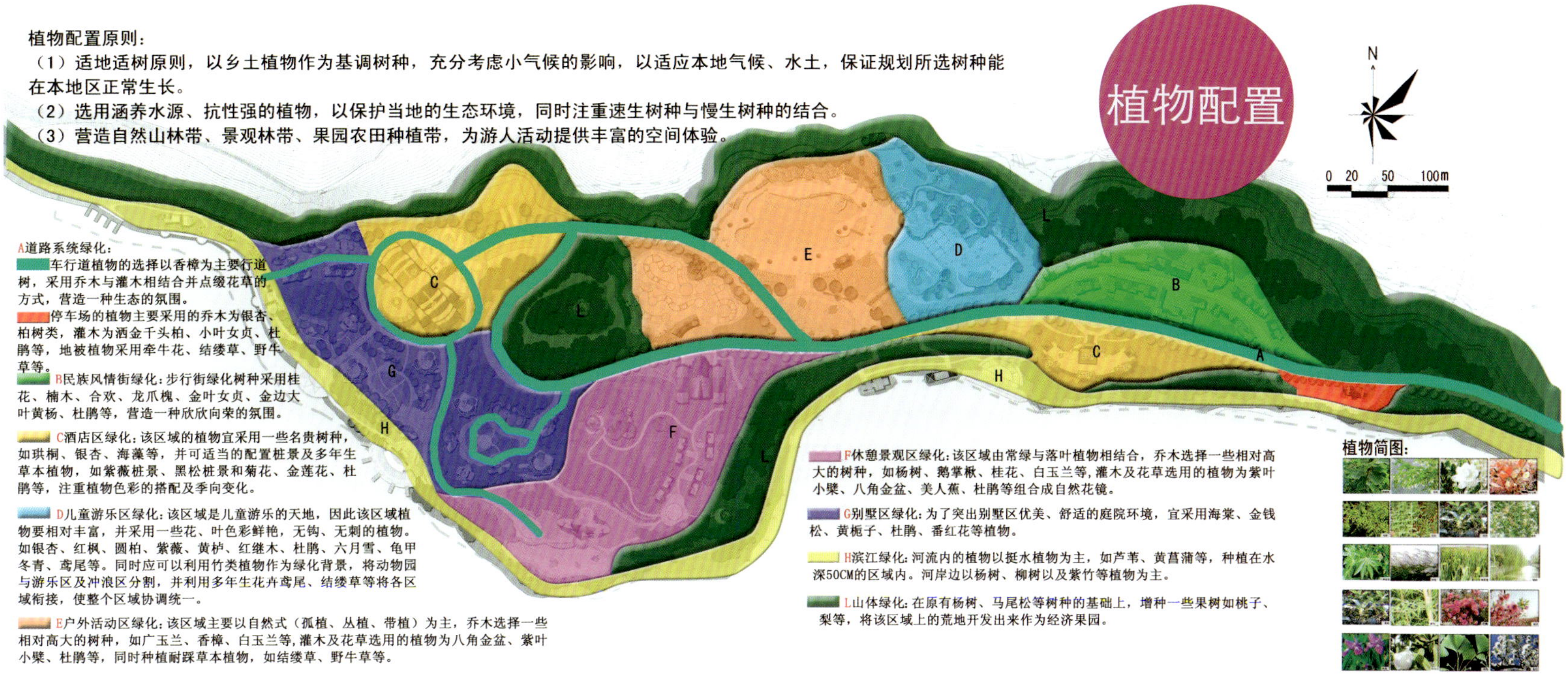

景观节点：分别规划有2处主要景观节点和7处次要景观节点。主要景观节点为杜鹃花酒店及篝火广场，次要景观节点为以啤酒屋、钓鱼台、中心广场、快艇码头、大众码头、儿童游乐场、冲浪区所形成的点状景观。

Landscape Nodes: There are two main landscape nodes and seven secondary nodes in the scheme. The main landscape nodes: Azalea Hotel and Bonfire Square; Secondary landscape nodes: Beer House, Fishing Platform, Center Square, Swift Boat Wharf, Public Wharf, Children's Play Ground, Surfing Area.

二、植物配置 II Plants

绿化是形成优美环境的重要条件，在保护现有绿地景观的前提下，注重植物高矮顺序与游人视线的关系，以及植物色彩与季节的关系和颜色相配的协调性，运用层次丰富的树种，营造良好的景观环境。

Afforesting is an important fundament of beautiful landscape environment. So we should preserve the current green land, pay attention to relation between height of plants and visitors' eyesight, and adjust the accordance of plants color and seasons to create excellent landscape.

景点布置

9 旅游市场分析及营销方案 Tourism Market Analysis and Marketing

第一节 化屋旅游市场分析

一、化屋旅游业发展现状分析

化屋旅游区是黔西县的一块净土，地处县域南部，从2006年起被列为黔西县“十一五”规划重点打造的旅游区之一。近年来，它良好的生态环境、独特的自然景观和人文资源吸引着周边县市甚至省外的游客流连忘返，游客数量逐年增长，知名度不断提高。为了满足当地不断增加的游客的需求，相关管理部门兴建了化屋度假酒店，提供旅游接待、餐饮、住宿等服务，营业额不断增长，旅游业发展前景可观。

Section Ⅰ Analysis of Tourism Market of Huawu

Ⅰ Current State of Tourism Development of Huawu

Dongfeng Lake Resort, located in the south of the Qianxi County, is a purified place of this county. It has been appointed to be one of the tourism region developed in priority of Qianxi County from 2006. In the recent year, with the excellent ecological environment, unique natural landscape and cultural resources, it has attracted constant streams of visitors from surrounding counties and prefectures even the other provinces. In order to meet the demands of more and more tourists, the relevant administration department has constructed Huawu Holiday Hotel for visitors' board and lodging. Tourism development in Dongfeng Lake Resort is prosperous.

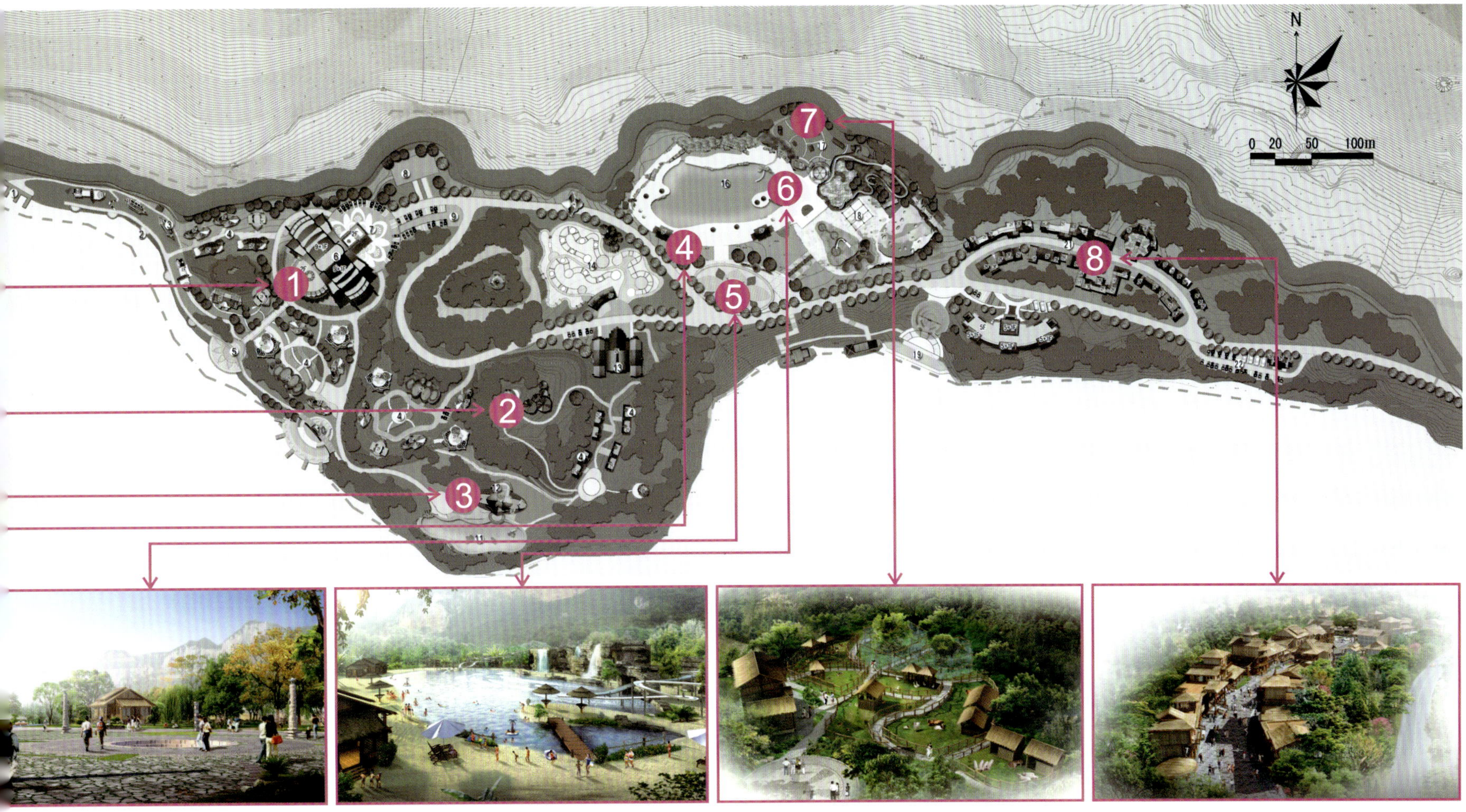

但目前化屋旅游项目开发深度不够，东风湖的旅游活动仅限于水上观光游览，水上及临水游乐活动尚未开发，旅游项目较为单调，而且游船码头不在黔西县境内，而是位于贵阳清镇市境内，这大大制约了旅游区的发展。因此，总体上说，东风湖旅游业发展还处于起步阶段。

二、化屋旅游市场分析

尽管化屋旅游区还在初期发展阶段，但是在度假旅游成为生活时尚、人们越来越追求高层次的旅游体验的今天，在化屋进行高端旅游深度开发的潜力十分巨大，它不仅对贵州省内的高端客源有巨大的吸引力，在贵广高速铁路、成贵铁路和厦蓉高速公路开通以后，还将吸引广东、福建、湖南等大量省外游客甚至国外客源。

However, Dongfeng Lake Resort has not been developed as it should be so far that only sightseeing in the lake without other entertainment activities for visitors. Also, there's wharf in the region of Qianxi County, but in Qingzhen, Guiyang, which, along with the monotonous tourism program, has greatly blocked tourism development of Dongfeng Lake Resort. In all, Dongfeng Lake Resort has developed primarily in tourism.

II Analysis of Tourism Market of Huawu

Although Dongfeng Lake Resort has developed primarily in tourism, it has great potential to develop high-end tourism at the time when holiday travel has become fashion and people have pursued high-level tour day by day. It attracts not only high-end visitors in Guizhou, but also the ones from Guangdong, Fujian and Hunan and even foreign countries after Guiguang Railway, Chenggui Railway and Xiarong Railway completed.

1. 一级市场

化屋旅游区的一级目标市场是以贵阳市、毕节市、黔西县、织金县为重点的周边市场。其中，贵阳作为全省经济、政治、文化的中心，是全省的旅游集散地。人均 GDP 居全省之首，大大高出其他地州市。清镇到黔西的高速公路建成通车后，贵阳到黔西的交通条件将得到很大改善，贵阳对黔西化屋的旅游集散作用也将随之增强。由于地缘优势，毕节市、黔西县、织金县的高收入人群也是化屋东风湖度假区的重要客源市场。

2. 二级市场

化屋旅游区的二级目标市场是以成贵铁路、贵广高速铁路和厦蓉高速公路为交通轴线的周边各省市客源市场，包括重庆、成都、粤、港、澳、广西、湖南、福建等地区。贵广高速铁路将于 2010 年建成通车，这条铁路连接省会贵阳与中国经济发展的桥头堡——珠三角地区，届时从广州乘火车到贵阳只需 6h，到黔西化屋只需 7h。

3. 机会市场

化屋旅游区的机会市场主要来源于对贵州省国内国际旅游市场的分流，以及长三角、京津唐等远程市场。长三角和京津唐是中国三大客源输出地之一，其居民可自由支配收入在国内首屈一指。只要做好营销宣传，将东风湖的高端度假旅游品牌打出一定的知名度，使其具有全国影响力，定能吸引这一市场的高端客源的注意力。

第二节 营销总体战略与目标

一、营销总体战略

化屋旅游区应围绕“集观光、会议、康体健身、民族风情于一体的中高端休闲度假集散地”的旅游价值定位来制定其营销总体战略和目标、营销方案和手段，开发包装并逐渐完善东风湖旅游区的主题品牌及产品系列，优化产品或服务的分销渠道，拓展市场空间。

根据上述价值定位及营销思路，化屋旅游区营销整体战略定位为以贵广高速铁路和厦蓉高速公路为交通轴线的周边各省市的一级客源市场为重点，以贵阳市和毕节市为核心的周边市场为次重点，以贵州省国内国际分流旅游市场以及长三角、京津唐等远程市场为补充，坚持不同市场不同策略的营销战略思路，充分利用广告直销、推介会分销、网络营销等手段，努力打造化屋景区“花舞蝶飞，梦境天堂”的旅游形象，提高景区的知名度和美誉度，塑造化屋高端休闲度假的旅游品牌，从而达到扩大景区游客规模、增加旅游收入的营销目标。

二、营销总体目标

1. 提高景区的知名度和美誉度，塑造化屋高端休闲度假的旅游品牌。

2. 积极扩大重点高端客源市场，发展二级市场，争取机会市场，培育以化屋为定期休闲度假目的地，以中高收入阶层为主要目标的高端旅游客源市场。

1. Primary Market

The primary market of Dongfeng Lake Resort is the surrounding markets of Guiyang City, Bijie City, Qianxi City and Zhijin City. Among them, Guiyang, as the center economy, politics, and culture, is the distribution of tour of the province. The traffic conditions will be great improved from Guiyang to Qianxi after the highway from Qingzhen to Qianxi completed, which will strengthen distribution market of tourists in Guiyang visiting Huawu, Qianxi. Also, people with high income of Bijie, Qianxi and Zhijin are the important customers of Dongfeng Lake Resort.

2. Secondary Market

Dongfeng Lake Resort has the secondary target market in every surrounding cities and provinces along Chenggui Railway, Guiguang Railway and Xiarong Highway, including Chongqing City, Chengdu City, Guizhou Province, Hongkong, Macao, Guangxi Province, Hunan Province and Fujian Province, and so on. Guiguang Railway, linking Guiyang City, the capital of Guizhou Province and the Pearl River Delta, the bridgehead of economic development of China, will be completed in 2010, when visitors can spend only six hours from Guangzhou to Guiyang and seven hours to Huawu, Qianxi by train.

3. Opportunity Market

The opportunity market resources of Dongfeng Lake Resort are mainly the division of domestic and foreign tourists of Guizhou Province, and the distant market of Yangtze River Delta and Beijing, Tianjin and Tangshan which is one of the main markets in China, and of which people's incomes is the NO. 1. As long as we publicize the product well to make Dongfeng Lake Resort be famous in China, we must draw attention of the high-end customers to this tourism product.

Section Ⅱ Overall Marketing Strategy and Objection

Ⅰ Overall Marketing Strategy

Dongfeng Lake Resort should be positioned as "the medium and high-end holiday travelling distribution with sightseeing, meeting, healthy activities and ethnic flavor", which is the fundament to make overall strategy and objection, and marketing plan and methods, and also to develop and improve gradually the thematic brand of Dongfeng Lake Resort so as to expand tourists market.

According to the above value orientation and marketing concept, the overall marketing strategy of Dongfeng Lake Resort of Huawu is determined as follow:

Ⅱ Objection

1. To promote reputation, and create brand of Dongfeng Lake Resort.

2. To expand high-end market, develop secondary market, and strive to win opportunity market so as to develop Huawu to be the destination of resort and the high and medium classes as the main target market.

第三节 市场营销方案

一、一级市场营销方案

化屋旅游区的一级目标市场是以贵阳市、毕节市、黔西县、织金县为重点的周边市场。针对这一市场的特点，可以采取以下营销手段来进行旅游推广：

选择类别	目标市场及媒介	行动策略
推介会	以贵阳、毕节及周边中心城镇为重点	每年在旅游旺季到来之前，以“花舞蝶飞，梦境化屋”为主题，分别在贵阳市和毕节市主要城市广场策划 1~2 次的旅游推介活动
户外广告	选择在贵阳到黔西省道的出入口等处	布设化屋景区的形象宣传画，以及景区形象口号，力求强烈的视觉冲击力
报纸	选择贵阳、毕节的主要平面媒体进行广告投放，如《贵阳晚报》、《毕节日报》等	主要宣传化屋的高端休闲旅游，或者定期发表化屋旅游专题文章，时间尽量选择在周末时间投放
宣传册	机场、酒店、旅行社	①在龙洞堡机场候机厅，向乘客免费发放化屋旅游宣传册和旅游形象宣传光盘 ②登机牌上印制化屋旅游形象广告，提升其旅游形象
电视	选择省内、毕节地区的主要电视栏目，贵州电视台的百姓关注节目、贵州大众生活频道等	尽量选择在重点客源市场有针对性地投放
旅行社	在贵阳、毕节的各大旅行社进行促销	主要采取人员推广和销售激励措施，建立起完善通畅的旅行社销售网络，促使旅行社开辟新线路
单位推广	贵阳市和毕节市各知名企业、大型国家企业和政府部门	组织专业营销队伍，深入到这些单位，针对离退休职工进行宣传，邀请其到化屋进行商务会议旅游，力争以折扣及增值服务来打动贵阳的潜在旅游客源市场

二、二级市场营销方案

化屋旅游区的二级目标市场是以贵广高速铁路和厦蓉高速公路为交通轴线的周边各省市客源市场，包括重庆、成都、粤、港、澳、广西、湖南、福建等地区。针对这一市场的特点，可以采取以下营销手段来进行旅游推广：

选择类别	目标市场及媒介	行动策略
推介会	主要参加泛珠三角地区的旅游推介会，省内重点城市为贵阳、都匀、荔波、凯里；省外重点城市为成都、重庆、广州、深圳、珠海、桂林、南宁、厦门、成都等地	推广自身旅游产品，提高景区知名度，借机组织、丰富、拓展自身在客源市场的销售网络
户外广告	选择在成贵高速公路、贵广高速铁路和厦蓉高速公路的出入口、交接口，深圳罗湖口岸	布设化屋景区的形象宣传画，以及景区形象口号，力求强烈的视觉冲击力，对粤港澳游客留下化屋的第一印象提升化屋知名度
报纸	选择客源地主要平面媒体进行广告投放，如《中国旅游报》、《南方周末》、《深圳晚报》等	主要宣传化屋的高端休闲旅游，或者定期发表化屋旅游专题文章，时间尽量选择在周末时间投放
电视	选择重要客源地（重庆、成都、广州、深圳等）电视旅游栏目，如重庆卫视、成都卫视、广州卫视、深圳卫视等	电视具有时间短、资金费用大等特点，选择应慎重，尽量选择在重点客源市场有针对性地投放
旅行社	在泛珠三角地区针对重点城市旅行社进行促销，如广州、深圳、珠海、桂林、南宁、厦门、成都等地	主要采取人员推广和销售激励措施，建立起完善通畅的旅行社销售网络，促使旅行社开辟新线路
差异化营销	珠三角市场在 7、8 月份出游比例高，青少年市场活跃，暑期前对泛珠三角各目标市场进行重点营销	观光、民族文化、休闲度假等不同需求进行分层促销

Section Ⅲ Marketing Plan

Ⅰ Primary Marketing Plan

The primary market of Dongfeng Lake Resort is Guiyang City, Qianxi City, Zhijin City and Bijie City, and the surrounding markets of these cities. Based on this feature of market, we adopt the following marketing strategies:

Category	Target Market and Media	Strategy
Promotion Shows	Stressing central cities and towns around Guiyang and Bijie	To hold one or two tourism promoting activities in Guiyang and Bijie with the theme of "flying flowers and butterflies, a dreamlike Huawu" before the busy seasons
Outdoor Advertisements	To set publicity posters and slogans of Huawu Scenic Spot with strong vision impact.	To set publicity posters and slogans of Huawu Scenic Spot with strong vision impact
Newspapers	To advertise through major plane major in Guiyang and Bijie, such as "Guiyang Night", "Bijie Daily"	To publicize high-end leisure travel of Huawu or present thematic articles about tourism of Huawu in the weekends
brochures	Airport, hotel, and travel agency	① To hand out tourism brochures and disks to passengers in waiting rooms of Longdongbao Airport ② To print image advertisement of Huawu scenic area in order to enhance tourism image of Huawu
Television	To choose main TV programs of Guizhou province and Bijie Prefecture such as the Concernment of Ordinary People of Guizhou TV, and the channel of public life of Guizhou	To advertise towards key tourism market
Travel Agencies	To promote in every great travel agency of Guiyang and Bijie	To adopt mainly methods of personnel popularizations and motivating sales and set up perfect sale nets, so as to accelerate travel agencies to open new lines
Units Promoting	Every famous enterprise, grand national corporations and government departments of Guiyang City and Bijie City	To organize professional sale group penetrating these unites to make propaganda especially to the retired people and invite them to Huawu and have tour of business meeting, striving to crack the potential market of Guiyang by discount and value-added services

Ⅱ Secondary Marketing Plan

Dongfeng Lake Resort has the secondary target market in every surrounding cities and provinces along Chenggui Railway, Guiguang Railway and Xiarong Highway, including Chongqing City, Chengdu City, Guizhou Province, Hongkong, Macao, Guangxi Province, Hunan Province and Fujian Province, and so on. Based on this feature, we can adopt the following marketing strategies:

Category	Target Market and Media	Strategy
Promotion Fair	To take part in tourism promotion fairs of Peal River Delta Area, especially the key cities such as Guiyang, Duyun, Libo, Kaili of Guizhou Province, and Chengdu, Chongqing, Guangzhou, Shenzhen, Zhuhai, Guilin, Nanning, Xiamen and so on	To publicize tourism products and enhance reputation of the spots so as to organize, enrich and expand the market
Outdoor Advertisement	To choose the entrances and exits of the highways of Chenggui, Guiguang and Xiarong, and Luohu Port of Shenzhen	To set image posters and slogans of Huawu beauty spots to realize strong vision impact and leave good impression to tourists of Hong Kong and Macao
Newspapers	To insert advertisements in the main plane media in the target markets such as "China Traveling", "South Weekly", and "Shenzhen Night", and so on	To publicize the high-end leisure traveling or publish articles about tourism in Huawu at stated times in weekends
TV	To choose the TV programs of tourism in the target markets such as Qingcheng TV, Chengdu TV, Guangzhou TV, and Shenzhen TV and so on	TV has the features of short time and much cost that we should choose it cautiously and in the key cities as possible as we can
Travel Agency	To make promotion in the key cities such as Guangzhou, Shenzhen, Zhuhai, Guilin, Nanning, Xiamen and Chengdu and so on	To take the solutions of popularization and sales motivation and establish perfect sale webs in travel agencies
Different sale	Traveling rate in the Pearl River Delta Area is relevantly high in July and August as the young's activities are frequently	Sightseeing, national culture, leisure vacation

三、机会市场营销方案

化屋旅游区的机会市场主要来源于对贵州省国内国际旅游市场的分流，以及长三角、京津唐等远程市场。针对这一市场的特点，可以采取以下营销手段来进行旅游推广：

1. 对贵州省国内国际旅游市场的分流方案

以贵州为旅游目的地的游客，一般都会以贵阳为首选目的地，因此贵阳是省内外游客的重要聚集地。随着“两高”的开通，黔西县将被纳入贵阳“一小时经济圈”，直接接受贵阳的客源辐射，化屋景区自然也在这一客源辐射区的范围内。由于这一客源市场接受营销宣传的方式与二级市场类似，不同的仅是以化屋为目的地的游客数量较二级市场少，所以，对这一客源市场的营销策略可以参照二级市场营销方案。

2. 长三角、京津唐等远程旅游市场营销方案

长三角、京津唐等远程旅游市场，特别是北京市场，具有很强的领头作用，居民整体文化素质较高，出游频率较多。主要的细分市场有企业员工、家庭市场、中老年人市场、大学生市场、公务员带薪度假市场等。目前，对这一市场的营销还处于品牌导入阶段，具体营销行动如下：

营销媒介	营销方式及内容
地铁 公交 酒店	在地铁、公交车身等处投放化屋景区的宣传海报，在星级酒店宾馆投放化屋自助旅游手册，详细介绍化屋高端休闲旅游活动
旅行社	加强与北京的旅行社合作，将化屋纳入贵州旅游精品线路中，提高化屋景区在贵州旅游资源中的地位
网站	与携程旅游网形成战略伙伴关系，直接开辟网络媒体宣传渠道
重点高校	在北大、清华、外国语大学、民族大学分别做景区推介，“先推销景区，再推销产品”，邀请专家学者和在校学生到化屋考察，提升其知名度，吸引高收入群体关注化屋旅游区
俱乐部	邀请知名的户外运动俱乐部会员到化屋考察，设计旅游产品

Ⅲ Opportunity Marketing Plan

The opportunity market resources of Dongfeng Lake Resort are mainly the division of domestic and foreign tourists of Guizhou Province, and the distant market of Yangtze River Delta and Beijing, Tianjin and Tangshan. Based on this feature, we can adopt the following marketing strategies:

1. Plan of marketing division of international and national tourism resources in Guizhou Province

Generally, visitors planning to tour in Guizhou will come to Guiyang first, so Guiyang is an important hinge of visitors from different places. With the two highways completed, Qianxi will be in the "one-hour economic cycle" of Guiyang to have the visitors of Guiyang, so will Huawu. As this market has the similar ways to make marketing promotion, the market strategy could take the one of the secondary market as example.

2. Marketing plan of distant tourism market

The target markets of Yangtze River Delta, Beijing, Tianjin and Tangshan, especially Bejing, is in the leading position as residents in these cities are of high quality and tourism rate. The main segmental markets are employees, families, colleger, and paid leaves of public servants. At present, this marketing is still in the stage of introducing brand. Following is the specification of marketing,

Promote Media	Promote means and contents
Subways Buss Hotels	To put posters of Huawu scenic spot on subways and buses, and place travel brochures elaborate high-end leisure and travel in Huawu in star hotels
Travel Agencies	To strengthen cooperation with travel agencies of Beijing so as to list Huawu into exquisite travel of Guizhou and enhance the position of Huawu scenic spot in tourism resources of Guizhou
Websites	To form strategic partnership with Xiecheng Travel Web and open directly propagation means of internet media
Key universities	To publicize the scenic spot in Peking University, Tsinghua University, University of Foreign Languages and National University with the principle of "promoting scenic areas first and then products", and invite experts and learners and students to visit Huawu so as to promote its fame and absorb high-income groups' attention on Huawu Scenic spot
Clubs	To invite members of well-know clubs of outdoor sports to visit Huawu and design tourism products for Huawu

总规划用地面积		247161.82m²
建筑总占地面积		14420.50m²
其中	民俗风情街	3204.80m²
	风情酒店	1720.84m²
	六星度假酒店	3257.88m²
	度假别墅	2892.48m²
	影视电玩城	1278.00m²
	康体会所	983.49m²
	啤酒屋	261.27m²
	管理用房	821.74m²
总建筑面积		45684.64m²
其中	民俗风情街	6782.31m²
	风情酒店	8762.41m²
	六星度假酒店	15987.27m²
	度假别墅	7307.60m²
	影视电玩城	2556.00m²
	康体会所	2950.47m²
	啤酒屋	261.27m²
	管理用房	1077.31m²
建筑密度		5.83%
容积率		0.18
绿地率		42.5%
停车位		108 个
其中	酒店停车位	58 个
	社会停车位	50 个

Total land use area in the plan		247161.82m²
Total occupied area of buildings		14420.50m²
including	Folk Street	3204.80m²
	Ethnic Hotel	1720.84m²
	Six-star Hotel	3257.88m²
	Resort villa	2892.48m²
	Electronic game center	1278.00m²
	Fitness club	983.49m²
	Beer house	261.27m²
	Management house	821.74m²
Total construction area		45684.64m²
including	Folk streets	6782.31m²
	Folk hotels	8762.41m²
	Six-star resort hotels	15987.27m²
	Resort villas	7307.60m²
	Electronic game center	2556.00m²
	Fitness club	2950.47m²
	Beer houses	261.27m²
	Manage houses	1077.31m²
Building density		5.83%
Net volume		0.18
Greening rate		42.5%
Parking lots		108
including	Parking lots for hotels	58
	Public parking lots	50

HuawuResort

效果图件 Effective Pictures

化屋度假区鸟瞰

现场鸟瞰

①

民族风情街

该片区为展示苗族、彝族特色民族文化并提供休闲购物、欣赏、体验的平台，通过民族技艺、民族手工艺品销售等民俗民风商业手段，展现浓郁的黔西民族特色，满足旅客的文化消费需求。

Folk Street

This area mainly demonstrated special ethnic culture of Miao which visitors can shop leisurely, enjoy and experience. With the ethnic business means of selling ethnic skills and crafts, the thick ethnic flavor of Qianxi can be showed to meet the tourists' needs of cultural consumption.

建筑设计以苗族、彝族文化为基础，它结合了化屋当地“歪梳苗”和黔西彝族古朴别致的原生态民居建筑风格，提炼其代表性元素和艺术语言，通过现代设计理念和艺术手法赋予其新的生命和时代气息，形成特色民族民居建筑群，弘扬民族地域文化。此一线基础建筑体以苗族干栏木楼为主，外凸出装饰以苗族“美人靠”为形，翘角、飞檐都取彝族居舍之形状，屋脊中部及两端有简单的起翘及起拱，门窗扇的装饰图案及小花格窗等极富建筑装饰效果。顶盖茅草及柱捆草绳等原始古朴的原生态元素让自然生态、民族风情和文化融合为一体，提升街区品质。

Combining with the local simple and fine architectural style of Yi and "Waishu" Miao of Huawu, buildings are designed based on Miao and Yi culture, constructed with local representative architectural elements and artistic language, and were endowed with new life and spirit of time with modern design concept and artistic technic, so as to form ethnic architectural complexes to carry forward ethnic regional culture. Most of the fundamental architectural complexes of this line are wooden houses on stilts with decoration of Yi style like "beauty settees", rake angles and upturned eaves etc. and simple arch in the middle ridge of the roof, decorated patterns on the doors and widows and small square windows, and so on.

①步行街鸟瞰 ②步行街局部效果 ③步行街局部效果

风情酒店

极具民族特色和古典韵味的设计，让游客有一种恍若隔世的感觉；民族与时尚的完美交融，宣告了古老化屋与现代化屋同在。

将歪梳苗的民族元素——分解，再融合进这座建筑当中，合理穿插现代元素，力求凸显民族格调和地域风情，特别讲究简约、实用、美观。屋顶是苗族牛角的造型，瓦片呈茅草的颜色，别致、古朴；木质门窗框的大量使用，更把原生态的风格体现得淋漓尽致，与民族风情街的文化建筑相互呼应。

Folk Hotel

With great ethnic flavor and ancient charming, visitors can feel living in the ancient time; the perfect integrity of ethnic and fashion has claimed the coexistence of ancient and modern Huawu.

We aim to highlight the hotel in ethnic style and regional flavor, especially simplicity, practicability and beauty by decomposing the ethnic elements of Waishu Miao and integrating them into this building and larded with modern elements reasonably. The roof is the horn shape of Miao and covered with tiles in the color of thatch, fine and simple; two huge reed-pipes in the front facing each other like dancing and playing harmony and prosperity; wooden doors and window frames are used commonly to embody thoroughly the original style and reflect with the cultural architectures along the folk street.

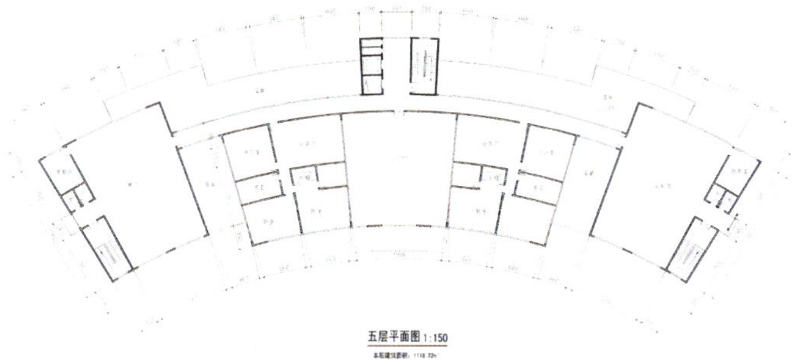

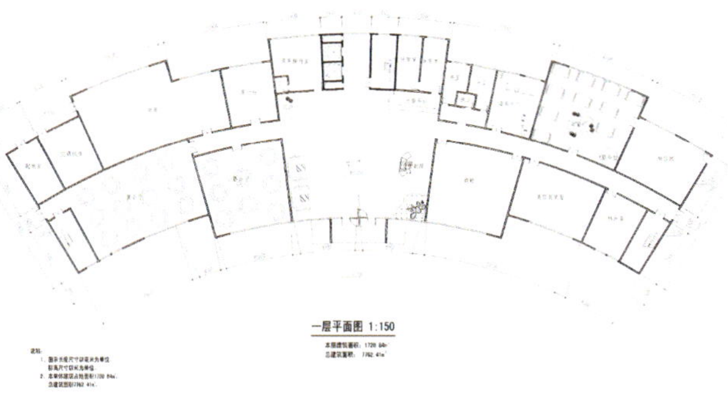

①～③风情酒店效果图

儿童游乐园

儿童游乐园入口大门设计主要采用苗族农耕文化元素为主，两侧门柱为苗族乐器芦笙形状并辅以苗族木鼓及舞蹈动作造型，展现了化屋歌舞之乡的风采。门楣上刻有日、月、鸟、兽等一些简单古朴的图案且悬挂木环，木环下串联着如链的铃铛，极具趣味性，微风吹过，悦耳动听的铃声引人入胜。

Children's Playground

Children's playground is with a gate at the entrance designed on the basis of farming culture. On the two pillars, standing two sides of the gate, are in the shape of the reed-pipe, a kind of Miao instrument, and wooden drum and some dancing gestures to reveal the charming of the hometown of dances and songs of Hua Wu; on the lintel of the gate, some simple patterns of the sun, moon, birds and animals are caved, and wooden circles are hung linking with bells in chainlike shapes, whistling in the wind.

游乐园集乡土气息、智力开发、童真童趣为一体，是针对小朋友回归自然、亲近乡土的主题乐园。在游乐园中为小朋友准备了弯月飞艇、旋转木马、飞天弹跳、疯狂飞碟、卡通城堡和摩天轮等丰富多彩的游乐设施，小朋友们游乐其中，找到了久别的快乐，在快乐中展现无瑕的童心，其乐无穷，同时也增加了知识、开阔了视野。

与游乐园相邻的迷你动物园里喂养着许多可爱的小动物，小朋友在观赏小动物的同时，能够增加对动物的了解，学习到关于动物的一些基本知识；还可以通过喂养小动物，培养了他们的爱心，为促进健康成长提供了很好的条件。

Children's playground, with flavor of rural life and childhood interesting, is a theme park for children to feel nature and rural environment. A plenty of amusement equipments such as airships, carousels, flying saucers, cartoon castles and so on are prepared for children.

Near the Children's Playground, it is the mini zoo with many cute small animals. Kids can watch them and learn some knowledge about them, and also feed them, which can cultivate their kindness to accelerate their health growth in both mind and physique.

①儿童游乐园、冲浪池、篝火广场鸟瞰 ②儿童游乐园局部鸟瞰 ③儿童游乐园跷跷板 ④儿童游乐园局部鸟瞰

①

冲浪池

冲浪一直是一项属于勇敢者的运动，胆量不够大的人只有望洋兴叹的份儿。不过，在这个陆上冲浪池，即使胆小的人也可以尝试浪头弄潮的刺激。冲浪池一端竖起幅度不同、起伏的石壁，多个水泵向这面弯弯曲曲的墙面上喷射水柱，强劲的水柱在投射到墙面后又反向溅射，从而形成了浪锋参差、几可乱真的人造浪。在冲浪池的人造浪峰上冲浪既不会像在海上冲浪那样有丧生的危险，又不像在封闭无景的室内冲浪池一样索然无味，而是既能够享受自然景色，又有着活力无限、激情四射的体验。

Surfing Pool

Surfing has been always a sport for the courage; no one can try without enough boldness. However, in this surfing pool even the cowards can try the thrilling of surfing. A rock wall with rough surface is standing on one side of the pool, water from pumps spurting against the wall and splashed by the rough surface to form artificial waves. Surfing here won't be in danger, but can enjoy natural scenery and thrilling experience.

①冲浪区鸟瞰 ②冲浪区局部效果图 ③冲浪区局部效果图

①

②

迷你动物园

与游乐园相邻的迷你动物园里喂养着许多可爱的小动物，小朋友在观赏小动物的同时，能够增加对动物的了解，学习到关于动物的一些基本知识；还可以通过喂养小动物，培养了他们的爱心，为促进健康成长提供了很好的条件。

Mini Zoo

Increase more effect on green, beauty of environment, let the architecture hidden into deep as well as lush forests mountains, avoid to the appearance of expanse of architecture in the view. beauty of environment, let the architecture hidden into deep as well as lush forests.

③

④

⑤

篝火广场

在冲浪池南边有一个篝火广场，整个广场的造型设计遵循生态学和民族美学理论，充分展示了人与自然地融合，突出彝族苗族文化内涵。广场地面的整体形状为一朵盛开的杜鹃花，地面铺装用青色鹅卵石镶嵌而成，周边结合地形以绿篱围合，适当用花池、盆景绿化点缀。游客在此可以欣赏婀娜多姿的彝苗舞蹈，倾听高昂的彝苗歌声，这些民俗演艺项目填补了度假区的民俗游览和晚间室外娱乐活动的不足。游客还可以进行各种竞技活动比赛和篝火晚会等活动，通过互动参与，在交融民族情感、凝聚民族精神的同时，展示传承民族文化，是民族文化事业的一处独特风景。

Bonfire Square

In the south of the surfing pool, there is a bonfire square designed with the theory of ecology and national esthetics to fully elaborate the harmony of human and nurture and highlight cultural connotation of Yi and Miao ethnic groups. The square is in the shape of a blooming azalea, paved with cobblestones, surrounded with green hedge and embellished by potted landscapes and flower beds. Visitors can enjoy graceful dance and joyous songs of Miao and Yi ethnic groups. In addition, visitors can participate the competitions and bonfire parties.

①迷你动物园鸟瞰 ②迷你动物园管理房效果图 ③～⑤篝火广场效果图

电玩城 Electronic Game Center

针对以寻求刺激、发泄为目的的目标客人而设计的娱乐型电玩城，其建筑外观设计就似一朵怒放的杜鹃花，圆形或弧形的流线体现随意性，其空间设计和布置也给人们随意感。

The Electronic Game Center, likes a blooming azalea in the appearance, is designed for the customers seeking for thrilling or pouring out emotion. Its smooth lines of cycle or semi-cycle embody casualness, as well as the spatial design and layout.

在创意设计中注重消费者与空间环境之间的感受体验、并运用电玩设备结合杜鹃花瓣主题环境包装的概念，使整个游电玩城具有强烈的游乐氛围。以健康向上的游乐项目引导大众娱乐、面向大众消费，满足顾客健康的需求，是释放激情、增加快乐、缓解压力的快乐发源地。电玩城同时还兼具了迷你剧院的功能。

The whole center will be of strong amusement atmosphere with the design emphasizing on consumers' feelings of space environment the concept of packing equipments with the image of azalea petals. It is a birthplace of merriness for customers to release enthusiasm, get fun and relieve stress with healthy entertainments.

①

②

③

④

⑤

⑥

户外拓展区

户外拓展区依山而建，拥有优美、怡静的自然景色，人于此更有回归自然的感觉。该区具有完善的配套设施，包括高空区、地面区和水上活动区域。人们在远离都市喧闹的自然环境里，以自然为舞台，以活动为道具，追寻和找回人与自然共为一体的天性，征服和超越自我。通过体验攀岩等拓展训练项目，并结合工作和生活进行交流、总结，增加了彼此之间的沟通和信任，挖掘了各自的潜力，促进抗挫折能力的养成，营造了良好的团队氛围和创新意识，对以后的工作和生活大有裨益。

Expanding Area

This area is built depending on mountains with peaceful and beautiful natural scenery and completed matched facilities including high-altitude area, ground area and water activities area which give people feelings of returning to nature. With nature as platform and activities as properties, people can pursue and find the concordance of human and nature to conquer and surpass themselves in this natural environment far away from noisy cities. Participators can communicate with each other and conclude combining with work and life, which is good to tap their potentials and foster their abilities resisting frustration and create good group atmosphere and creative connotation.

①～②电玩城效果图 ③拓展中心管理房 ④～⑥拓展中心局部效果

化屋度假区鸟瞰

夜景效果

①～④沿江效果图

①

商务会所 Business Club

休闲会所采用原木本色与白色作为主体基调，外观以帆船和杜鹃花瓣为造型，采用了简洁而统一的设计手法，体现了设计的自然气息和人文精神，原木的运用以及民族元素的现代处理，焕发出建筑新的生命力，表达了对民族传统文化的尊敬和理解。利用建筑空间的形式营造和谐、安逸、舒适、宁静的休闲环境，是具桑拿、饮食、娱乐、保健、休闲于一身的综合性的、全方位一站式休闲娱乐服务场所，为众多追求健康、钟情美丽的高雅人士搭建一个自然、放松、温馨、宁静的港湾。

Designed simply and uniformly, the club is mainly in the original color of wood and white and in the shape of sailing boat and petals of azalea to embody natural air and cultural spirit and express our respect and understanding to ethnic traditional culture by using woods and modernizing ethnic elements. It is a comprehensive and all-dimensional leisure and entertainment center with harmonious, comfortable, peaceful and relaxing environment created by using the form of architectural space, the service programs including sauna, diet, entertainment, health care and leisure, which is a natural, fresh, worm and tranquil bay for the elegant who pursue health and beauty.

①商务会所效果图 ②会所户外泳池效果图 ③会所户外泳池效果图

六星级酒店 Six-star Hotel

规划设计以杜鹃花为灵魂，与民族特色元素相辅相成。酒店外观是一朵含苞待放的杜鹃花，中间为花蕊，左右两侧是两片花瓣。酒店大门、窗框、台阶等都融合了歪梳苗的图腾、刺绣元素，以点带面地刻画歪梳苗的文化特征。五彩斑斓的彩色玻璃拼嵌成色彩艳丽的苗族“条带裙”，流光溢彩的景象形成了两个不同时代的对比，新世纪的新发展和民族文化的继承在此呼应彰显。穿行于世俗尘世里的游人，拖着疲惫的心绪来到大山深处寻找栖息之所，杜鹃花造型的酒店，一定会让他有一个如花蕾般绽放的假期。

The hotel is designed with azalea as soul and ethnic elements as counterparts. Specifically, the appearance of the hotel is like an azalea in bug with "stamens" in center and two "petals" on right and left sides. The elements of totem and embroidery of Waishu Miao are added in the design of gates, window frames and steps to unfold the cultural features of Waishu Miao ethnic group. Colorful grasses are pieced together to form a beautiful "strip dress" of Miao, the bright scene discloses the comparison of two different times of the new development of new century and the inheritance of ethnic culture. Visitors, passing through the world with a tired mind and physique must have a holiday like a blooming azalea living in such hotel.

①六星级酒店效果图 ②贵宾区入口效果图 ③六星级酒店夜景

①

②

度假别墅

在酒店的周围，星罗棋布地分布着以杜鹃花形或杜鹃花叶为顶，形态各异的别墅群落，在这样一种建筑环境里，温馨雅致的感觉填满心头。人们的身心得以从喧嚣的都市中解脱出来，享受生活所带来的欢愉，无论从感观还是精神，都令人流连忘返。

③

④

⑤

Holiday Villas

Villas in different appearances and with roofs like azalea or the leaf of azalea scattered around the hotel like stars in the sky. Living in such environment, people must be relaxing and joyous like releasing from noisiness.

①～⑨别墅效果图

滨湖游览区 Lake Shore

游客在闲暇时光漫步于滨湖步道欣赏湖光山色，亦可在码头登船欣赏沿江美景，望山远眺，极目舒展，重峦叠嶂，烟云包裹滨水。沿滨湖岸边设置服务设施，水榭、茶舍、亭子、啤酒屋等建筑小品均有适宜的尺度，其小巧、简洁、清新的建筑风格与山水相得益彰。

Visitors can walk along the lakeshore and admire scenery, and also enjoy boating in the lake and imaging as distant as the row upon row of verdant mountains in the distance. Service facilities along the bank like tea houses, pavilions, beer bars etc should be in proper size so that the smart, elegant and fresh constructions can reflect the magnificent scenery of lake and mountains.

①啤酒屋效果图 ②啤酒屋效果图 ③钓鱼台效果图 ④贵宾区码头 ⑤大众区码头

滨湖游览区 Lake Shore

游客在闲暇时光漫步于滨湖步道欣赏湖光山色，亦可在码头登船欣赏沿水域与湖岸结合设置内容多样、可供选择性大的游览活动项目，有观光游艇、脚踏船、手划船、皮划艇、水上单车、垂钓等，使其成为人们远离都市喧嚣、享受清静、自然的滨水休闲空间。

Many activities are available both on the bank and in the lake: sightseeing on ships, boating by feet, boating by hands, riding bicycle on water and fishing, which contribute to the leisure space for people leaving away from noisiness of cities and enjoying peace and nature.

①

②

③
步行街
游乐场
拓展中心
步行街
游乐场
码头
拓展区
会所
大酒店

④

⑤

⑥

①观景亭效果图 ②观景亭效果图 ③～⑥小品

Xiangzhigou Spring Resort
香纸沟温泉度假村

Tourism
旅游度假区 Resort

Xiangzhigou Spring Resort

项目规划 Item Planning

1 综合现状 Comprehensive Current State

一、区域概况

贵阳香纸沟风景名胜区，位于贵阳市东北部的乌当区新堡布依族乡，地处东经 106° 50' ~ 106° 56'、北纬 26° 45' ~ 26° 48' 之间。该乡南接水田镇、下坝乡；东接百宜乡；西邻白云区及新场乡；北抵羊昌镇。面积 58.6km^2。有贵（阳）开（阳）公路经过，距贵阳 42km，距开阳 48km。

全景区属扬子地台黔北台隆的贵阳复杂构造变形区，地貌特征为地势隆起的山原台地，地势高，起伏大，总的是西北高，东南低，最高 1565m，最低 960m，相对高差 300 ~ 500m。有极为发达的沟谷流水地貌，喀斯特地貌和溶洞广泛发育，地下水与地表水交替频繁。

I Regional Overview

Guiyang Xiangzhigou scenic spot, situated in Xinbao Bouyei rural areas in Wudang district in Northeast Guiyang, is located at 26° 45'~26° 48'N and l06° 50'~106° 56'E. The countryside, with an area of 58.6km2, links to Shuitian town to its South, Baiyi countryside to its East, Baiyun district and Xinchang countryside to its West and Yangchang town to its North. The countryside, with a Guiyang–Kaiyang road running through it, is 42km away from Guyang and 48km away from Kaiyang.

The whole area belongs to a Guiyang complicated structural deformation zone in Tailong, North of Guizhou in Yangtze platform, featuring rising highland and tableland with high relief. The overall situation is northwest high and southeast low with the highest 1 565 meters, the lowest 960 meters and the relative elevation 300~500 meters. The area is endowed with developed ravine and running waters, Karst landform and cavern, as well as frequently alternated ground water and surface water.

区位分析图

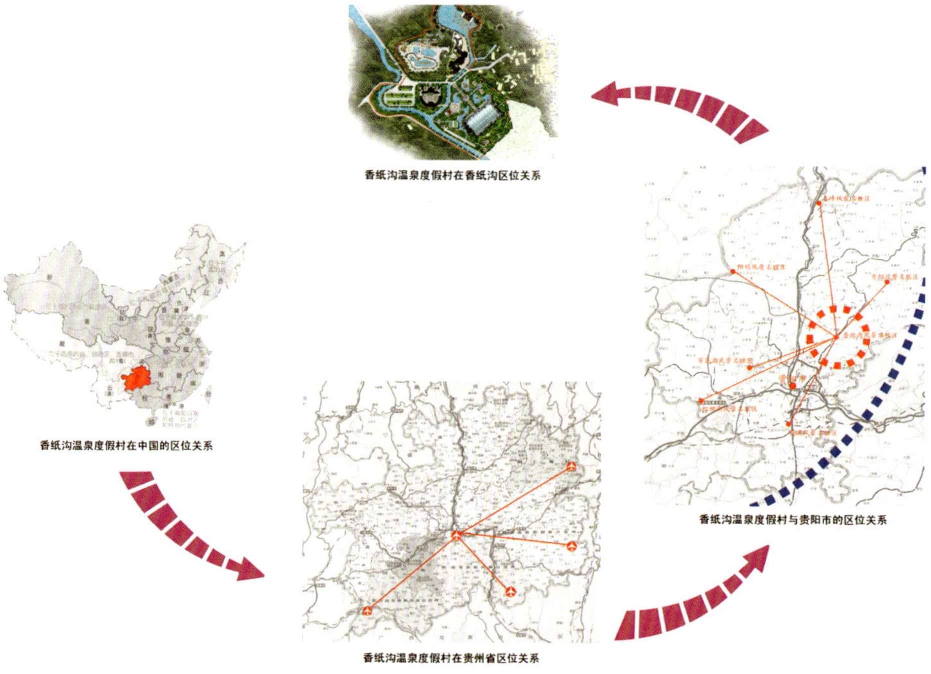

景区内河流属乌江水系，南明河支流，由阿渡河、白水河、龙泉河三条支流组成。区内降水丰富，年均降水量约为 1180mm，而且出露地层透水性较差，地表水系发达。地下水与地表水交替频繁，泉水出露众多。几乎所有河流的源流和支流都来自地下河或喀斯特大泉。

全景区属亚热带季风湿润气候，气候特点是冬无严寒，夏无酷暑，水热同季，无霜期长，雨量充沛，垂直气候差异比较明显，景区内相对湿度 81% 以上，阴天多，日照少。

景区属亚热带常绿阔叶林带，植被以常绿林为主，树种多为拷槠类植物，间有枫香、响叶杨、花揪、麻栎等树种。区内计有森林 3300hm^2，树径 lm 以上大古树 22 株，树径 50 ~ 100cm 的古树 120 多株，珍稀濒危植物 10 种，观赏植物 130 多种，野菜 50 多种，野果 40 多种。

景区所在地新堡布依族乡境，1949 年前属贵筑县羊昌区的新堡和陇上两个保。1952 年，建立新堡乡政权，下设新堡村和陇上村，属贵筑县。1983 年，建立新堡布依族乡人民政府。居住着汉族、布依族、苗族、彝族等四个民族。现辖 7 个行政村，35 个村民小组，总人口 1274 户，5449 人，其中农业户 1133 户，5280 人，少数民族人口 2843 人，占总人口的 53.8%。风景区内有 6 个村，27 个村民组，农户 919 户，4304 人。该乡造纸小区已形成规模，主要生产瓦楞纸、平板纸、餐巾纸。旅游业已为全乡龙头产业，并被视为“富民兴乡”的有效之路。

高程分析图

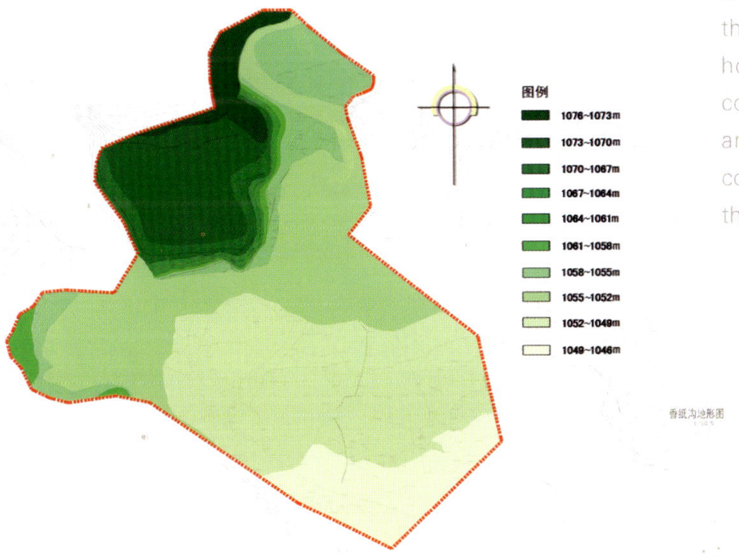

The rivers within the scenic area belong to Wujiang River Water. Nanming river tributaries consist of Adu River, White River and Longquan River. The area has abundant rainfall with an average annual rainfall at about 1,180mm, low emergence stratums hydraulic permeability and developed surface drainage pattern. The underground water and surface water alternate frequently, and spring water emerges from time to time. Almost all the sources and tributaries of the river come from underground river and Karstic springs.

The whole area enjoys subtropical monsoon climate characterized by no cold in winter, summer without heat, hot water in the same quarter, long frost-free period, and abundant rainfall with relatively distinct vertical climate differences. The relative humidity in the area is above 81% with many cloudy days and few sunny days.

The spot is subtropical evergreen broad-leaves forest with evergreen forests as the main vegetative cover. Most of the trees belong to castanopsis. The forest coverage is 3 300 hectares, including 22 big ancient trees with above one meter tree-walk, 120 ancient trees with 50~100 tree-walk, 10 species rare and endangered plants, more than 130 species ornamental plants, more than 50 species edible wild herb and more than 40 species wild fruit.

The scenic spot which is located at Xinbao Bouyei countryside is part of Xinbao and Longshang of Yangchang district in Guizhu County before 1949. In 1952, the countryside established Xinbao regime and set up Xinbao village and Longshang village under it, which belongs to Guizhou County. In 1983, the people's government was established in Xinbao Bouyei countryside. There are four nationalities live in the countryside—Han people, Bouyei people, Miao people and Yi people. There are 7 administrative villages, 35 village groups, and 1 274 households with total population of 5 449, of which agricultural households 1 133 with 5 280 and ethnic minorities 2 843, taking up 53.8% of total population. Within the scenic spots, there are 6 villages, 27 village groups, 919 agricultural households and 4,304 people. The papermaking communities in the countryside have taken shape, mainly producing fluting paper, flat paper and napkins. Tourism has become the leading industry of the whole countryside and has been regarded as a promising road towards "enriching the people and rejuvenating the countryside".

二、景区资源

在新堡乡 58.6km^2 的土地上，分布有丰富的风景名胜资源。山、水、林、瀑布、峡谷、岩溶景观、文化古迹、民族风情，都有独特之处及较高价值。特别是天然林覆盖率达 60% 以上，水资源最为丰富。2000 年 2 月，经贵州省人民政府批准，贵阳香纸沟列为贵州省第四批省级风景名胜区之一。

1. 资源类别

贵阳香纸沟风景名胜区的风景名胜资源，根据《总体规划》总计自然景源和人文景源两大类、7 分类、21 种类，124 个景物。

2. 资源等级

在香纸沟风景名胜区两大类 7 分类 21 种类不同特征的景物景观中，《总体规划》的一级景物有 6 个，二级景物有 21 个，三级景物有 32 个。 一至三级景物景观，共有 59 个（处）；若将天然林按片计算，总共 71 个（处）。

II Scenic Resources

The rich resources of scenic spots and historical interests are distributed on 58.6km^2 of territory in Xinbao countryside. Mountains, rivers, forests, waterfalls, canyons, karst landscapes, cultural relics, customs—all these resources boast uniqueness and high value. In particular, the forest coverage reaches over 60% and water resources are the most abundant resources. In February, 2000, with the approval of the people's government of Guizhou province, Guiyang Xiangzhigou was listed as one of the fourth scenic spots and historical interests at the provincial level.

1. Resource Class

According to General Rules, Guiyang Xiangzhigou scenic spots are divided into two large types of natural scenic and human scenic, 7 categories, 21 species and 124 sceneries.

2. Resource Level

Among the sceneries and landscapes with varied features in Xiangzhigou scenic area mentioned above, and in line with General Rules, there are 6 first-level sceneries, 21 second-level sceneries and 32 third-level sceneries—all together 59 sceneries from first-level to third-level. If natural forests are calculated by patches, then there are altogether 71 sceneries.

现状分析图

水体分析图

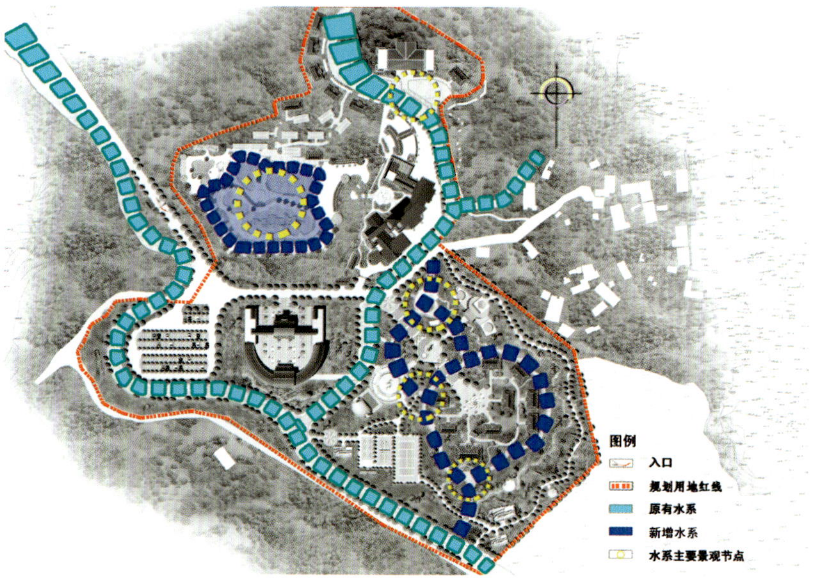

2 建设条件分析 Analysis of Construction Conditions

一、优势

1. 黔中秘境，自然乌当

乌当区地处黔中，山清水秀，气候宜人，人杰地灵，文化丰厚，产业结构，调整更新，构成了城乡结合、古中有今、传统文化与现代文明兼容、自然秘境的泉城形象。“黔中秘境，自然乌当”具有一定知名度。

2. 旅游区位优势明显、交通条件优越

乌当区是省会卫星城市贵阳市区的北大门，是连接我省东北部区县的必经之地，乌当区处在贵阳、遵义两座一级中心城市之间，受其辐射影响较大。

3. 可整合的旅游资源多，产品的多样性和丰富性强

贵阳香纸沟风景名胜区物种丰富，生态良好，环境优美，涵养了丰富的水资源，形成了千姿百态的水景。还包括了悬崖绝壁，奇峰异石，峡谷，瀑布，古树名木，民族风情，更有杜寨农民画乡和古代造纸作坊，资源丰富，品位高。

4. 适游性强

由于可利用的布依族文化资源丰富、可开发产品多种多样、适于全年发展旅游。

Ⅰ Strengths

1. Fairyland in Guizhou and Natural World in Wudang

Wudang district, located in the middle part of Guizhou province, is a good place for outstanding people with beautiful mountains and clear waters as well as rich cultures. The place has also witnessed the adjustment and updating of industrial structures. All these characteristics have constituted the image of spring cities of combining the city with the countryside, ancient with modern and traditional culture with modern culture. "Fairyland in Guizhou and Natural World in Wudang" enjoys a certain degree of good fame.

2. Distinct Advantages for Tourist Locations and Transportations

Wudang District is the Northern Gate of Guiyang city, the capital and satellite city of Guizhou province. It is a must path for connecting counties and districts of northeast parts of Guizhou province. Wudang district is situated between the two first-class central cities of Guiyang and Zunyi, which is greatly affected by radiation.

3. Excessive Integrative Tourist Resources and Strong Diversity and Richness of Products

Guiyang Xiangzhigou scenic spots and historical interests are endowed with rich species, pleasant ecological environment, which has nurtured rich water resources and formed kaleidoscopic waterscape. Others include sheer precipices and overhanging cliffs, picturesque peaks and rocks, canyons, waterfalls, ancient and rare trees, customs, and in particular rural peasant paintings in Du village and ancient papermaking workshop, which indicate abundant resources and high taste.

4. Favorable Conditions for Tourism in All-Round Year

Owing to rich Bouyei cultural resources available and a great variety of exploitable products, the place is favorable for travelling in four seasons.

二、劣势

1. 生产生活与环境保护的矛盾

大量家庭造纸作坊的存在，不利于环境的保护。村寨环境状况较差，给水排水系统不完善。

2. 开发保护资金缺乏

新堡乡以农业为主要经济来源，地方财力有限。

3. 服务设施不完善

公路等级低，基础设施差，服务设施不完善，需进一步改进。

4. 管理机制不健全

村民利用村寨民房作服务接待，对房屋进行不适当的改造，大大破坏村寨风貌，缺乏有效地控制与整治。

Ⅱ Weaknesses

1. Contradictions Between Life – production and Environmental Protection

A great number of family papermaking workshops are unfavorable to environmental protection, and bad environmental conditions hinder drainage system.

2. Shortage of Fund for Development and Protection

Xinbao countryside takes agriculture as its main resource of funding, so there is a limit to its local financial capacity.

3. Imperfect Service Facilities

Low-level roads, bad infrastructure and imperfect service facilities need further improvement.

4. Unsound Management Mechanism

The villagers take advantages of civilian buildings for service reception and inappropriately transform the houses, which has damaged the appearance of the village.

三、机遇

1. 政策机遇

贵州省国民经济“十一五”规划及 2010 年长远规划中明确了旅游是作为我省支柱产业之一，作为新的经济增长点加以大力发展与培育。全省各地、州、市也都先后将旅游业列为当地大力发展的优势产业。

贵阳市委、市政府、乌当区委、区政府都把旅游业确定为新兴的支柱产业来加以培育和发展。这是香纸沟风景名胜区发展前景有利的大气候、大环境。

2. 产业结构调整机遇

随着国务院批准《贵阳市城市总体规划（修编）》所确定“两点、两翼、三通道”的总体布局，使乌当区经济和社会发展面临重大转折，现实生产力重新布局为“一点一翼一通道”，成为贵阳市城市城区规划九个组团之一。目前，正在规划以贵阳为龙头的“1 + 10 经济圈”。市、区、乡（镇）政府及有关部门极为重视风景名胜区的事业，正在采取各项政策和措施促进旅游及相关产业的发展。

四、挑战

旅游产品面临着激烈的市场竞争。

Ⅲ Opportunities

1. Policy Opportunities

Both national economic "eleventh five-year" plan and 2010 long-term plan of Guizhou province state clearly that tourism should be regarded as pillar industry and new economic growth point for massive development and cultivation. The regions, states and cities in the whole province also list tourism as advantageous industry in the region.

Guiyang Municipal Committee and Municipal Government, Wudang District Committee and District Government all nurture and develop tourism as new pillar industry. These are favorable climate and environment for the future development of Xiangzhigou scenic spots and historical sites.

2. Restructuring Industries

With the approval of the State Council for the general layout of "two points, two wings and three channels" identified in Urban Master Planning in Guiyang city (revised edition), the economic and social development in Wudang has experienced a major turning point. However, Practical productive forces rearrange the layout as" one point, one wing and one channel", which has become one of the nine groups in urban and district planning in Guiyang city.

Ⅳ Threats

Tourism products face intensive market competition.

3 规划总则 Planning Principle

一、规划依据、原则和指导思想

1. 规划依据

(1) 法律法规

(2) 规划参考材料

2. 规划原则

(1) 坚持“严格保护、统一管理、合理开发、永续利用”的基本方针。

(2) 规划突出本区资源特色，风景区建设与资源环境、地方风格协调一致。

(3) 生态效益、社会效益和经济效益并重，注重协调发展。

(4) 保护生态环境，严格控制用地，对用地进行用途管制。

(5) 与贵阳市、乌当区及新堡乡的有关规划相协调。

Ⅰ Planning Basis, Principles and Guidance

1. Planning Basis

(1) Rules and Regulations

(2) Planning references and materials

2. Planning Principles

(1) Adhere to the basic guidelines of "strict protection, unified management, reasonable development and sustainable Utilization".

(2) Plan and highlight characteristics of resources in the region as well as coordinate scenic spot construction and resource environment with local style.

(3) Attach equal importance to eco-efficiency, social efficiency and economic efficiency and lay stress on coordinated development.

(4) Protect eco-environment, control land and control land use strictly.

(5) Coordinate with Guiyang city, Wudang district and Xinbao countryside for related planning.

二、规划范围

天池温泉度假村，地处香纸沟风景名胜区六个景区中的主景区——香纸沟景区。该景区属陇脚村，位于白水河上游。该景区面积 13.9km^2，是香纸沟风景名胜区的精华所在。

三、性质与目标

1. 度假村的性质

以物种丰富的天然林和清秀多姿的水文景观为依托，以田园温泉、布依族文化、布依族美食、古代造纸作坊为特色，适宜于都市郊区的休闲度假、生态审美、观光游赏的温泉度假村。

2. 规划目标

本规划的总体目标是 ：在“规划是前提，保护是核心，管理是关键”的思想指导下，把风景名胜区的规划建设纳入乌当区国民经济与社会发展之中，协调社会、经济、技术等诸多因素，使温泉度假村的建设与当地国民经济建设同步发展，创造一个交通便捷、布局和功能合理，生态、经济、社会相协调的复合生态系统，满足现代人生活、文化娱乐、旅游的需求。

本规划的具体目标是：将建设休闲度假、生态审美、观光游赏的都市郊区的温泉度假村和建设社会主义新农村的陇脚村有机地结合起来，互利互惠，共同繁荣，共同发展。

Ⅱ Scope of Planning

Heaven pool Spa resort is located at the main scenic spot of six scenic spots in Xiangzhigou scenic spots and historical sites—Xiangzhigou landscape which is situated at upstream of White River in Longjiao village. Its area is 13.9km^2 and it is the essence in Xiangzhigou scenic spots and historical sites.

Ⅲ Properties for Resorts

1. Properties for Resorts

Silhouetted by natural forests with rich species as well as clear and charming waterscape, characterized by rural hot-spring, Bouyei culture and food as well as ancient papermaking workshop, the resort is suitable for relaxation, ecological aesthetics and sightseeing in the suburbs.

2. Planning Target

The overall planning object is: under the guidance of "planning is precondition, protection is the core and management is the key", the planning and construction of scenic spots and resorts will be absorbed into national economic and social development in Wudang district, harmonizing society, economy, technology and factors so as to synchronize the development of construction of hot-spring resort and local national economy, create convenient transportation, reasonable layout and function as well as complex of ecosystems of coordinated ecology, economy and society, and meet the needs of modern people for life, cultural entertainment and traveling.

The specific objective of the plan: to organically combine the construction of hot-spring resort for holidays, ecological aesthetics and sightseeing in metropolitan suburbs with the construction of Longjiao village in socialist new rural areas on the basis of mutual benefit, reciprocity, common prosperity and common development.

4 总体布局 Overall Layout

一、规划布局

香纸沟风景区，是一个集中连片的风景区。《总体规划》根据天然地势所形成环境风景名胜区，布局为 6 个景区（大寨田园风光景区、龙箐沟峡谷景区、杜寨画乡景区、龙泉河景区、香纸沟景区、白水河景区）、1 个休闲区（新堡村避暑休闲区）、1 个管理服务区（设于陇脚村）。

其中，由于历史原因，管理服务区设于陇脚村，在上、下陇脚寨及基本农田的周围，已建成用 300 床位的财政宾馆一幢，造纸博物馆一幢。两个寨子中有餐饮服务户 28 家，农家床位 868 床，还有一个约 2000 平方米的停车场及民族风情表演场。在此集散的骑马观光马匹，有 150 多匹。已经形成了一个风景区的服务中心，面积约 0.4 平方公里，其中大部分是农田。香纸沟风景名胜区管理处也在停车场边建有一排办公用平房。按《总体规划》，将香纸沟风景名胜区管理处从景区中划出，成为风景名胜区中的管理服务区，设主要停车场及各种服务设施及行政管理设施。

本《详细规划》，就是在改造利用原管理服务区的占地范围的基础上进行布局。

Ⅰ Planning Layout

Xiangzhigou scenic area is a concentrated blanket scenic spots. According to environmental scenic spots formed by natural topography, General Rules divides six sceneries (Dazhai rural landscape, Longqinggou canyon scenery, rural peasant paintings in Du village, Longquan River landscape, Xiangzhigou landscape and White River landscape), and one Recreational Areas (Xinbao village summer resort), and one management service zone (in Longjiao village).

Due to historical reasons, the management service zone is located at Longjiao village. Around Longjiao village both up and down and basic farmland, the Finance Hotel with 300 beds and a papermaking museum were built. Within these two villages, there are 28 households of Food and Beverage Services, 868 beds in peasant family, a 2 000 m^2 parking lot, an arena with ethnic customs and 150 horses for sightseeing. A scenic spot service centre has been formed with an area of 0.4 km^2. of which a majority of part is farmland. Xiangzhigou scenic spots management office has also established a row of office bungalows along the parking lot. According to General Rules, Xiangzhigou scenic spot management office is divided out from the scenic spots and has become the management service zone for historical interests with main parking lot and various kinds of services and administrative management facilities.

The Detailed Planning is to conduct layout on the basis of transforming and utilizing the occupied land for original management service zone.

1. 规划用地原则

(1) 突出景区土地利用的重点与特点，合理确定景点用地范围及环境保护范围。

(2) 保护风景游览地、优良耕地、林地和水源地。

(3) 根据风景区的要求，合理调整土地利用，发展符合风景区特征的土地利用方式与结构。

(4) 建设用地坚持少占或不占耕地，坚决不占高产耕地和基本农田保护区，不破坏森林植被。

2. 土地利用协调规划

按照土地使用的主导性质，温泉度假村用地以改造利用原管理服务区的占地为主，严格控制交通与工程用地及农民建设用地，积极培育风景游赏用地，合理布局旅游服务用地，综合规划农林生产用地。

3. 主要经济技术指标

1	规划用地面积	73296 m^2	8	游泳池面积	600 m^2
2	建筑占地面积	10220 m^2	9	儿童戏水池面积	250 m^2
3	建筑面积	20620 m^2	10	道路及硬地面积	6780 m^2
4	建筑密度	13.94%	11	绿化面积	42622 m^2
5	容积率	0.3	12	绿化率	58.15%
6	小泡池面积(13 个)	110 m^2	13	停车场面积	4070 m^2
7	大泡池面积	720 m^2	14	停车位	135 个

1. Planning Principles for Land Use

(1) Highlight the key points and characteristics for land use, and reasonably identify the scope of land use and the scope for environmental protection.

(2) Protect resorts, good arable lands, forests and water source areas.

(3) Appropriately adjust land use and develop ways and structure for land use that fits the characteristics of sceneries in line with the requirements of scenic spots.

(4) The land for construction will not or try to not take up the arable land, never take up arable lands with high yields and basic farmland reserves, and not damage the forest cover.

2. Coordinated Planning for Land Use

According to the dominating character of land use, land use for hot-spring resort gives priority to transforming and utilizing original management service zone, strictly control land use for transportation and engineering as well as for peasant construction, actively cultivate sightseeing land use, properly layout tourist service land use and comprehensively plan land use for agricultural and forest production.

3. Major economic and technological targets

1	Planning covering area	73296m^2	8	Swimming pool area	600m^2
2	Building covering area	10220m^2	9	area of children's pool	250m^2
3	Total building area	20620m^2	10	Area of road and hard ground	6780m^2
4	Building density	13.94%	11	Green area	42622m^2
5	Plot ratio	0.3	12	Landscaping ratio	5815%
6	Area of small bathing pools (13)	110m^2	13	Parking lot area	4070m^2
7	Area of big bathing pools	720m^2	14	Parking space	135

总平面图

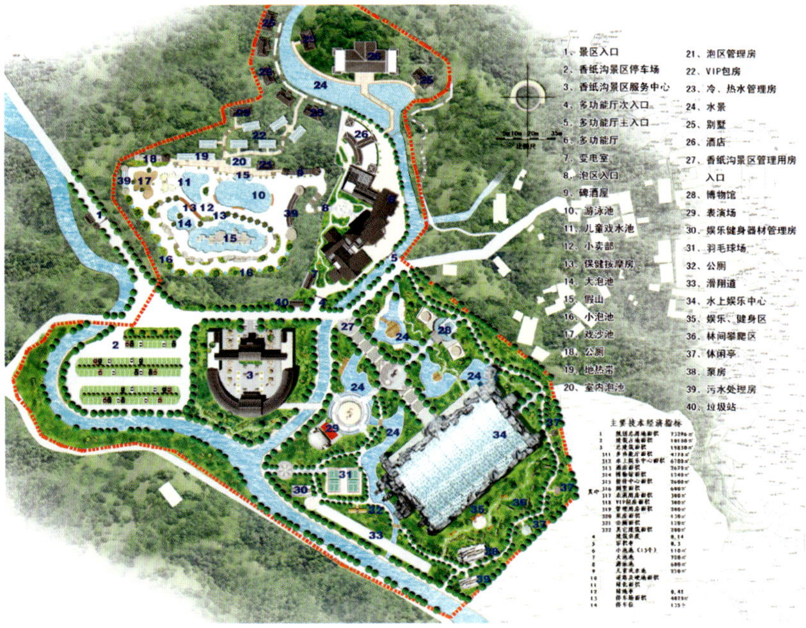

二、功能分区

按温泉度假村所具有的功能作用，及发挥功能作用中所必需的相关设施要求，划分下列功能区：

1. 入口、停车及餐饮购物区

该区位于温泉度假村的西南部，西与香纸沟景区的入口相接，包括停车场和购物中心。停车场为生态地面，占地面积 4070m²，设有大小 135 个停车位。购物中心采取前开后合的布局手法、并将布依族建筑与我国江南传统民居建筑融为一体，占地面积 1300m²，分为前后两个部分，前为接待休息等，后设商店购物。

2. 娱乐健身区

该区位于温泉度假村的东南部，包括博物馆、表演场、滑翔道、水上娱乐中心、健身区、攀爬区等。博物馆在原造纸博物馆的基础上进行改建，为当地布依族传统民居建筑样式，以当地布依族至今仍然使用的铜鼓及其图案为文化元素符号装饰，陈列展示当地布依族文化。表演场同为当地布依族传统民居建筑样式，舞台采取梯级移动手法处理，作为当地“汉灯、苗舞、布依歌”的表演场地。水上娱乐中心占地面积 6700m²，为游客提供水上娱乐活动。

3. 温泉洗泡区

该区位于温泉度假村的西北部，包括游泳池、儿童戏水区、大泡池、小泡池、地热带、室内泡区、VIP 包房等，为游客提供各种洗浴场地。其中，小泡池共有 13 个，分别建在天池洗浴区西南边缘的竹林丛中；VIP 包房共有 6 小幢，分别建于天池洗浴区北面的树林之中。

II Function Division of Tianchi Spa Resort

According to functions possessed by hot-spring resort and necessary related requirements for facilities, the place is divided into following function zones:

1. Entrance, Parking, Dinning and Shopping Zone

This zone, located at Southwest part of hot-spring resort, is connected with Xiangzhigou scenic spot entrance to its West, including parking lot and shopping centre. The parking lot is eco-ground with an area of 4 070m², setting up 135 parking spaces for cars. The shopping centre adopts the layout of front opening and back closing, and integrates Bouyei building with traditional residential building in the regions south of the Yangtze River of China. Its area is 1 300m², and divides into two parts of back and forth. The front is for services like reception and rest while the back is for shopping.

2. Entertainment and Fitness Areas

The area is situated in Southeast part of hot-spring resort, including museum, arena, glide path, overwater leisure and recreation center, fitness area and climbing area. The museum is reconstructed on the basis of the original papermaking museum with the style of local Bouyei traditional residence building. The museum is decorated with drums and their patterns as cultural elements and symbols which are still used by local Bouyei people and displays Bouyei culture. The arena is also the traditional residence buildings of local Bouyei. The stage adopts the method of moving steps and serves as the performance place for "Han lanterns, Miao dances and Bouyei songs". The overwater leisure and recreation center, with an area 6,700 m², provides overwater leisure and recreational activities for tourists.

3. Hot Spring Area

This area is located at the northwest of the hot spring resort, including swimming pools, children's swimming pools, big soaking pools, small soaking pools, geothermal belts, indoor pools, VIP compartments and so on for tourists. Among them, there are 13 small soaking pools, located in the bamboo forest in the southwest part of Tianchi bath region respectively; 6 VIP compartment buildings, situated respectively in the forest in the north part of Tianchi bathing area.

功能分析图

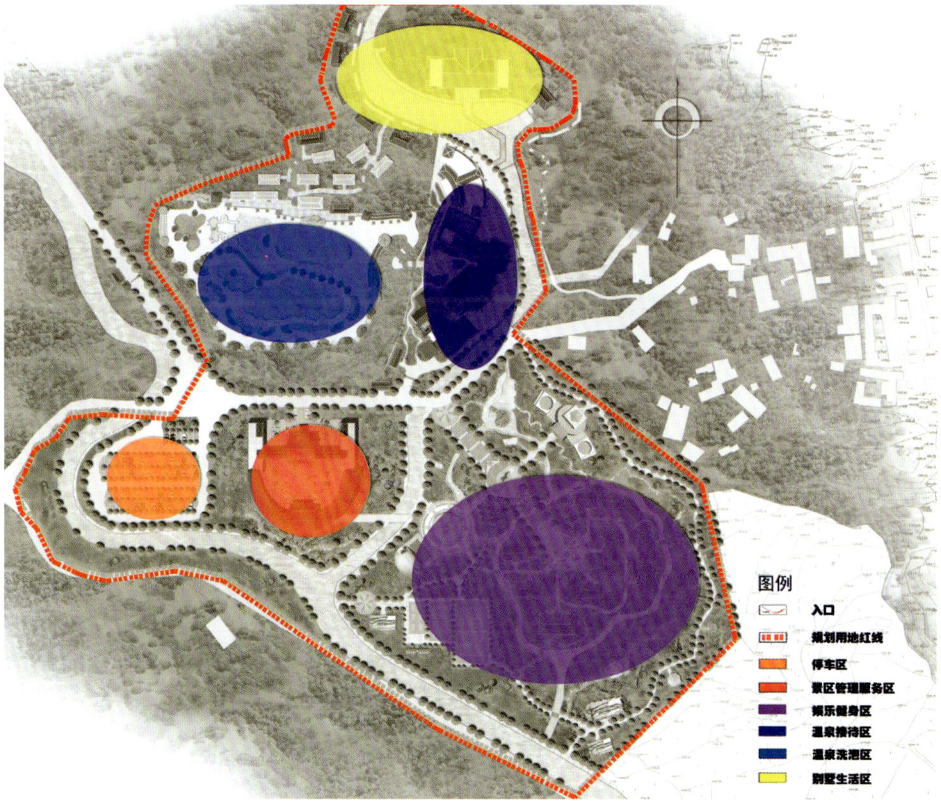

4. 酒店服务区

该区位于温泉度假村的东部，包括酒店、多功能厅和接待中心等服务设施，由原香纸沟财政宾馆等建筑物改建而成。对景区的保护、建设及开放营运，游人的吃、住、行、游、购、娱及安全等进行全面管理的工作区域。选择适当的位置，树立温泉度假村的文化标志——独特的牛头形象，既反映了布依族水田稻作文化中对牛的崇拜，也反映了“田园温泉”的自然属性，又适合于现代人以“牛”为强劲的竞争心理，并以当地簸箕画的形式加以塑造。

5. 别墅生活区

该区位于温泉度假村北部的沟壑两岸，北岸有 4 幢别墅，南岸有 3 幢别墅，均为当地布依族传统民居建筑样式。

在以上 5 个功能区中，温泉洗泡区和酒店服务区为一期建设工程，其他为二期建设工程。

三、游客容量分析

按《总体规划》的测算方法，游人所占面积指标定在下限 $5m^2$/ 人。全风景名胜区日容量 = 1800 + 1500 + 1000 + 1600 + 1650 + 1300 + 1 500 = 10350，风景区年容量 = 10350x 210d(7 个月)= 217.35 万人次。其中，香纸沟景区为：

游览步道长 12km，计容游人 2000 人。

游览车道长 3km，计容游人 150 人 (以马车计算，每 100m 一辆马车，计 30 辆马车，每车 5 人)。

骑马，计容游人 200 人；

上陇脚，计容游人 100 人；

下陇脚，计容游人 100 人；

水库水面，计容游人 150 人；

南静寺，计容游人 300 人；

篝火晚会，计容游人 200 人；

古造纸作坊，计容游人 100 人。

景区为二日游。

全日合计容人量 =（2000 + 150 + 200 + 50 + 100 + 150 + 50 + 300 + 200 + 100）÷ 2 = 1650 人。

温泉度假村的规划总用地面积为 $73296m^2$，按 5% 为游客活动空间计算，有 $3664.8m^2$；那么，游客容量可达 733 人。

4. Hotel Service Area

This area is reconstructed from the original Xiangzhigou Financial Hotel, located in the east part of the hot spring resort, consisting of hotels, multifunction halls and reception centers. It is the working area devoted to the overall management including the protection, construction and open operation of the resort, eating, lodging, transportation, touring, shopping, entertainment and safety of the tourists. Select a suitable location so that to set up the cultural symbol of the hot spring resort—the unique image of a cow head, reflecting not only Bouyei nationality's worship towards cows in its rice culture, but also its nature of "the idyllic hot spring". What's more, this symbol also plays up to the modern people's competition psychology that cow represents robustness. This symbol should be presented in the form of dustpan painting which is a kind of local art.

5. Villa Living Area

This quarter is located on the banks of the gully in the north part of the hot spring resort, with 4 villas on the north bank and 3 on the south. All the villas are built according to the traditional Bouyei residents' architecture style.

The phase I project includes the hot spring area and hotel service area and the phase II the others of the above 5 functional areas.

Ⅲ Tourists Capacity Analysis

According to the calculating method in the《general plan》, the target for prescribed minimum area per tourist is $5m^2$ per person. The daily total tourists capacity of the resort=1800+1500+1000+1600+1650+1300+1500=10 350, annual capacity=10 350+210 days(7months)=217.35 million people. Among them, the Xiangzhigou Resort includes:

Walking road for sight-seeing is 12km long, with capacity of 2 000 persons.

Driving distance for sight-seeing bus is 3 km, with capacity of 150 persons (to calculate by carriages, one carriage every 100 meters, all together 30 carriages, 5 people on each carriage.)

Horse riding, with capacity of 200 persons;

Shanglongjiao, with capacity of 100 persons;

Xialongjiao, with capacity of 100 persons;

The surface of reservoir, with capacity of 150 persons;

Nanjing Temple, with capacity of 300 persons;

Campfire party, with capacity of 200 persons;

The ancient paper-making workshop, with capacity of 100 persons;

The scenic area is for two-day tour.

Total amount of daily population=(2000+150+200+50+100+150+50+300+200+100) ÷ 2=1650 persons

The total planned land of the hot spring resort is 73 296 square meters. If tourists take up 5% of the whole space, that is 3 664.8 square meters, then it can accommodate all together 733 tourists.

5 专项规划设计 Special Planning and Design

一、道路交通系统规划设计

1. 外部交通现状

香纸沟风景名胜区现在唯一的对境外联系道路为省道贵开公路。

景区内部交通以新堡——香纸沟为主干线，计 8km，有贵阳——香纸沟，新堡——香纸沟两条营运路线。新堡——香纸沟达陇脚村的乡村公路正在改造提高。

新堡——香纸沟达陇脚村的乡村公路，从温泉度假村中穿过，成为游客和当地居民的公共交通道路。为了适应温泉度假村的需要，只规划新建停车场和游览步道。

I Design Plan of the Road Traffic System

1. External Traffic System

The only outbound route of the Xiangzhigou Resort is the provincial road—Guiyang Highway.

Inside the resort, Xinbao—Xiangzhigou road is the main line, 8 kilometers in total; another two lines are the Guiyang—Xiangzhigou road and Xinbao—Xiangzhigou road. The Xinbao—Xiangzhigou village highway to the Longjiao Village is now under reconstruction.

Xinbao—Xiangzhigou village highway to the Longjiao Village goes through the hot spring resort, serving as the public transportation route for tourists and local residents. To meet the need of the hot spring resort, only now parking lots and sightseeing pedestrians will be built.

2. 设计依据

①《城市道路设计规范》
②《城市道路交通规划设计规范》
③《城市用地竖向规划规范》
④《停车场规划设计规则（试行）》

2. Design Basis

1 Code of Designing Urban Roads
2 Code of Designing Urban Road Construction Plan
3 Code of Vertical Planning Urban Land
4 Code of Designing Parking Lots (trial implementation)

交通流线图

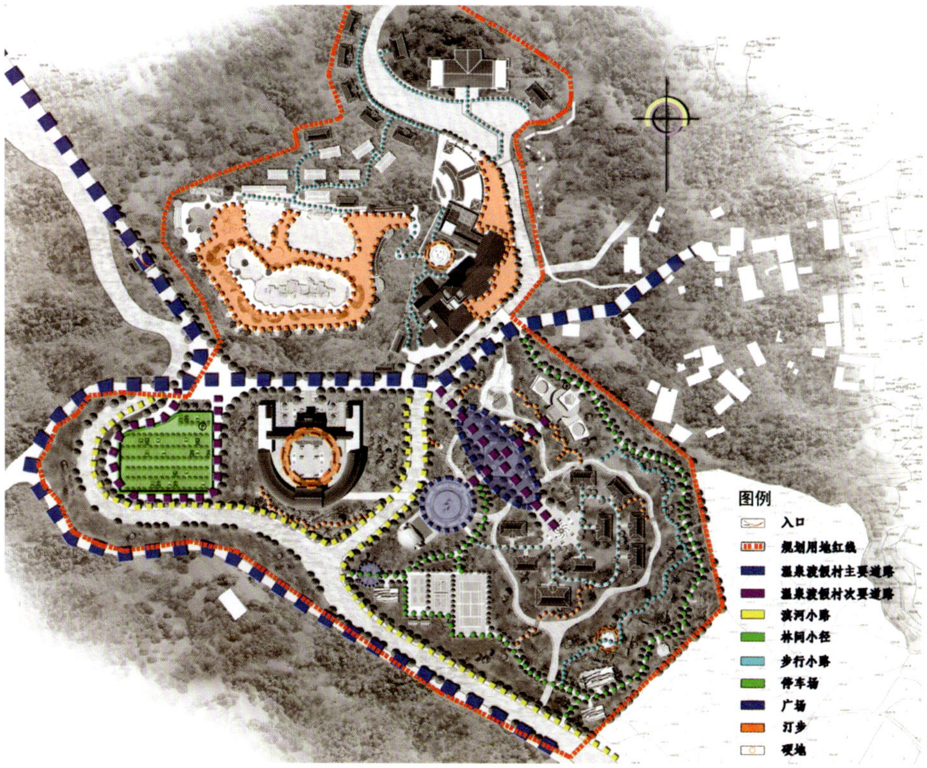

3. 道路分级

主要道路宽 6m，次要道路宽 4m，步行道宽 1.5m。

4. 道路横断面设计及坐标网系统

道路横断面形式根据风景区道路红线要求综合确定。考虑了道路的性质、等级和功能，同时也考虑了环境和工程设施等方面的要求。做到节约用地，满足交通畅通和安全，布置好道路绿化。坐标网系统采用 1954 年北京坐标系，1985 年黄海高程基准，标注精度为 1mm。

5. 交叉口设计

为确保一定的行车安全和行车车速，根据道路设计规范要求，对道路交叉口路缘石转弯半径及视距三角形加以控制。同时也考虑它与地下管线、绿化、照明、排水及交叉口建筑的配合和协调。

6. 静态交通设施规划

度假村停车场面积 4070m^2，每个停车位按 30m^2 计算（含通道），共 135 个。

3. Road Classification

Main road is 6 meters wide, subordinate road is 4 meters wide and pedestrian is 1.5 meters wide.

4. Design of the Road Cross Section and Coordinate Net System

The form of the road cross section is comprehensively determined by the scenic resorts red line requirements, taking the properties, classification and function into consideration, as well as the requirements in terms of environments and project facilities. The construction should save land, to meet the goal of smooth and safe traffic with virescence along the road. We apply the 1954 Beijing coordinate net system and 1985 Huanghai height datum, and the marked accuracy is 1mm.

5. Design of the Intersection

In order to insure safe driving and driving speed, according to the code of road designing, the curb turning radius and stadia triangle of the intersection should be controlled. At the same time, the coordination and coherence with the underground pipelines, virescence, illumination, drainage and buildings around the intersection should also be considered.

6. Static Traffic Facility Plan

The parking lot in the resort has an area of 4 070m^2, if one parking space takes up 30m^2 (passage included), altogether 135 parking spaces.

二、竖向规划设计

本规划设计场地的地势有一定起伏。规划中依据原场地的自然状况，结合建筑、道路、排水等要求标高相协调的原则，在满足使用要求的同时，又满足经济、安全和景观等方面的要求。确定了道路交叉口与控制点以及场地的标高，并保证道路最小纵坡 > 0.3%，场地最小坡度 > 0.5%，道路横坡取 1%。

Ⅱ Design of Vertical Plan

This plan requires the land to be uneven. In the plan, according to the natural conditions of the original land, taking into consideration the requirement that the elevations of architecture, road and drainage should be coherent, the construction should also meet the need in use, budget, safety, sightseeing and so on. We should determine the elevation of the intersection, the control point and the land and make sure that the minimal longitudinal slope of the road is>0.3%, the minimal slope is>0.5%, and the transverse slope should be set at 1%.

三、生态环境保护规划

1. 生态保护规划

(1) 生态现状

贵阳香纸沟风景区总面积 50.0km^2，人口 4304 人，村民以养殖、土法造纸、旅游服务为主。风景区森林面积大，物种丰富，空气清新，水质清澈，整个生态环境良好，水土流失不严重。

Ⅲ Eco-Environment Conservation Plan

1. Ecological Conservation Plan

(1) Current condition of eco-system

The Xiangzhigou Resort in Guiyang has an area of 50.0 square kilometers with a population of 4303. Villagers mainly live on cultivation, papermaking through indigenous method, tourism service. The resort boasts a large area, a rich bio-diversity, fresh air and clean water, enjoying a good overall eco-environment with no serious problem in soil and water loss.

(2) 保护规划原则

① 有利于资源的综合保护和永续利用。

② 保护与培育风景区内的自然生态环境。

③ 保护与恢复风景区中的景观及其环境。

④ 开发与保护培育相协调。

⑤ 建设项目与风景环境相协调。

⑥ 运用有关法律、法规、条例加强保护培育规划的法律效力。

(3) 生态保护规划

① 特级生态保护区

香纸沟风景区处在高山密林之中，风景名胜区连片的森林生态环境形成了需要绝对保护的核心区，山高林密人也难进。因而，形成了生态保护和特级保护相一致。这部分区域面积为 25.0km^2，占风景名胜区总面积的 50%。

② 一级自然景观游览保护区

本保护区主要指峡谷和水景的一部分，由于景观美，是人们主要游览之地。这部分区域约 7.6km^2，占风景名胜区总面积 15%。本区尽量保持原有生态面貌，原则上只修步道、休息设施，供游人游览休息，并加强防火及固体废弃物管理，不修建其他设施。

③ 二级人文景观保护区

香纸沟风景名胜区是以布依族为主的民族地区，有着浓郁的民族风情。这些区域要控制机动车的进入，最好是采用马车，电瓶车等游览运输工具或骑马游览。

④ 三级风景恢复及发展控制区

主要是指陇脚村原划出的管理服务区，有一定的景观，但由于位置和环境的关系，要进行必要的服务设施的建设。也是由于服务设施的建设而影响了景观和环境，因此要加以有效的管理与控制。

(4) 生态保护措施

① 建设景观路，加强旅游沿线道路的绿化建设；

② 进入风景区车辆实行分级限制，具体为：新香路通行风景区的专用车、小轿车、小型面包车、消防车；景区内部道路通行电瓶车、马车、自行车、消防车或外事专用车。

(2) Principles of the conservation plan

① To be beneficial to the comprehensive conservation and sustainable utilization of the resources.

② To conserve and cultivate the natural ecosystem in the resort.

③ To conserve and recover the sights and related environment in the resort.

④ To coordinate exploration, conservation and cultivation.

⑤ To coordinate project construction with the natural environment.

⑥ To strengthen the legal validity of the conservation and cultivation plan through the application of related laws, regulations and rules

(3) Ecological conservation plan

① Special grade reserve

Xiangzhigou Resort is located in the dense forests of the high mountains, the forest eco-environment with a series of famous resorts form the core area that needs absolute conservation. It's very difficult for people to get into the thick forests. Therefore, ecological conservation is in accordance with special grade reserve conservation. This region takes up an area of 25.0 square kilometers, occupying 50% of the whole resort.

② First grade natural sightseeing spot reserve

This reserve mainly refers to a part of the valley and waterscape. Due to its natural beauty, it's a popular destination of tourists. This region has an area of approximately 7.6 square kilometers, taking up 15% of the overall resort. The original ecological features should be maintained with the most efforts. In principle, only pedestrians, facilities for rest can be built, which facilitate the sightseeing and resting of the tourists. Fire prevention and solid waste management should be strengthened. No other facilities are allowed.

③ Second grade human landscape reserve

Xiangzhigou Natural Resort is a minority concentrated region with a strong tinge of folk customs. Most of the people living here are Bouyei. The access of motor vehicles should be restricted in these regions. Carriage, battery cars and other sightseeing transportation tools or horse riding are preferred.

④ Third grade recovery and development management reserve

This grade mainly refers to the originally planned management and service area in Longjiao Village. There is some scenery there, but due to its location and surroundings necessary service facilities should be built there. Just because the construction of service facilities affect the scenery and surroundings, effective management and control are required.

(4) Ecological conservation measures

① To build sightseeing pedestrian, and to strengthen the virescence along the sightseeing routes.

② Vehicles in the resort should be restricted through classification, specifically, special vehicles, sedan cars, minivans, fire engines which are allowed on the Xinxiang road; battery cars, carriages, bicycles fire engines and special cars for external affairs which are allowed on the roads inside the resort.

③ 停车场用植草砖作表，禁止用水泥地面。

④ 按“以林为主”的原则，植树造林，培植牧草，尽快修复开山采石、修路造成的植被破坏、山体裸露。

⑤ 建设用地控制

a. 陇脚村可在原有基础上增加20%建设用地，村寨建设用地控制在人均90m² 以内。需新建建筑必须报有关部门同意，符合规划要求。

b. 香纸沟影响景观的建筑应拆除，拆除及新建筑物按编制的详细规划实施。

c. 在陇脚村，凡影响交通、乱搭乱建、贴面砖的现代建筑予以拆除或改建，按详细规划进行控制。

(5) 水体保护

① 陇上村的造纸厂近期内治理污水，远期搬迁。规划保留香纸沟沿线六处，作参观、科普、教育之用；其余作景观，供游人观景。

② 建立生活污水、垃圾收集处理系统，设环卫监督管理保洁队伍，随时处理脏、乱、差问题。

③ 风景区河流、小溪两岸不少于50m范围划定水体保护带，保护带以植物景为主。对现有排污口进行检查、清理、控制排放标准。

④ 河流两侧山脊以内农田限制使用农药、化肥，提倡科学种田，改进耕作技术。

⑤ 维持合理的环境容量，降低环境损坏度。

⑥ 提倡对环境影响小的“生态旅游”，发展生态交通，能设步行道的不设车行道。

⑦ 为协调景区生态和旅游环境，将景区内农田、菜地的田园风光作为生态观光农业园，将村庄、农庄作新的观光旅游元素。

⑧ 加强村寨绿化，改变能源结构，多建沼气池，节能灶或使用电热，加强固体废弃物的清除管理。

2. 环境保护和环境卫生系统规划

(1) 环境保护规划

①规划依据

a.《环境空气质量标准》GB 3095—1996

b.《地表水环境质量标准》GHZB 1—1999

c.《城市区域环境噪声标准》GB 3096—93

d.《贵州省地面水域水环境功能划类规定》(黔府发(1994)22号)

e. 香纸沟风景名胜区环境质量基础资料。

③ The parking lot covers with grass and bricks, no cement ground.

④ Plant trees and forage grass on the forest-based principle and restore the destroyed plant cover and the bare parts of the mountain as soon as possible.

⑤ Restriction on construction land

a. Longjiao Village can have 20% more of construction land on the original basis, and village construction land should be limited within 90m² per person. The construction of new buildings requires the permission of the relevant authorities and should be in line with the plan.

b. Buildings that affect the scenery of Xiangzhigou should be dismantled. The process of dismantling and reconstructing should be implemented according to the detailed plan.

c. In Longjiao Village, any modern buildings that affect traffic, violate the rules or face bricks should be dismantled or reconstructed, and the process should be controlled according to the detailed plan.

(5) Water body conservation

① The papermaking workshop needs to deal with its waste water in the short term, and it will move to a new place in a specified future. The plan requires that 6 of them along the Xiangzhigou be remained for the purpose of visiting, disseminating science and education; the rest of them are left to be sightseeing spots for tourists.

② To build a collecting and disposal system for domestic sewage and garbage. To set up a environment and management supervision cleaner team, dealing with dirty, disorderly and bad problems at any time.

③ To designate a water body conservation belts along the banks of rivers and streams within no less than 50 meters, mainly aimed at protecting the flora. To examine, clean current sewage outlets and control the discharge standards.

④ Within the mountain ridges besides the rivers, the use of pesticides and chemical fertilizer is restricted, and to promote scientific farming, and improve farming skills.

⑤ To maintain suitable environment capacity, and minimize the damage to environment.

⑥ To promote eco-tourism which exerts less damage to the environment, develop eco-transportation, build pedestrians and avoid roadways.

⑦ In order to coordinate the ecosystem and environment in the resort, the farmland, idyllic vegetable field serve as eco sightseeing Agriculture Park while the village and farmstead the new sightseeing elements.

⑧ To strengthen village virescence, transform the energy structure; build more methane tanks, energy saving stoves or use electric heating; and reinforce the management of disposing solid wastes.

2. Environmental Protection and Sanitation System Planning

(1) Environmental protection planning

① Planning basis

a. "Ambient Air Quality Standard" (GB 3 095-l 996)

b. "Surface Water Quality Standard" (GHZB l-l 999)

c. "Urban Regional Environmental Noise Standard" (GB 3 096-93)

d. "Guizhou Province Ground Water Environment Functional Category" [(Qianfufa (1994) 22)]

e. Basic Materials on the Environmental Quality of Xiangzhigou Scenic Resort

② 环境保护工程规划

a. 生态环境建设：主要交通干线、河流沿岸，服务设施群落四周建设缓冲林地。对荒山、坡地植树造林、种草。对成林地带进行封山育林，提高内外景区的绿化率。

b. 工程管线规划：风景区内新建的工程管线宜采用地下敷设，对原有架空，出地工程管线逐步改为地下管线。

c. 风景区的垃圾、生活废水处理：根据风景区内景点分散，地形复杂，人口相对较少的特点，因地制宜，建设污水处理站，处理生活污水，生活粪便先进入化粪池处理后才能进入排水系统，各景区选择次要隐蔽地段挖坑掩埋，各景区建设规范的垃圾转运站和堆存点，建设水冲式厕所或生态式、旅游式公厕，确保景区内水质及旅游环境的质量。

③ 空气环境保护措施

控制煤烟型污染，推广液化气、沼气池及农村电气化，应用优质清洁能源。

④ 水环境保护措施

控制水污染，改善水质，对主要河流段城镇生活污水进行净化处理，并对上游工业排污口进行清理，对排污总量进行控制。保护水质，保土固水，要从上游做起，在景区内大搞植树造林，退耕还林种草，实施绿色工程计划，提高森林覆盖率，限制使用农药、化肥，减少水土流失，改善水质。

⑤ 噪声环境保护措施

重点是降低交通噪声，出入风景区的车辆，不准鸣笛，基建及卫生用车宜夜间行驶。不准用高音喇叭在景区内呼叫，严格控制人为噪声。

(2) 环境卫生系统规划

建立健全环卫机构，设置环卫管理站，负责风景区的环卫工作。配置必需的环卫设施，垃圾运输、清洁车辆及洒水车等。

①垃圾的产量及处理

按照风景区规划的总人数，每日产生垃圾以1kg/人计算，则日产垃圾为15t/d，垃圾清运率为100%，规划在陇脚设一座垃圾转运站，面积50 ～ 100m²/座。垃圾箱、果皮箱，达到国家规定的标准，并由专人管理收集清运。风景区的道路，场站每日清扫两次，白天保洁。

② 公厕

在香纸沟温泉度假村各功能区和陇脚上寨出入口等处，设公厕。

厕所的粪便，必须经化粪池处理后才能进入排水系统。

② Environmental engineering planning

a.Ecological environment construction: Buffering woods will be constructed around the major roads, river banks, and service facilities. Trees and grass will be planted on the barren hills and slopes. For mature wood, hillsides will be closed to facilitate afforestation and raise the green rates within and outside the scenic spots.

b.Engineering pipeline planning: the new engineering pipelines in the scenic area should be laid underground, and the exposed and overhead pipelines will be gradually replaced by underground ones.

Garbage and domestic wastewater treatment in the scenic spots: with the characteristics of scattered spots, complicated terrain and a relatively small population, the sewage treatment station will be built to treat domestic wastewater according to the local conditions. Excrement is first treated in the septic tanks and then enters the drainage system. Garbage is buried in the hidden places of scenic spots. Normative transfer stations and deposit sites, flush or ecological and tourist toilets are built in the scenic spots to ensure the quality of the water and tourist environment in the scenic spots.

③ Air environmental protection measures

Control coal and smoke pollution, promote liquid propane gas, biogas and rural electrification, and use quality clean energy.

④ Water environmental protection measures

Control water pollution to improve water quality. The domestic wastewater is purified in the main section of the river and the industrial sewage outlets in the upper river are checked up to control the total amount of sewage. Water quality protection and water and soil conservation should be done from the upper river. By vigorous afforestation in the scenic spots and returning farmland to forest and grass, the green project is implemented to increase forest coverage, restrict the use of pesticides, fertilizer, reduce soil erosion, and improve water quality.

⑤ Noise environmental protection measures

The key point is to reduce traffic noise. Vehicles to the scenic area are not allowed to whistle and the vehicles for infrastructure and sanitation drive at night. Loudspeakers are not allowed to use in the scenic area and the manmade noise is strictly controlled.

(2) Environmental sanitation planning

Establish and improve the sanitation institute, and build sanitation management station in charge of scenic sanitation work. Configure the necessary sanitation facilities, such as garbage transport, clean vehicles and sprinklers, etc.

① Waste production and disposal

According to the total number of Scenic Spots and the daily garbage production of 1kg/person, the daily output of garbage is 15T/Day, with a garbage collection rate100%. planning set a foot in Gansu a garbage transfer station is planned to be built in Longjiao, with an area of 50~100m². Litter bins and garbage bins meet the national standards and are managed, collected and cleaned up by professional staff. The roads and stations in the scenic spots are cleaned up twice daily and kept clean during the daytime.

② Public toilets

Public toilets are built near the entrances of the functional areas of the Xiangzhigou Spa Resorts and the Longjiaoshang Village.

Toilet stool must be treated in septic tanks and then enter the drainage system.

③ 健全环卫机构，完善环卫设施

随着游人的增加，旅游业的发展，人口的自然增长，建立专门的环卫机构，统一管理环境卫生势在必行。增加环卫专用车辆，建设洗车场，制定相应的管理措施，并监督执行。

3. 绿地系统规划设计

为加强生态环境的保护，特将温泉度假村的绿化面积定为 42622m^2，绿地率达 58%，建筑密度 14%，容积率 0.3，并对建筑物进行必要的美化措施。

四、旅游服务设施及附属设施系统规划设计

香纸沟温泉度假村既拥有自己的服务设施，又依托陇脚上寨来发展服务，利用农房，利用农村剩余劳动力，既办好服务，又少兴土木，少增加人口，农民还增加收入，一举数得。根据香纸沟的情况，只设接待中心和服务村。

温泉度假村接待中心

香纸沟温泉度假村设接待中心，满足旅游度假活动的要求，取代原财政宾馆为三星级或四星级酒店，保持原 300 床位以上，可接待高、中档游客的休闲度假。

陇脚服务村

陇脚村现有 28 户农家乐型的食宿户，可住 800 多个游客，并提供马车、马匹等游览工具。

以上合计 1100 床位。预测游客每天平均约 800 人，五一、十一、节假日的日游客超过 1000 人，而进入淡季则游人较少。因此，发展农家住宿是适应淡旺季需要的较好选择。

③ Improve sanitation institutes and sanitation facilities

With the increased amount of tourists, the development of tourism, the natural growth of population, it is imperative to establish specialized sanitation agency and need unified management of environmental sanitation. We will increase specialized sanitation vehicles, build car washing stations, and work out and implement appropriate management measures.

3. Green land system planning and design

To strengthen the ecological environment protection, the green land area of spa resort is set as 42 622 square meters, green land rate 58%, building density14%, plot ratio 0.3, and the necessary landscaping measures will be taken for the buildings.

IV The Planning and Design of Tourist Service Facilities and Ancillary Facilities and Systems

The Xiangzhigou Spa Resort not only has its own service facilities, but also relies on the Longjiaoshang Village to develop services. To make use farming rooms and rural surplus labor can offer good service with small-scale building, small amount of employees and increase farmers' income which can achieve several purposes at the same time. According to the conditions of Xiangzhigou, only the reception center and service village are set up.

Spa Resort Reception Centre

The Xiangzhigou Spa Resort establishes a reception center to meet the requirements of travelling activities. The former Finance Hotel is replaced to a three-star or four-star hotel, maintaining more than 300 beds which can accommodate high, medium and leisure tourists.

Longjiao Service Village

There are 28 joyous farmer's houses to accommodate more than 800 tourists and the carriages, horses and other tourist vehicles can be provided.

The total of the above is more than 1 100 beds. It is estimated that there are about 800 tourists per day in average and over 1 000 on May 1, October 1, and holidays, but fewer in the off season. Therefore, the development of farm accommodation is a better choice to meet the needs in off and peak season.

绿地系统

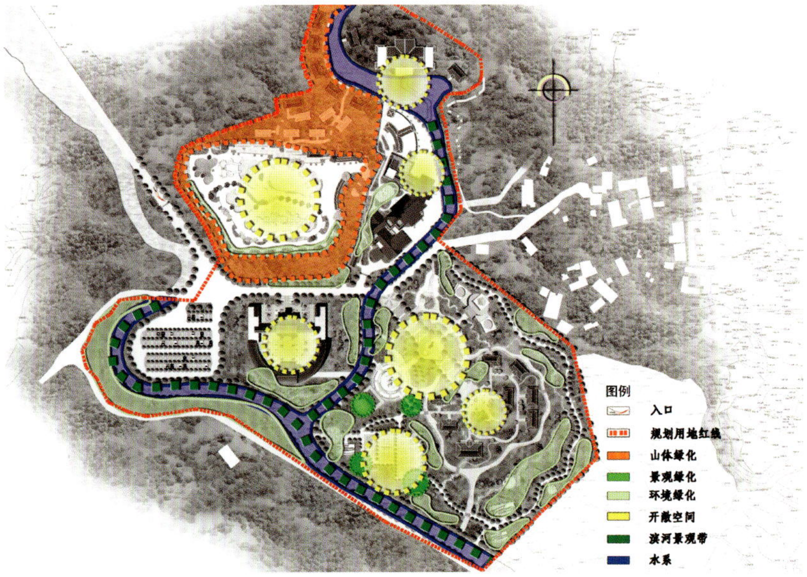

景观视线

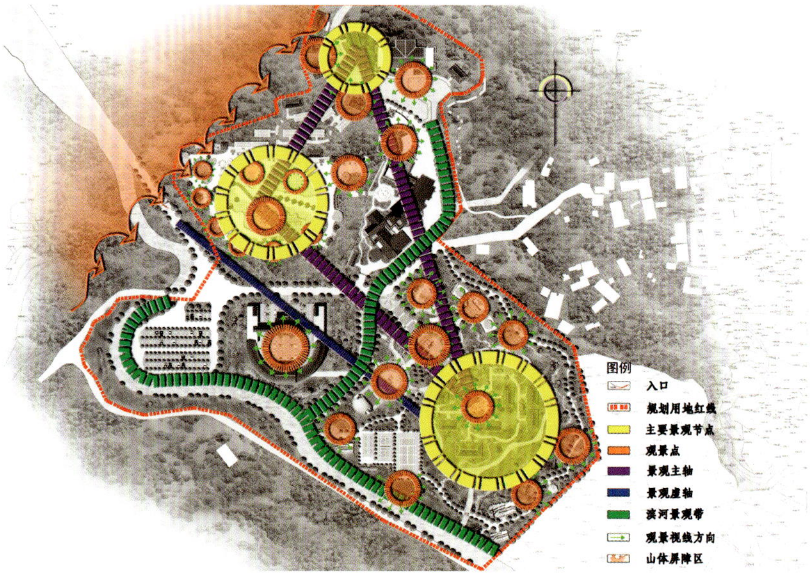

五、工程管线系统规划设计

1. 给水排水规划

(1)给水排水现状

现景区内用水户，主要是当地村民。水源为山泉水，重力自流供水，由于自然植被良好，水质通常较为清澈，源水经管道输送后直接作为生活用水。

景区内缺乏完善的排水系统，原财政宾馆、村民的生活污水，均未经处理而直接沿山坡自然排放，粪便污水经旱厕粪坑蓄存后用于农田施肥

(2)给水规划

① 用水量定额

根据现行国家有关规范及规定，结合当地气候条件及生活习惯，最大日用水量定额采用如下：

a. 旅游散客用水量定额为 15L/(cap・d)。

b. 旅游接待床位用水量定额为 200L/(cap・d)。

c. 景区服务人员用水量定额为 180L/(cap・d)。

d. 绿化用水量定额为 2L/(㎡・d)。

e. 未预见水量及管网漏失水量，按生活用水量的一定比例计入。

f. 同一时间内火灾次数为一次，一次灭火用水量 10L/s。

② 用水量

《总体规划》将香纸沟景区的用水量定为 60m^3/d，原管理服务区（陇脚村）的用水量定为 320m^3/d。现规划温泉度假村的用水量，预测将超过以上的总和，初步定为 550m^3/d。

③水源

景区内绿化植被完好，泉水和河水均较为充沛。应就近选用优质山泉水作为给水水源，尽可能采用重力自流供水节省供水能耗。其次，可采用河水作为供水水源。除此以外，温泉度假村给水水源，主要是抽取地下温泉，每天出水量可达 600t。

④ 水处理

景区内源水仅需过滤和消毒处理。

⑤ 输配水

管道沿规划区内道路敷设，并尽量呈环状布置，以提高供水的安全可靠性。主干管为 DN100 球墨铸铁管，支管为 DN40 ~ 80 的 UPVC – U 给水塑料管。给水管应敷设在地面以下，其中在车行道下的覆土深度不小于 0.8m。

消防采用常高压系统，室外地上式消火栓管道直径为 DN100，消火栓间距不超过 120m。消火栓距建筑物的外墙距离不小于 5m，距道路边的距离不大于 2m。不通行车的道路应在建筑室外墙上设置 DN 50 室内消火栓，连接管道管径为 DN70。室内消火栓的间距不大于 50m。

V Planning and Design of Engineering Pipeline System

1. Water Supply and Drainage Planning

(1) Current situation of water supply and drainage

Water users in the scenic spots are mainly the local villagers. Water resource is spring water, supplying by gravity. Due to good natural vegetation, the water is usually very clear and it can be directly used as drinking water through pipes.

The drainage system in the scenic area is imperfect. The domestic sewage of the former Finance Hotel and the villagers drain along the slopes without any treatment and the stool water is deposited in cesspit and then fertilize the farmland.

(2) Water supply planning

① Water consumption quota

Under the existing provisions of the relevant state regulations and the local climatic conditions and living habits, the maximum daily water consumption quota is as follows:

a. Water quota of foreign independent tourists is 15 L/(cap・d).

b. Water quota of tourist reception beds is 200 L/(cap・d).

c. Water quota of scenic staff is 180 L/(cap・d).

d. Water quota of afforesting and greening is 2 L/(m^2・d).

e. Unforeseen water consumption and leakage is counted by a certain percentage.

f. One fire disaster at the same time, the fire–fighting water consumption is 10L/s.

② Water consumption

"Master plan" designates the water consumption in the Xiangzhigou scenic area as 60m^3/d, the original service area (Longjiao village) as 320m^3/d. The planning water consumption of the Spa resort is predicted to exceed the total of the above and be supposed to be 550m^3/d.

③ Water sources

The nearest high–quality spring water are selected as the water sources, supplying by gravity as much as possible to save energy consumption and the river water can be another source of water. In addition, the underground hot spring, 600T per day, is the main water source to supply the spa resorts.

④ Water treatment

The water inside the scenic areas only needs filtration and disinfection.

⑤ Transmission and distribution of water

Pipelines are laid along the roads in planning areas and arranged like a ring as much as possible in order to improve the safety and reliability of water supply. The main pipes are DN100 ductile iron pipes, and branches are DN40 to DN80 UPVC–U plastic pipes. The pipes should be laid underground, with the depth of over 0.8m under the roadway.

High pressure system is adopted for fire control. The diameter of outdoor overground fire hydrant pipe is DN100, and the space between fire hydrants is not more than 120 meters. The fire hydrant is not less than 5 meters far from the building's exterior wall, and 2 meters from the road side. The DN50 indoor fire hydrant is installed on the exterior wall of the building along the impassable road, and the diameter of the connecting pipe is DN70. The space between the indoor fire hydrants is not more than 50 meters.

(3) 排水规划

① 排水体制

规划采用不完全分流制排水系统，生活污水经收集接纳输送后进行处理，雨水沿山地坡度及道路过沟自然排放。

② 污水量

污水量按用水量的 80%确定，《总体规划》将香纸沟景区的污水量预测为 50m³/d，原管理服务区（陇脚村）的污水量预测为 260m³/d。现规划温泉度假村的污水量，预测将超过以上的总合，初步定为 440m³/d。

③ 污水管网

污水管道沿规划内道路及山地坡度布置，管材为 UPVC 排水管。污水管道直径不小于 DN200。在管道转弯处、管道交汇处、管径或坡度改变处以及直线管段上每隔一定距离处设置污水检查井。

(3) Drainage planning

① Drainage system

The incomplete split–flow drainage system is employed. The domestic sewage is collected and delivered for treatment and the rain water flows along the mountain slopes and the road ditches.

② Sewage flow

The wastewater flow is calculated by 80% of total water consumption. It is estimated that the wastewater flow of Xiangzhigou scenic area will be 50m³/d and the original service area (Longjiao village) 260m³/d in the "master plan", and the wastewater flow of the planning spa resort will exceed the total of the above and be supposed to be 440m³/d.

③ Sewer network

The pipes are UPVC drainage pipes, with the diameter of not less than DN200, and are laid out along the roads and mountain slopes within the planning area. The manholes of sewage pipe will be located at the turns and junctions of the pipes, the parts changed of the pipe diameter or the slope, and the straight pipes at intervals.

2. 污水处理及排放要求 Sewage Treatment and Discharge Requirements

(1) 污水处理 Sewage treatment

粪便污水需经化粪池预处理，餐厅污水需经隔油池预处理。

Fecal sewage is pretreated in the septic tank, and restaurant wastewater in the grease trap.

(2) 污水排放 Sewage discharge

经稳定塘的污水出水用于农灌或绿化，或用于厕所及卫生间的冲洗水，污水必须 100%回用，做到零排放。严禁直接排入溪河及溶洞内。沿河两岸不得新建排污口。

The wastewater from the stabilization pond is reused for agricultural irrigation or urban greening, or for flushing washrooms and toilets. The wastewater must be recycled by 100% and achieve zero release. Direct discharge into rivers and caves is strictly prohibited. The discharge outlets are not allowed to build along the river banks.

给水排水规划图

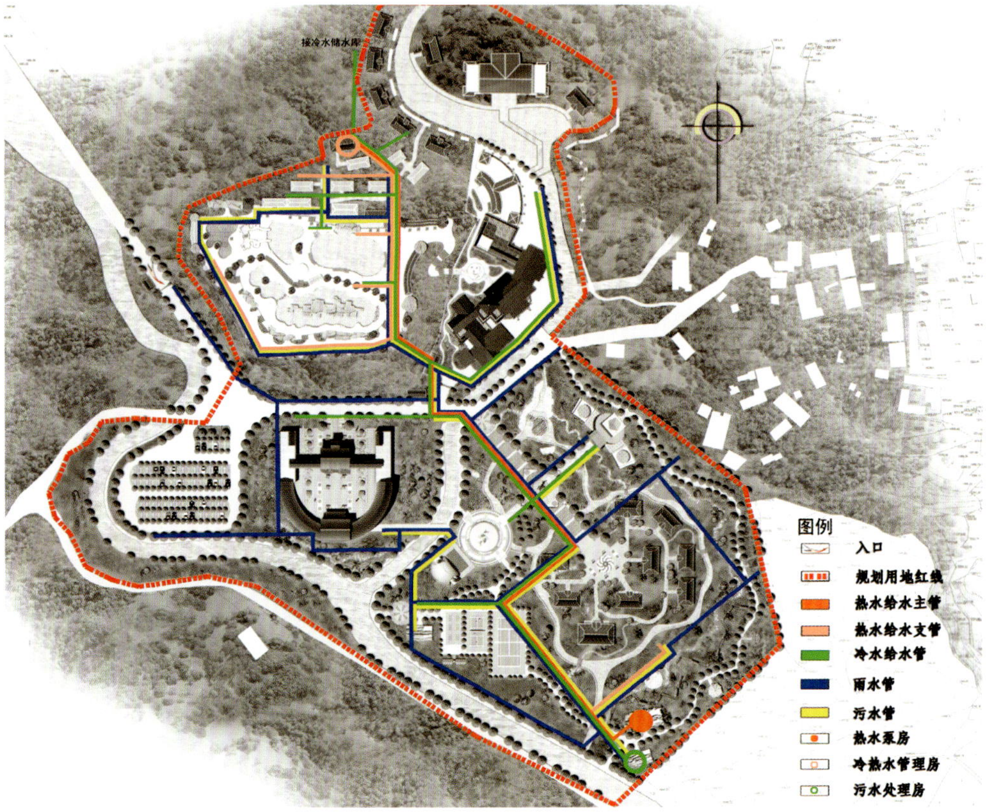

六、电力电信规划

1. 电力

(1) 现状

香纸沟风景名胜区上级电源为新天 110kV 变电站，经 10kV 线路引入至水田村，全乡各村组均已通电，有充足的供电能力。

(2) 负荷预测

《总体规划》考虑同时系数后，预测风景区共需用电 2000kW，其中原陇脚管理服务区（含原财政宾馆），用电 930kW。本规划温泉度假村用电需增加到 1550kW，风景区总用电也要相应增加。

(3) 电网规划

《总体规划》按照贵阳市农村电网建设与规划，2005 年全区的供电能力已大幅度提高。香纸沟风景名胜区的各景点、景区分布较为集中，规划各景区原则上由就近的农网 10kV 线路供电。

供电电源采用 10kV，在农村电网建设改造的同时应兼顾各景区服务设施的用电负荷，引入 10kV 专线供电。

近期内，各景区已全部解决其供电，完善供电设施，提高供电能力。远期，对各景区供电线路进行改造，有条件的景区、景点可增加一回 10kV 线路供电，以提高供电的可靠性和供电质量。

各景区、景点的 10kV 变配电所，位置由详规或在设计中确定，原则上规定变配电所设置于负荷中心，变压器的负荷率约为 70%为宜。负荷相对集中的景区服务点，其低压供电半径≤ 250m，负荷较小且分散的景区服务点，其低压供电半径宜按导线电压降选择。架空线路应尽可能避开主要景观，有条件时，局部采用电力电缆埋地敷设，以利于景观保护。

本详规将温泉度假村的变电室设在多功能厅次入口处的左侧，按《总体规划》的有关要求施行。

VI Power and Telecom Planning

1. Power

(1) Current situation

The power of the Xiangzhigou scenic area is supplied from the Xitian 110kV transformer substation to the Shuitian Village by 10kV line and all the villages of the towns are electrified with adequate power supply capacity.

(2) Load prediction

The simultaneity factor is considered in the "Master plan". The total power consumption of the scenic spots will be 2 000kW which includes 930kW for the former Longjiao management and service area (including the former Finance Hotel). The power demand of the planned spa resort will increase to 1 550 kW and the total electricity consumption in the scenic area will be increased accordingly.

(3) Power grid planning

The "Master plan" was put forward in accordance with the Guiyang rural power grid construction and planning. The region's power supply capacity increased significantly in 2005. Because the distribution of all scenic spots in the Xiangzhigou scenic area is comparatively concentrated, the power of the planned scenic area is supplied by the nearest 10kV rural grid.

The Power is supplied by 10kV power source with 10kV electric wires. While the rural grid is constructed, the electricity load of scenic services must be considered.

The power will be completely supplied to the scenic spots, power supply facilities will improve, and the supply capacity will enhance in the short run. In the long run, the power supply line of all the scenic spots will be reconstructed and another 10kV line will be installed in the scenic spots where conditions permit to increase the reliability and quality of power supply.

The locations of 10kV substations of all scenic spots are specified in the detailed plan or design. In principle, the substations should be installed in load center, with transformer load rate of 70%. The low-voltage power supply radius is ≤ 250 meters in the scenic services with relative concentrated load and the low-voltage power supply radius is chosen according to wire voltage in the scattered scenic services with relative small load. Overhead lines should avoid the main landscapes as much as possible. If the conditions permit, underground power cables will be used to protect landscapes.

The transformer room of the spa resort is located at the entrance to the multifunctional hall in the detailed planning, which is implemented by the related requirements of the "Master Plan".

2. 电信

(1) 现状

目前，香纸沟风景名胜区附近的村组均已开通了程控电话，以新堡乡为中心的电信线路呈辐射形遍及各村组，在移动通信方面，新堡乡现已开通移动电话基站一处，位于香纸沟，移动电话覆盖全乡的 95%以上区域。现有移动用户 500余户。

(2) 电信设施规划

根据风景区总体规划要求，各景区服务设施预测共需电话 1200 门左右。景区服务设施的通信系统就近接入新堡乡本地电信网络，较为集中的香纸沟已设置程控电话交换点，少数边远景区线路不便敷设或用户较少较分散的，已采用 ETS（无线接入网）方式运行。

香纸沟风景名胜区各景区、景点电话均已与贵阳市等位拨号，开通国内及国际长途直拨，满足游客的多种需求。

温泉度假村服务设施的通信，要求开通综合业务数字网（ISDN）及宽带网，满足语音、图像、数据通信等要求，建立酒店管理系统、智能卡电话、数据通信、旅游电子商务等。规划要求通信设施的功能和能力，应与香纸沟风景名胜区的开发相适应，使香纸沟风景名胜区的通信更加完善，为游客提供优质便捷的服务。

2. Telecommunication

(1) Current situation

The program-controlled telephones have been equipped in all villages near the Xiangzhigou scenic area now. The telecommunication lines radiate from the Xinbao Town to all other villages. As for mobile communications, a mobile phone base station covering over 95% of the region has been built in Xiangzhigou of Xinbao Town. There are more than 500 mobile users at present.

(2) Telecommunications facilities planning

According to the requirements of the general planning, about 1 200 telephones should be equipped in the scenic spots. The communication system of Scenic service facilities links to the nearby local telecommunications networks of Xinbao Town. The program-controlled telephone exchange has been installed in Xiangzhigou with more telephones and ETS (Radio Access Network) has been adopted in remote scenic areas where the communication lines are hard to be equipped and the users are scattered.

The telephones can be used in all scenic spots of Xiangzhigou as equally as in Guiyang City. IDD and DDD phone services are provided to meet the various needs of tourists.

For the spa resort facilities for communications, Integrated Services Digital Network (ISDN) and Broadband Networks are equipped to meet the voice, video, and data communication requirements and to establish hotel management systems, smart card telephones, data communications, tourist e-commerce, etc. According to the planning, the function and capability of the communication facilities should adapt to the development of the Xiangzhigou scenic area to make its communications more perfect and to provide quality and convenient services for tourists.

电力电信

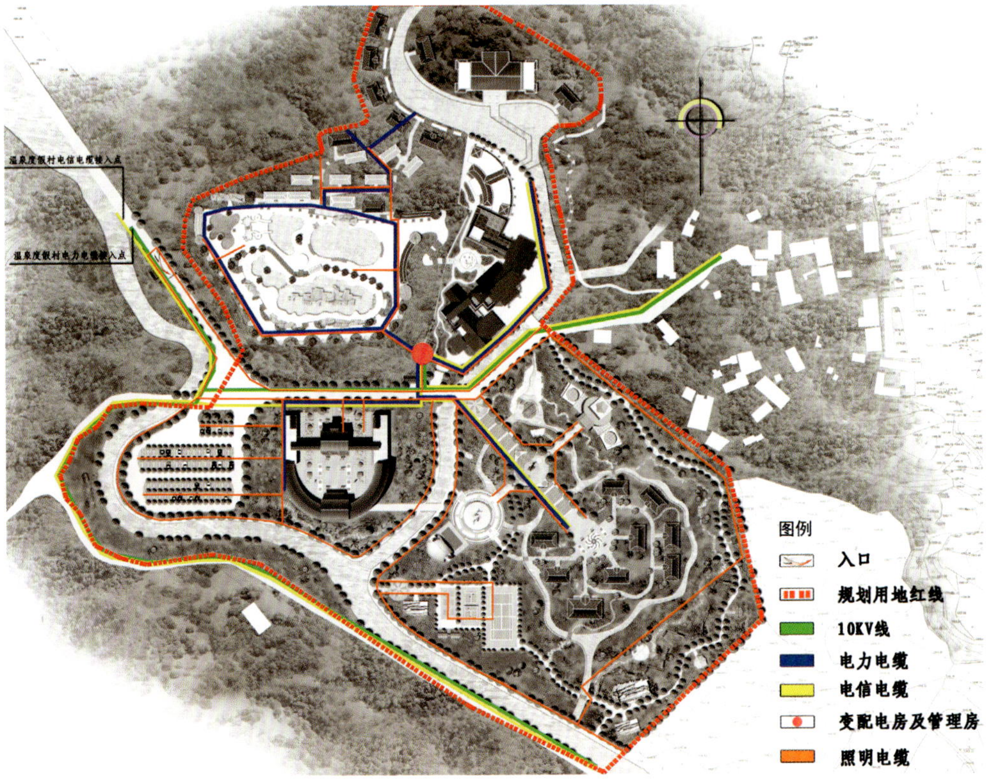

七、防灾系统规划

1. 防火

香纸沟风景区的天然林护林防火，纳入乌当区森林防火体系建设，严禁乱设烧烤点，在各重要地段设置防火标志，森林消防队伍应专业化，适当位置设置防火瞭望台。

2. 森林病虫害防治

风景区建立防治标准站，配备专业人员，定期检查并实施一定保护措施。

3. 防洪

风景区河流为山区性河流，容易形成暴涨暴落，在峡谷和溶洞，要特别注意泄洪通畅，应该疏浚河道，预留足够的排洪滩涂，严禁各类建筑物侵占河滩，故在风景区内不建防护坝，已建应拆除，并作自然式斜坡处理。

4.防地质灾害

在峡谷或溶洞内，应检查是否有松动的石块，如有及时清除，在陡峭疏松现积地段，避免游人攀登，严禁放炮建房，以免振动动岩石，形成危岩或发生滑坡。

Ⅶ Disaster Prevention System Planning

1. Fire Prevention

The protection and fire prevention of the natural forests in Xiangzhigou scenic area has been integrated into the forest fire prevention system of Wudang District. It is strictly prohibited to set up BBQ sites arbitrarily. Put up fire control symbols in key areas, set up professional forest fire control team, and establish an observation station in an appropriate place.

2. Forest Pest Prevention

Establish standard control stations with professional staff to regularly inspect and implement some protective measures in the scenic area.

3. Flood Control

Rivers in the scenic area are mountainous rivers, so it is easy to rise and fall suddenly. Special attention should be paid to keep discharge unobstructed in the valleys and caves, dredge the rivers, set aside adequate drainage beach and prohibit all kinds of buildings occupied the beach. The protective dam is not allowed to build in the scenic area do not build, and those built should be dismantled as natural slopes.

4. Geological Disaster Prevention

Check whether there are loose stones in the canyon or cave and remove in time if there are. Prevent anyone from climbing, blasting and building in cliffy and loose slopes so as not to shock rocks to become unstable rocks or landslide.

6 近期建设规划 Recent Construction and Planning

一、规划目标

加强香纸沟温泉度假村的基础设施建设，首先完成温泉度假村的一期工程建设，完善服务设施的配套，争取在 2007 年 4 月接待游客。紧接着开展二期工程建设，争取在 2009 年 4 月全面接待游客。

二、建设项目一览表

期限	序号	项目	所在功能区	规模	投资（万元）	备注
一期	01	游泳池	所在功能区	600m^2×1000 元 /m^2	60	
	02	儿童戏水区	天池洗浴区	250m^2×1000 元 /m^2	25	
	03	大泡池	天池洗浴区	720m^2×2000 元 /m^2	144	含室内泡池
	04	小泡池	天池洗浴区	13 个，110m^2	40	
	05	地热带	天池洗浴区		5	
	06	室内泡池	天池洗浴区		15	
	07	VIP 包房	天池洗浴区	6 幢 ×10 万	60	
	08	多功能厅		改造 3500m^2×1500 元 /m^2	525	含装修
				新建 2370m^2×3000 元 /m^2	711	含装修
	09	厕所		3 幢，180m^2	18	
	10	泵房		130m^2	50	含水处理
	11	管理房			10	
	12	大门		2 处	15	
	13	泡区入口			25	
	14	长廊			20	
	15	观景亭		4 个	10	
	16	保健按摩房			10	
	17	别墅	别墅生活区	7 幢，600m^2	120	含装修

期限	序号	项目	所在功能区	规模	投资（万元）	备注
二期	01	停车场	停车及购物区	4070m^2	40	
	02	博物馆	休闲娱乐区	1240m^2×2000 元 /m^2	248	含装修
	03	酒店	酒店服务区	2670m^2×3000 元 /m^2	801	含装修
	04	接待中心	酒店服务区	2600m^2×2000 元 /m^2	520	含装修
	05	表演场	休闲娱乐区	300m^2	15	
	06	羽毛球场	休闲娱乐区		3	
	07	滑翔道	休闲娱乐区		3	
	08	水上娱乐中心	休闲娱乐区	6700m^2	1800	
	09	道路及硬地		6780	50	
	10	健身区	休闲娱乐区		2	
	11	管理房			10	
	12	攀爬区	休闲娱乐区		20	

I Planning Objectives

Strengthen infrastructure construction of the Xiangzhigou Spa Resort. At first, complete the Phase I project of the spa resort and improve relevant service facilities to receive tourists from April 2007, and then followed the Phase II construction project to receive tourists from April 2009.

II List of Construction Projects

Phase	No.	Project	Functional Area	Scale	Investment (Ten Thousand Yuan)	Remaek
I	01	Swimming pool	Tianchi bathing area	$600m^2 \times 1000yuan/m^2$	60	
	02	Dabbling pool for children	Tianchi bathing area	$250m^2 \times 1000yuan/m^2$	25	
	03	Big bathing pool	Tianchi bathing area	$720m^2 \times 2000yuan/m^2$	144	Including indoor poor
	04	Small bathing pool	Tianchi bathing area	13pools, $110m^2$	40	
	05	Geothermal zone	Tianchi bathing area		5	
	06	Indoor bathing pool	Tianchi bathing area		15	
	07	VIP private room	Tianchi bathing area	6buildings × 100000yuan	60	
	08	Multifunctional hall		Reconstruction: $3500m^2 \times 1500$ yuan/m^2	525	Including fitment
				Construction: $2370m^2 \times 3000$ yuan/m^2	711	Including fitment
	09	Toilet		3Toilets，$180m^2$	18	
	10	Pumping house		$130m^2$	50	Including water treatment
	11	Management house			10	
	12	Gates		2	15	
	13	Entrance to bathing area			25	
	14	Gallery			20	
	15	Pavilions		4	10	
	16	Massage room			10	
	17	Villa	Villa living area	7villas, $600m^2$	120	Including fitment

Phase	No.	Project	Functional Area	Scale	Investment (Ten Thousand Yuan)	Remaek
II	01	Parking Area	Parking and shopping areas	$4070m^2$	40	
	02	Museum	Recreation area	$1240m^2 \times 2000$ yuan/m^2	248	Including fitment
	03	Hotels	Hotel service area	$2670m^2 \times 3000$ yuan/m^2	801	Including fitment
	04	Reception center	Hotel service area	$2600m^2 \times 2000yuan/m^2$	520	Including fitment
	05	Arena	Recreation area	$300m^2$	15	
	06	Badminton area	Recreation area		3	
	07	Glide road	Recreation area		3	
	08	Water recreation center	Recreation area	$6700m^2$	1800	
	09	Roads and hard ground		6780	50	
	10	Fitness area	Recreation area		2	
	11	Management house			10	
	12	Climbing area	Recreation area		20	

Xiangzhigou Spring Resort

效果图件 Effective Pictures

温泉度假区整体鸟瞰

温泉度假区一期鸟瞰

入口、停车及餐饮购物区

该区位于温泉度假村的西南部，西与香纸沟景区的入口相接，包括停车场和购物中心。停车场为生态地面，占地面积 4070m²，设有大小 135 辆停车位。购物中心采取前开后合的布局手法，并将布依族建筑与我国江南传统民居建筑融为一体，占地面积 1300m²，分为前后两个部分，前为接待休息等，后设商店购物。

③

④

Entrance, Parking, Dinning and Shopping Zone

This zone, located at Southwest part of hot-spring resort, is connected with Xiangzhigou scenic spot entrance to its West, including parking lot and shopping centre. The parking lot is eco-ground with an area of 4 070m^2, setting up 135 parking spaces for cars. The shopping centre adopts the layout of front opening and back closing, and integrates Bouyei building with traditional residential building in the regions south of the Yangtze River of China. Its area is 1 300m^2, and divides into two parts of back and forth. The front is for services like reception and rest while the back is for shopping.

①多功能厅效果图 ②多功能厅局部效果图 ③～④温泉入口景观

温泉洗泡区

该区位于温泉度假村的西北部，包括游泳池、儿童戏水区、大泡池、小泡池、地热带、室内泡区、VIP 包房等，为游客提供各种洗浴场地。其中，小泡池共有 13 个，分别建在天池洗浴区西南边沿的竹林丛中；VIP 包房共有 6 小幢，分别建于天池洗浴区北面的树林之中。

Hot Spring Area

This area is located at the northwest of the hot spring resort, including swimming pools, children's swimming pools, big soaking pools, small soaking pools, geothermal belts, indoor pools, VIP compartments and so on for tourists. Among them, there are 13 small soaking pools, located in the bamboo forest in the southwest part of Tianchi bath region respectively; 6 VIP compartment buildings, situated respectively in the forest in the north part of Tianchi bathing area.

①大泡池效果图 ②大泡池室内效果图 ③小泡池效果图 ④小泡池效果图

①

②

①室内泡池效果图 ②沙池效果图 ③～⑤ VIP 包房

①

②

③

④

①泳池效果图 ②保健按摩房 ③公厕 ④小卖部

①

酒店服务区 Hotel Service Area

该区位于温泉度假村的东部，包括酒店、多功能厅和接待中心等服务设施，由原香纸沟财政宾馆等建筑物改建而成。对景区的保护、建设及开放营运，游人的吃、住、行、游、购、娱及安全等进行全面管理的工作区域。选择适当的位置，树立温泉度假村的文化标志——独特的牛头形象，既反映了布依族水田稻作文化中对牛的崇拜，也反映了“田园温泉”的自然属性，又适合于现代人以“牛”为强劲的竞争心理，并以当地簸箕画的形式加以塑造。

This area is reconstructed from the original Xiangzhigou Financial Hotel, located in the east part of the hot spring resort, consisting of hotels, multifunction halls and reception centers. It is the working area devoted to the overall management including the protection, construction and open operation of the resort, eating, lodging, transportation, touring, shopping, entertainment and safety of the tourists. Select a suitable location so that to set up the cultural symbol of the hot spring resort—the unique image of a cow head, reflecting not only Bouyei nationality's worship towards cows in its rice culture, but also its nature of "the idyllic hot spring". What's more, this symbol also plays up to the modern people's competition psychology that cow represents robustness. This symbol should be presented in the form of dustpan painting which is a kind of local art.

别墅生活区 Villa Living Area

该区位于温泉度假村北部的沟壑两岸，北岸有 4 幢别墅，南岸有 3 幢别墅，均为当地布依族传统民居建筑样式。

在以上 5 个功能区中，温泉洗泡区和酒店服务区为一期建设工程，其他为二期建设工程。

This quarter is located on the banks of the gully in the north part of the hot spring resort, with 4 villas on the north bank and 3 on the south. All the villas are built according to the traditional Bouyei residents' architecture style.

The phase I project includes the hot spring area and hotel service area and the phase II the others of the above 5 functional areas.

①酒店效果图 ②～④别墅效果图

①

娱乐健身区

该区位于温泉度假村的东南部，包括博物馆、表演场、滑翔道、水上娱乐中心、健身区、攀爬区等。博物馆在原造纸博物馆的基础上进行改建，为当地布依族传统民居建筑样式，以当地布依族至今仍然使用的铜鼓及其图案为文化元素符号装饰，陈列展示当地布依族文化。表演场同为当地布依族传统民居建筑样式，舞台采取梯级移动手法处理，作为当地“汉灯、苗舞、布依歌”的表演场地。水上娱乐中心占地面积6700m^2，为游客提供水上娱乐活动。

Entertainment and Fitness Areas

The area is situated in Southeast part of hot-spring resort, including museum, arena, glide path, overwater leisure and recreation center, fitness area and climbing area. The museum is reconstructed on the basis of the original papermaking museum with the style of local Bouyei traditional residence building. The museum is decorated with drums and their patterns as cultural elements and symbols which are still used by local Bouyei people and displays Bouyei culture. The arena is also the traditional residence buildings of local Bouyei. The stage adopts the method of moving steps and serves as the performance place for "Han lanterns, Miao dances and Bouyei songs". The overwater leisure and recreation center, with an area 6 700m^2, provides overwater leisure and recreational activities for tourists.

①表演广场效果图 ②主题雕塑效果图

①

②

①水上娱乐中心效果图 ②水上娱乐中心室内鸟瞰

1

2

3

①～⑤水上娱乐中心室内局部效果

①～③景观效果图 ④公共服务设施及小品

④

名酒池

名花池

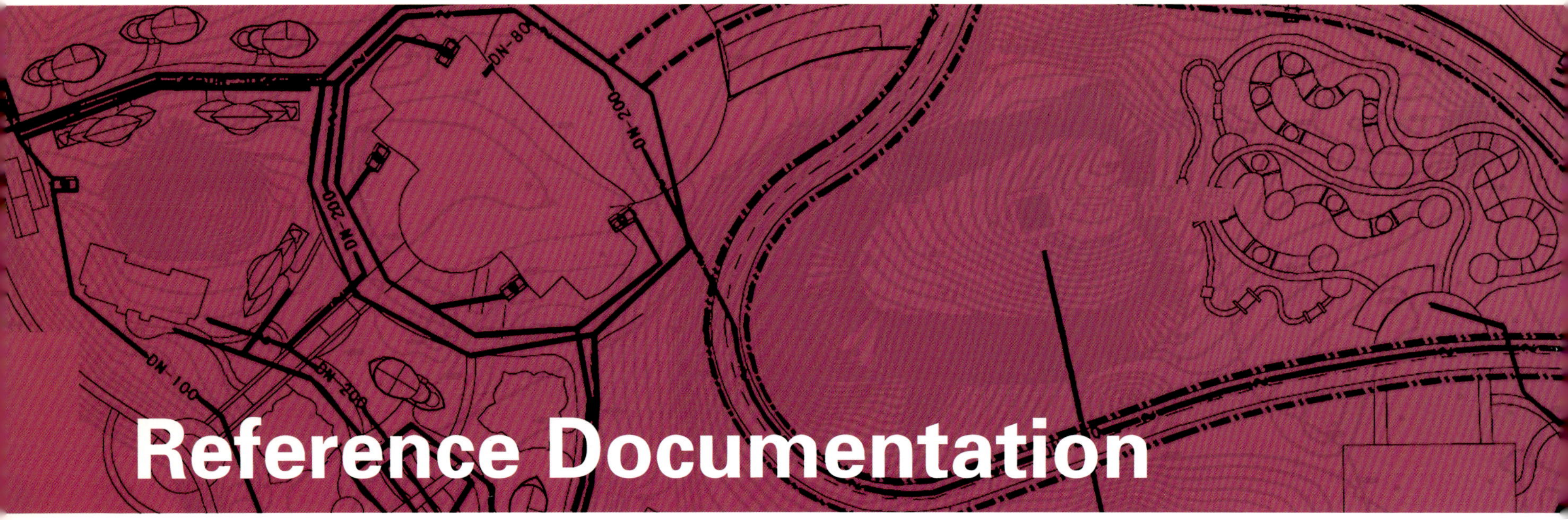

Reference Documentation

参考文献

[1] 陈南江 . 滨水度假区旅游规划创新研究［D］. 上海：华东师范大学 ,2005.
[2] 张建 . 城市主题公园的创新与发展——环球嘉年华现象的启示［J］. 价格月刊 . 2004,2：22–23.
[3] 赵平 . 从概念和功能看我国旅游度假区的发展［J］. 太原大学学报 . 2006,3：81-83.
[4] 熊瑛 . 大型主题公园策划与规划研究［D］. 北京：北京工业大学 ,2001.
[5] 熊瑛，王珊，宛素春 . 大型主题公园评析［J］. 北京规划建设 . 2003,5：16–19.
[6] 郑宝华 . 迪斯尼对我国主题公园发展的启示［J］. 产业经济 . 2006,6：241.
[7] 米冰 . 都市旅游发展中的主题公园开发研究［D］. 长春：东北师范大学 ,2006.
[8] 赵海霞 . 度假区景观规划研究初探［D］. 广州：华南师范大学 ,2005.
[9] 汪芳，侯安扬 . 对旅游区详细规划的探讨［C］. 和谐城市规划——2007 中国城市规划年会论文集 .749–752.
[10] 姚昆遗 . 对上海主题公园的评析和思考［J］. 上海大学学报 . 1998,5：87–90.
[11] 周小棣 . 关于主题公园的思索［J］. 东南大学学报 . 2001,11：120–123.
[12] 芦宝英 . 国内主题公园发展研究再思考［J］. 成都行政学院学报 . 2004,11：66–68.
[13] 芦宝英 . 国内主题公园开发存在的缺憾与反思［J］. 西华师范大学学报 . 2005,1：75–79.
[14] 周建明 . 旅游度假区的发展趋势与规划特点［J］. 国外城市规划 . 2003,18：25–29.
[15] 刘家明 . 旅游度假区发展演化规律的初步探讨［J］. 地理科学进展 . 2003,22：211–218.
[16] 邵颖莹 . 南京金牛湖旅游度假区规划实践［D］. 南京：东南大学 ,2004.
[17] 肖春娟 . 旅游度假区市政基础设施规划研究［D］. 西安：西安建筑科技大学 ,2007.
[18] 王惠建 . 旅游主题公园核心竞争力的识别和培育研究［D］. 湘潭：湘潭大学 ,2003.
[19] 王宁 . 论主题公园的开发操作［J］. 城市规划汇刊 . 1997,4,27–28.
[20] 李永文 . 论主题公园的区域经济影响、建设与发展［J］. 经济地理 . 2005,25：694–697.
[21] 王忠丽 . 论主题公园建设与城市旅游竞争力的提升［D］. 郑州：河南大学 ,2005.
[22] 夏海燕 . 浅析主题公园的主题策划［J］. 城乡规划 . 2003,5：61–62.
[23] 张怡，吴骥良 . 谈度假区项目可行性研究［J］. 规划师 . 2004,5：100–102.
[24] 李春生 . 我国城市主题公园的发展与创新［J］. 城市问题 . 2005,6：30–33.
[25] 王宇智 . 我国度假地发展的制约因素和对策［J］. 桂林旅游高等专科学校学报 . 2002,13：35–38.
[26] 周翀燕 . 我国旅游度假地开发研究［D］. 厦门：华侨大学 ,2005.

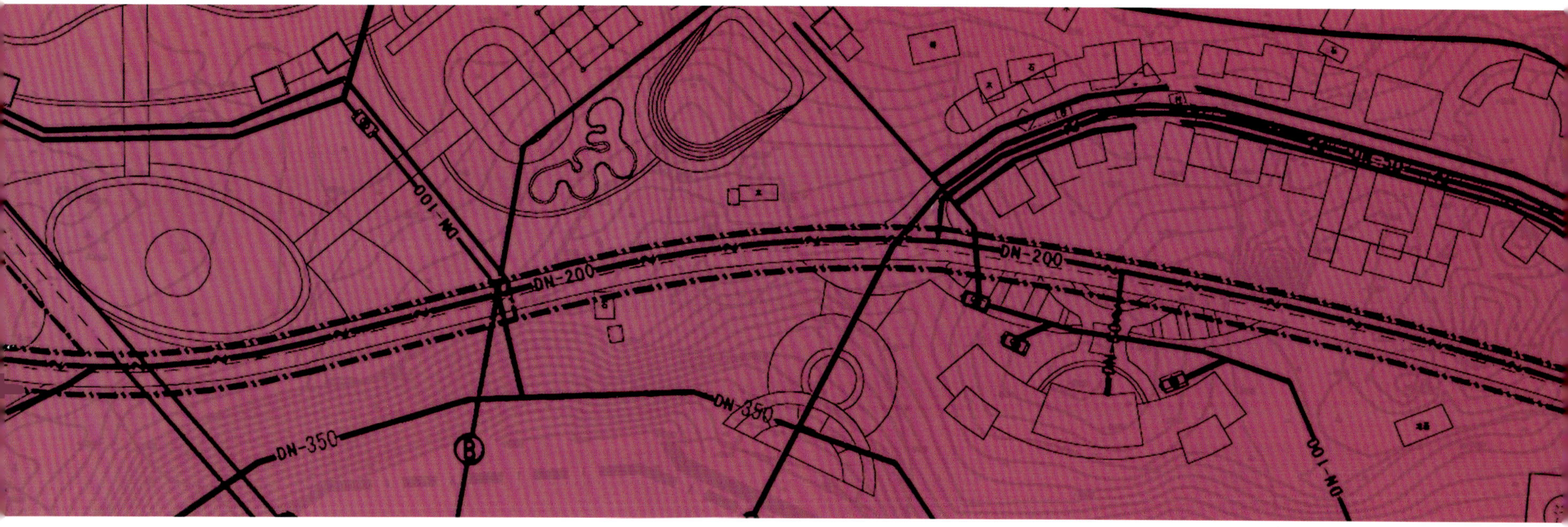

[27] 唐继刚 . 我国旅游度假区的开发现状、问题及发展构想——以苏南地区为例［D］. 南京：南京师范大学 ,2002.
[28] 毕斗斗 . 我国旅游度假区开发模式研究［J］. 计划与市场 . 2002,3：32–33.
[29] 张东祥 . 我国旅游度假区开发问题研究与对策［J］. 商场现代化 . 2007,7：349.
[30] 任一 . 我国主题公园差异化战略的构建［D］. 成都：西南财经大学 ,2005.
[31] 王钰 . 我国主题公园产业创新体系研究［D］. 长沙：中南大学 ,2007.
[32] 龚海鹏 . 我国主题公园的发展策略研究［D］. 长沙：中南林业科技大学 ,2006.
[33] 李春生 . 我国主题公园的发展、现状与创新［J］. 地域研究与开发 . 2007,26：71–74.
[34] 张沛 . 我国主题公园发展转换研究［D］. 昆明：昆明理工大学 ,2007.
[35] 彭红霞，李娟文 . 我国主题公园开发的误区与对策［J］. 湖北大学学报 . 2002,5：46–49.
[36] 李慧云 . 中国主题公园存在的问题和发展对策［J］. 发展战略 . 2004,4：54–56.
[37] 保继刚 . 中国主题公园的发展反思及国际主题公园进入中国的透视［J］. 风景园林 . 2005,2：26–31.
[38] 曹晓珍 . 中国主题公园的发展建设与规划研究［D］. 北京：北京林业大学 ,2006.
[39] 王欣 . 主题公园的发展现状 _ 问题与对策［J］. 商业经济与管理 . 2000,1：28–30.
[40] 陈准 . 主题公园发展、城市化战略及产业结构调整［J］. 北京规划建设 . 2003,5：11–12.
[41] 冯维波 . 主题公园规划设计探讨［J］. 地域研究与开发 . 2001,20：86–89.
[42] 朱美荣 . 主题公园如何才能走出困境［J］. 社会观察 . 2004,12：20–21.
[43] 李志强 . 主题公园文化的规划表达初探［D］. 重庆：西南大学 ,2007.
[44] 侯新冬 . 主题公园与历史文化名城互动发展研究［D］. 上海：华东师范大学 ,2006.
[45] 陈静 . 主题公园与区域经济的互动关系研究［D］. 西安：西北大学 ,2006.
[46] 龚映梅 . 主题公园原理与可持续发展模式研究［D］. 昆明：昆明理工大学 ,2002.
[47] 张广瑞，刘德谦等 .2008 年中国旅游发展分析与预测［M］. 北京：社会科学文献出版社 . 2008.
[48] 王德刚等 . 旅游开发学［M］. 济南：山东大学出版社 . 2007.
[49] 工旭科，赵黎明 . 旅游区规划的城市化问题及其对策研究 [J]. 人文地理 . 2007,6（6）：94–97.
[50] 马聪玲，张金山 . 对我国旅游规划现状的反思与评价——兼谈“十一五”期间的旅游规划思路［J］. 北京工商大学学报 (社会科学版).2007，3（2）：88–92.
[51] 杨建朝，朱菁菁 . 浅谈旅游规划中的误区 [J]. 市场论坛 . 2005,1：94–96.
[52] 谢彦君 . 关于中国旅游规划操作问题的几点思考［J］. 东北财经大学学报 . 2001,7（4）：40–42.
[53] 李立华 . 论中国旅游可持续发展的规划策略［J］. 山地学报 . 2000,2（1）：76–83.